Contemporary Trends and Issues in Science Education

Volume 62

The book series Contemporary Trends and Issues in Science Education provides a forum for innovative trends and issues impacting science education. Scholarship that focuses on advancing new visions, understanding, and is at the forefront of the field is found in this series. Authoritative works based on empirical research and/or conceptual theory from disciplines including historical, philosophical, psychological and sociological traditions are represented here. Our goal is to advance the field of science education by testing and pushing the prevailing sociocultural norms about teaching, learning, research and policy. Book proposals for this series may be submitted to the Publishing Editor: Claudia Acuna E-mail: Claudia.Acuna@springer.com

Cristiano B. Moura
Editor

A Sociopolitical Turn in Science Education

Towards Post-pandemic Worlds

Springer

Editor
Cristiano B. Moura
Simon Fraser University
Burnaby, BC, Canada

ISSN 1878-0482 ISSN 1878-0784 (electronic)
Contemporary Trends and Issues in Science Education
ISBN 978-3-031-78585-6 ISBN 978-3-031-78586-3 (eBook)
https://doi.org/10.1007/978-3-031-78586-3

This Springer imprint is published by the registered company Springer Nature Switzerland AG
The registered company address is: Gewerbestrasse 11, 6330 Cham, Switzerland

Foreword

This volume solidifies, for me, a sense of belonging that I have been seeking for a long time. Don't get me wrong, I have been fortunate enough to participate in collectives of more-than-science-teachers who have provided a deep sense of community in and beyond my professional life. But in the polarization of post-pandemic educational politics, broader conversations about the possibilities for science education are frequently stifled by the rise of repressive policies. In the United States, the question that diverts an imaginative conversation is often something like: "But what about Florida?" This question refers to a particular state that has become infamous for banning from schools any discussion of equality or oppression along multiple dimensions. Of course, addressing regressive political circumstances, wherever they exist, is important. But there is also a risk in allowing the far right to pull discourse in their direction even in locales where they do not hold political power.

Many educators who resist these rightward tugs are increasingly made to feel as if our ideas are crazy and unrealistic outside of our "pockets of resistance" (Frausto Aceves et al., 2022). But this book is an affirmation that a community of scholars is coalescing between those pockets and across the world to confront, study, and change the inherent politics of science education. Moreover, it provides a brilliant reminder of how much we can learn through dialogue with each other. For me, the sociopolitical turn in science education documented by this book feels like opening a door to an unfamiliar building, immediately feeling the warmth inside, and then being struck by ironic graffiti on the wall. It feels like arriving with trepidations at a social gathering—only to be pulled into rich conversations that are simultaneously welcoming and provocative. Even as I feel a sense of belonging in these pages, I am challenged by terms, concepts, and even deities that are completely new to me.

In this volume, Cristiano Moura has curated works by a group of authors I am humbled to call colleagues and, in some cases, friends. Together the sociopolitical turn advocated by the editor and authors is a gentle hand on our collective chin. This metaphorical hand is directing the attention of the field toward issues that have existed as long as the field itself but have typically been ignored.

Much of the wisdom herein comes from the Global South and diasporic or Indigenous communities in the North. And we know that the scientific enterprise has always borrowed, stolen, and incorporated (i.e., exploited) these kinds of wisdom (Smith, 1999). This extractive relationship mirrors the material relationship between these worlds (Gandolfi, this volume). That latter extractive relationship has thrust the various communities of Earth into positions that are increasingly precarious for all of us, but to different degrees. The authors are telling us, with tension in their voices and hope in their eyes, "you might want to look over here." The sociopolitical turn advocated in this book is therefore an effort to confront reality in order to form different, horizontal, and reciprocal relationships around how human communities produce and share knowledge with each other—beginning with what and how we teach young people about the sciences.

My use of the plural of science and my emphasis on beginning with young people in the previous sentence are both intentional word choices. Notice that above, I referred to a singular scientific enterprise, but here to multiple sciences. This is an acknowledgment that a conglomeration of corporations, universities, foundations, and nation-states has claimed a monopoly on knowledge production in a process that has been part-and-parcel of European colonialism, US imperialism, and racial capitalism (Roy, 2018; Wallenberg et al., 2021). Yet, even as hegemonic as that process has been, referring to a singular or universal science belies substantial epistemic heterogeneity and contestation. This heterogeneity exists even within and among the sciences under the umbrella of the scientific enterprise—to say nothing of the sciences nourished by the rain falling outside (Bang et al., 2018).

The chapters of this book show us the problems of the enterprise and the possibilities of the sciences. In so doing, they are aligned with the kinds of engagement we should encourage in science classrooms. Of course, science education does not happen only in classrooms nor are young people the only participants. In chapters throughout this book, all ages and groups of people and even other Earth beings are considered to be both learners and teachers. Still, even if we reject romantic or cliché notions about youth, I contend that young people have a special place in any discussion about social change or possible futures. As several chapters note, the pandemic briefly shook, but then ultimately reinscribed, the role of schools and other educational institutions in our society. Still, this book also documents alternatives that survived or emerged from the pandemic.

I write this foreword while on sabbatical in Santiago de Chile, just 100 km or so from a glacier that is described in this book as a science educator itself (Torres-Olave & Guerrero, this volume). The authors critique how so-called "sustainable" development is both accelerating the glacier's shrinking and threatening the water supply of 7 million people in the metropolitan area. This makes my temporary home in Santiago feel precarious compared with my permanent home in a similar-sized metropolitan area: Chicago, Illinois. Indeed, the enterprise of science has played a central role in this disparate distribution of precarity between Global North and South. Chicago sits on Lake Michigan, one of the most abundant sources of drinking water on Earth. But then again, in a northern suburb of Chicago, a decommissioned nuclear power plant stores radioactive waste just off the lakeshore. That spent fuel represents an ever-present threat of contamination to a massive ecosystem, including millions of humans, with no technological solution in sight. And few people in the region are aware of its existence.

This comparison might make many teachers and students rethink the role of science education in both locations. Here in Chile, there is a new requirement for secondary students that engages with the most pressing issues in local contexts and in the world (Cabello, 2024). Back in the United States, the push to prepare more compliant workers for the scientific enterprise remains the myopic focus of most politicians and policy makers who mention science education. The reform in Chile is far from perfect (Guerrero & Torres-Olave, 2021). But it nonetheless demonstrates one of the lessons of this book: the Global North would be smart to turn toward the Global South for leadership in science education. And this book also shows us that such a turn is well underway, indeed it may be a different way of describing the sociopolitical turn.

I was recently asked after a presentation why I think sociopolitical approaches and conversations of social justice are increasingly popular in science education. Reflecting on that question made it clear to me that the world is demanding this shift. We saw this demand in the political battles about vaccines—whether they were trustworthy, but also which communities and which parts of the world got them first. We saw this demand in the outrage sparked by the murder of George Floyd by Minneapolis police coinciding with the height of the pandemic. We continue to see this demand in social movements for environmental and climate justice. We see it in wildfires, flooding, blizzards, and hurricanes. In other words, the increasing popularity of sociopolitical themes in science education is not a trend or a fad. For those of us who have long been fascinated by the politics of science education, it is not "our turn," to use a different definition of that noun. And just as this turn is not owned by a particular group, neither is this a turn in a singular direction. This is a turn toward multiple possible futures, multiple possible worlds, as Moura (this volume) writes in the introduction. But then again, as described above, it is a turn towards the wisdom of the Global South. The turn described in the chapters of this book encourages the field to attune our senses and open our minds to consider: How might education about the sciences inform political projects that establish more reciprocity and balance on Earth and move toward more just and sustainable possible futures? I encourage you to find answers in this fascinating text.

College of Education,
University of Illinois Chicago,
Santiago de Chile
September 2024

Daniel Morales-Doyle

References

Bang, M., Marin, A., & Medin, D. (2018). If Indigenous Peoples stand with the sciences, will scientists stand with us? *Dædalus, the Journal of the American Academy of Arts and Sciences, 147*(2), 148–159.

Cabello, V., Zúñiga, C. G., Amador Valbuena, C., Manrique, F., Albarrán, M. J., & Moncada-Arce, A. (2024). "We are not being taught sustainable citizenship!": Podcasts for critical science literacy in teacher education. *LUMAT: International Journal on Math, Science and Technology Education, 12*(2), 18–49. https://doi.org/10.31129/LUMAT.12.2.2135

Frausto Aceves, A., Torres-Olave, B., & Tolbert, S. (2022). On love, becomings, and true generosity for science education: Honoring Paulo Freire. *Cultural Studies of Science Education, 17*(2), 217–230. https://doi.org/10.1007/s11422-021-10098-w

Gandolfi, H. (2024). (Re)considering science education in the face of environmental injustices in pre- and post-pandemic worlds: Insights from a critical-decolonial nature of science. In C. Moura (Ed.) *Science Education for post-pandemic worlds: Towards a socio-political turn.* Springer.

Moura, C. B. (2024). In the eye of the storm: For a sociopolitical turn in science education to build post-pandemic worlds. In C. Moura (Ed.) *Science Education for post-pandemic worlds: Towards a socio-political turn.* Springer.

Roy, R. D. (2018). Decolonise science – Time to end another imperial era. *The Conversation.* https://theconversation.com/decolonise-science-time-to-end-another-imperial-era-89189

Smith, L. T. (1999). *Decolonizing methodologies: Research and indigenous peoples.* Zed Books.

Torres Olave, B., & Guerrero, G. R. (2024). Critical imagination and conscientisation for just post-pandemic Science Education: Park Rangers as public science educators. In C. Moura (Ed.) *Science Education for post-pandemic worlds: Towards a socio-political turn.* Springer.

Wallenberg, E., Adams, A., Chow, C. S., Glaser, E., Hasan, N., Raghavan, M., Moss, M., Hetzner, E., Cho, S., Jett, E. Woolfson, B., Cox, B., & Hundahl, O. (2021). Letters from the editors. *Science for the People, 24*(1). https://magazine.scienceforthepeople.org/volume-24-number-1-racial-capitalism/

Daniel Morales-Doyle is an associate professor of science education and coordinator of the licensure strand of the MEd Science Education program at the University of Illinois, Chicago. His research examines the potential for science education to act as a catalyst for alternative futures. Specifically, his work focuses on the sociopolitical dimensions of science curriculum, teaching, and teacher education. He was a high school teacher in the Chicago Public Schools for more than a decade before coming to UIC. Dr. Morales-Doyle just released the book "Transformative Science Teaching," which reveals his vision for science education that supports meaningful learning in the sciences, focusing on incorporating practice-based teacher education, justice-centered science pedagogy, and youth participatory science, which could give all students opportunities to be engaged with the sciences, with their communities, and in the world.

Acknowledgments

As someone who studies the History of Science, I am not easily convinced by heroic narratives when it comes to producing science. Even though I am the solo editor of this book, it would not have been possible without the encouragement, generosity, and help of many people. First of all, I am indebted to Jesse Bazzul, for encouraging me at each step of my career, including when I was in doubt about taking on the challenge of organizing a book as the solo editor. On another unrelated occasion, Jesse told me to "say what is at your heart, and they will understand." It worked greatly. Since then, I have followed this advice and listened to my heart whenever I doubt where to go or what to do; the responses do come—thank you, Jesse. I must also thank Larry Bencze for the encouragement and essential mentorship at the beginning of this project. Your perspectives were pivotal for this project to take off the ground. Thank you to Steve Alsop, who organized with me the Symposium at the XX IOSTE conference that turned into this book; thank you also for the inspiration, good sense of humor, and wise perspectives on our research field, which prompted me to reflect on what kind of researcher I wanted to be and also my priorities when doing research. In this vein, I also owe Andreia Guerra and Tania Camel a huge thank you for the years of collaboration, friendship, and for everything I learned from you and the time, wine, and laughs we have shared so far. There is much more to come! My research group, NIEHCC (acronym in Portuguese for Research Group on History of Science, Culture and Education), is also responsible for my formation as a researcher and person, helping in my reflections on fundamental themes and continuous political formation from the exchanges I had with such an inspiring group of people. I will not mention all the names to avoid omitting anyone who contributed to my thinking over more than ten years of collaborative work. You are and will always be in a very special place in my heart. Thank you! My beloved wife, Olga, has also been a source of encouragement and understanding about all it takes to dedicate myself to academic projects. My energy comes mostly from the love we nurture daily with every tiny silly thing that makes us laugh together.

Coming from the south of the Global South, growing up in a *favela*, first gen, son of migrants from Northeastern Brazil, it comes with a great surprise that at any point in my life, I would have something to say to an international audience in academia. Or that I would be joined by such an amazing group of people from many corners of the world to say something as important as reaffirming that all education (including science education) is political. It takes courage and tenacity to walk all the way I (we) did, alongside such brave people, to "speak truth to the power." This courage I learned, no doubt, from my dad and mom, Maria and Linaldo, and my grandpas, grandmas, and family (some deceased, but still living in my heart), many of whom still live in the countryside of Northeastern Brazil, and, despite all the machinery that work every day to deny their rights to existence in that form, still enjoy working on the land with their bare hands and picking up fresh fruits from trees. Thanks to Daniel Morales-Doyle for a beautiful and truly inviting foreword! Thanks to Ana Maria Navas, my great friend, with whom I share the days at my University, for the support and encouragement, for hearing my stories, telling hers, and sharing ideas and ideals for the future!

I had an enjoyable and collaborative exchange all the way through with Claudia Acuña and her team, especially Karthika Menon. Without the hard work of the Springer team, this book would not have been possible.

It is always virtually impossible to recognize all those who have impacted our lives in decisive ways, especially as lucky as I am to have so many colleagues, friends, and great people around me who always have something special to share. Every human being is a rich entanglement of several threads one picks up from others along the way. I consider the multicolored fabric I am becoming to be due to the generosity I found in each of the rich, colorful and well-woven threads I was lucky to pick up in my path. *Oxalá* (I hope) I can continue this path and also serve to color someone's tapestries along the way as others did to me. Finally, I would like to thank my ancestors and Orixás, without whom nothing of that would be possible.

Contents

About the Editor

Cristiano B. Moura is an assistant professor in the Faculty of Education at Simon Fraser University, Canada. His work is informed mainly by History, Philosophy and Sociology of Science, post-colonial studies, and cultural and sociopolitical perspectives on science and education. His writings span from History of Science and Science Education under decolonial lenses, research methodologies in science education, and also had been reflecting on the boundaries of science education research and practices amid and beyond the COVID-19 pandemic. He was editor of Brazilian journals in the field of science education and was recently appointed editor-in-chief of the *Science & Education* journal from the IHPST (International History, Philosophy and Science Teaching) group.

Part I
Introduction

Chapter 1
Introduction—In the Eye of the Storm: For a Sociopolitical Turn in Science Education to Build Post-pandemic Worlds

Cristiano B. Moura

1.1 Introduction

> Good afternoon. In the past two weeks, the number of cases of COVID-19 outside China has increased 13-fold, and the number of affected countries has tripled. There are now more than 118,000 cases in 114 countries, and 4,291 people have lost their lives. Thousands more are fighting for their lives in hospitals. WHO has been assessing this outbreak around the clock and we are deeply concerned both by the alarming levels of spread and severity, and by the alarming levels of inaction. We have therefore made the assessment that COVID-19 can be characterized as a pandemic. (Tedros Adhanom Ghebreyesus, WHO director-general on March 11, 2020. (WHO, 2020))

It was March 11th, 2020, roughly 3 years before I started to write this introduction, when Tedros Adhanom, WHO (World Health Organization) director-general, went to public, with a frown on his face and a grave tone to declare COVID-19 a pandemic. By that time, we did not know the scenes we would witness worldwide in the coming months. As of March 10th, 2023,[1] roughly 3 years since the beginning of the pandemic, official numbers counted 6.8 mi deaths and 676.6 mi total cases, according to Johns Hopkins Coronavirus Resource Center (Johns Hopkins University, n.d.). Unprecedented and surreal scenes in the once busiest cities of the world became widespread in the news: city centers devoid of people, masks and face shields as part of daily attire, digital audiences in TV shows, and empty shelves in grocery stores due to a collective fear of running low on essential items. We also

[1] This was the date when the Johns Hopkins Coronavirus Resource Center stopped collecting data about the pandemic. It is worthy remember the pandemic was officially ended by WHO on May 5th, 2023.

C. B. Moura (✉)
Simon Fraser University, Burnaby, BC, Canada
e-mail: cb_moura@sfu.ca

C. B. Moura (ed.), *A Sociopolitical Turn in Science Education*, Contemporary Trends and Issues in Science Education 62,
https://doi.org/10.1007/978-3-031-78586-3_1

witnessed scenes of despair in several countries around the world with corpses piling on Equatorian streets, collective burial and overture of improvised cemeteries in Brazil, and hospitals in remote cities of the Amazon region running out of oxygen.[2] The same calamity state was reported in many corners of the world, some examples being India, Italia, and Peru, the last one being the highest death toll per capita reported. Those are just examples of the collective breakthrough that the world population went through during the pandemic.

If it is true to mention the "world population" when recalling the collective suffering of those times, it is also true that the experiences of each populational group differed greatly from each other. Research (Magesh et al., 2021; Rocha et al., 2021) shows that COVID-19 hit harder racialized populations and people with low socioeconomic status in general. This was commented on in previous analyses of COVID-19 impacts in science education circles (Alsop & Bencze, 2020). More than a public health emergency, the COVID-19 pandemic must be seen as part of a broader context. As Means and Slater (2021) analyze, the emergence of coronavirus is linked to socio-historical processes of incessant resource extraction, and environmental degradation. According to them, the sense of disorientation that haunts contemporary social life is shown by the general inability to locate the emergence of the pandemic within global capitalism. This inability is an obstacle to emancipatory articulations, as Means and Slater (2021) note.

Going to the broader conjuncture, a report by Oxfam (Ahmed et al., 2022) showed that during the pandemic, the wealth of the ten richest men has doubled, while the incomes of 99% of humanity are worse off. Globalization and ecological devastation have been associated with the emergence of new infectious diseases and the reappearance of diseases previously classified as under control, leading to what has been defined as a new epidemiological transition (Platto et al., 2020). The increase in human contact with previously isolated wildlife habitats and their disturbance by anthropogenic factors have increased the risk of transmission of diseases formerly kept in environments far from humans. As boundaries (and Earth's limits) have been pushed by capitalism, such diseases emerge in the human population. In other words, the political-economic system that regulates our modes of existence is closely linked to the emergence of infectious diseases, as already reported by Jones et al. (2008). If we further analyze the differences between the global North and global South (broadly conceived) with regard to the impacts of the pandemic, a strikingly unequal scenario will be even more evident (Rezende et al., 2021; Raveendran & Bazzul, 2021).

If we look closely enough, the pandemic also placed science and science education in what I call the "eye of the storm". According to The Oxford Dictionary of Idioms, being in the "eye of the storm" can have two opposite meanings: it can mean being in the calm region, at the centre of a storm or hurricane; or, this can mean staying at the most intense part of a tumultuous situation. A quick look at

[2] See, for example: https://www.cnn.com/interactive/2021/03/world/coronavirus-pandemic-cnn photos/

different dictionaries points out the same paradox. If we understand the pandemic as a big storm threatening human life, we can understand science and science education staying at this bewildering place during the pandemic: we saw an unprecedented collective effort from scientists to produce vaccines and several studies analyzing different aspects of the pandemic and nurturing public policies to mitigate the spread and the effects of the disease. Also, the general public was struck by several aspects of the science "in-the-making" that was unfolding in front of their eyes. Every week new scientific reports, an unending back-and-forth in recommendations to stop the virus spreading, and the political entanglements of public policies as they came up in the daylight (Moura et al., 2021).

When it comes to science education, there was also some turmoil. Some journals issued special calls for papers analyzing the COVID-19 pandemic: the first published collection of papers (in English and with international reach) was part of the *Journal of Activist Science & Technology Education*. Some months before, *Science & Education* journal issued a special call for papers analyzing the pandemic and its effects, which was followed also by a call from *Cultural Studies of Science Education*. It goes beyond the scope of this introductory piece to analyze the similarities and differences between the calls and the papers that responded to the calls, both in those journals and beyond them. It can be said, though, that the pandemic caused a great impact on the field, with many papers published on this theme. But, at the same time, despite the peak of papers analyzing the pandemic during the disruption, what followed seemed to be a comeback to a business-as-usual approach in science education. This is precisely a symptom Means and Slater (2021) noted—a failure to see the entanglements between the disturbances caused by the pandemic and capitalist forces. As Bazzul (Chap. 15, this volume) notes, the pandemic, on its own, is a meaningless event—unless it changes the very forms of social life we are enmeshed in. This is exactly what this book is about: if we are careful enough to understand that the COVID-19 pandemic was more than a pandemic—or if we want to make it more than a pandemic (when explicitly addressing the sociopolitical in this pandemic)—what are the consequences for science education?

Before jumping directly into this question to introduce this book's central themes, I shall briefly present some studies that dealt with the pandemic as an object of study, highlighting those inclined to see the pandemic as an event with broader imbrications and also implications—those emphasizing the sociopolitical dimensions of science education. Next, I will present how authors who cling to the so-called sociopolitical turn have been fixing its meaning and envisioning the future of research and practice under this "umbrella" definition, and I expand this definition based on a pluriversal worldview. Finally, I will present the chapters that are part of this volume.

1.2 COVID-19 and Science Education

With the surge of COVID-19, many journals in science education launched calls or special issues to debate not only what was happening during the pandemic but to gaze forward into the consequences for science education. In what might have been one of the first open calls for papers addressing COVID-19 as a topic for science education, Erduran (2020) drew from the history of other pandemics to reflect on the COVID-19 pandemic. From that reflection and building parallels with the present, she called for a series of papers reflecting on the pandemic under HPSS lenses as well as several related topics. This call yielded part of the papers commented upon in this section. Reiss (2020), departing from his background in biology, walks the readers through a brief history of some pandemics, then shifting to some indications of how the COVID-19 pandemic could be explored as a topic in science education under diverse lenses—from the history, philosophy of science and moral philosophy, to sociology and even other disciplines such as economics and psychology. Reiss (2020) provides a good example of how integrated HPS approaches can be valuable for science education but, at the same time, calls for temperance when proposing reforms in science education when science-related issues arise.

In his paper, Reiss (2020) also suggests that the pandemic could be seen through different theoretical lenses, and could be considered, for example, a socio-scientific issue. This perspective is taken by García-Carmona (2021) as a departure point to elaborate on how Nature of Science (NOS) aspects could be explored with students using digital media, with four very illustrative examples and a general guidance on organizing activities in schools. Çalık (2021) also explores some examples and key features that would allow one to position COVID-19 as a socio-scientific issue. Going beyond that classification are Braund (2021) who classifies the COVID-19 pandemic as a wicked problem, as it involved "high complexity, uncertainty and contested social values" (p. 351)—something previously suggested by Sharma (2020)—, and also Moura et al. (2021) who understand the pandemic as an event that would defy current classifications from science-in-context programs (Bencze et al., 2020) such as SSI and others, since it involved high politically charged (and messy) debates. I will come back to this argument in this section. The pandemic also meant that scientific literacy, as defined by many, has an intrinsic limitation when it does not address the wider socio-political milieu in a critical manner, as argued by Braund (2021).

In addition to those discussions, many papers were published during the pandemic reporting the shifts that happened in educational settings due to social distancing. Full online or hybrid courses and programs, and the use of digital educational resources and tools were soaring and represented major and unsettling changes for teachers and students around the world. Gerard et al. (2022), for instance, documented the use of open educational resources to implement remote science instruction by 12 teachers in the USA, analyzing the potential of those resources in facilitating self-directed learning and promoting knowledge integration. Studies like this one, exploring diverse instructional resources and strategies in a context of remote learning abounded during the pandemic, with examples of laboratory practices in an online fashion (Gamage et al., 2020; Babinčáková & Bernard, 2020), promotion of instructional strategies such as gamification (Rincon-Flores & Santos-Guevara, 2021) and others. Still in an online setting but coming back to a perspective of analyzing claims from teachers in a perspective of their understanding of scientific claims, Saribas and Çetinkaya (2021) studied how pre-service teachers analyzed claims containing fallacies, conspiracy theories, and scientific arguments related to the COVID-19 pandemic. After an intervention with the pre-service teachers, the authors found that the general quality of their analysis of the claims increased considerably. Still talking about online environments, social media was a central piece of the pandemic, for the good and for the bad. Bichara et al. (2021), recognizing the increase in the use of social media during the pandemic, proposed an analysis of tweets from the social media Twitter ® to understand what the mentions to "scientific methods" reveal about public engagement with NOS. As expected, authors found that this was a highly politically charged debate on social media, with many interactions involving politicians and also touching a ranging of NOS aspects, with some weight on non-epistemic NOS categories.

As mentioned in the introduction, even though science education research during the pandemic went in many different directions, as shown by the brief analysis of previous paragraphs, in this book, we are interested in those studies that used sociopolitical lenses or highlighted a justice-centred landscape when analyzing the pandemic. As a first example, Moura et al. (2021) analyze the unfolding of pandemic events in Brazil and the US. They found that the strategies used to disseminate disinformation were more sophisticated than simply the use of completely wrong pieces of information: some politician leaders would misleadingly use scientific studies to convince the broader public of their positions. This would call for revising how we address such controversies in science education, notably the explicit inclusion of politics in those controversies. Waight et al. (2022), exploring the problem of trust in science during the COVID-19 pandemic (and vaccine hesitancy), start by calling attention to what was interpreted as a supposedly bigger distrust in science by communities of colour in the US, deconstructing this argument and then exposing continued racism in scientific and technological research. The authors propose a closer examination of the nature of science and technology/knowledge-making processes in relation to social justice and equity issues. In the same critical vein, Raveendran and Bazzul (2021) take up questions of (bio)power and subjectivity in socialized health systems in a historical outlook, focusing on the Indian context and complicating the conversation around vaccination and social control/resistance.

The papers in the special issue of the Journal of Activist Science and Technology Education are an example of addressing the complexities of the pandemic under a sociopolitical lens. Alsop and Bencze (2020) set the tone by highlighting the social consequences of the pandemic, bringing data and analyses that go beyond the walls erected by traditional communities of science education: concerns around care, marginalized populations, the relationships with hypercapitalism and so on. Those concerns and analyses are joined by the authors participating in the issue: Dillon and Avraamidou (2020) turn their attention to the science education field and its scholarly professional associations to ask seven questions enticing the community to act in the face of the problems posed by the pandemic. Sharma (2020) and Carter (2020) focus on neoliberal ideologies in the current climate and point to recent reforms in science and technological education that can serve as starting points for a more profound engagement with sociopolitical issues in science education—what is likewise highlighted by Bencze (2020) when exploring STEPWISE framework as a response to sociopolitical challenges imposed by the pandemic.

Those are, of course, only examples of some initiatives that serve as inspiration and platform for this book and examples of a sociopolitical look into the pandemic (and beyond). Even though those are good examples, the question of what the so-called sociopolitical turn entails still remains. I turn to this question in the next section.

1.3 The Sociopolitical Turn in Science Education: Where Did This Idea Come from, and Where Can it Drive Us?

It might sound wearing or even pushy for some to hear about a new turn in (science) education. I remember when I was presenting a seminar, and at a certain point, I started to talk about the sociopolitical turn in science education. In the end, someone in the audience asked very politely but also with a tired glance: "what was made of the discursive, sociocultural and other turns we had in science education?" Their intention was (from what I captured in the conversation that followed) to ask whether we need a new turn in science education—and if so, why? What would one gain from pushing the field towards this direction? How would we do that, and what would be the fundamental differences from previous movements in our field? I probably will not be able to answer all those questions, but I understand their concerns. I consider, however that, before deciding whether this is a firm path forward, one should better understand what is understood by sociopolitical turn—for that, my task here is to review briefly what the field has been calling "the sociopolitical turn" in science education. This is not a comprehensive review of initiatives that could be located in the realm of sociopolitical perspectives in science education—for a broader review, I suggest the first part of Valladares' paper (Chap. 6, this volume)—which I believe is lacking in the current literature. Rather, it is an attempt to come up with a definition based on previous work dealing with this theme.

Among recent studies explicitly mentioning the sociopolitical in science education, Gandolfi (2023) sharply identifies the task associated with such a turn as a move that goes beyond recognizing cultural diversity, aiming at deconstructing and transforming science education against sociocultural homogenization and oppressions, a vision that is shared by Valladares (2021). Gandolfi (2023) remarks that the sociopolitical turn would entail a rethink of taken-for-granted notions like the relationship between science and society, the idea of science education for citizenship, and concepts such as "scientific literacy" and "socio-scientific issues." In this book, there are several examples of rethinking such notions, and many of the contributors to this book have done this work elsewhere for a while now (e.g. Bazzul, 2015; Sjöström & Eilks, 2018). Another example of authors who explored those ideas, Tolbert and Bazzul (2017) understood such a turn as imminent in science education, as social forces (for example, environmental in/justice, contradictions from capitalism, racism, etc) push us (or should push us) to rethink what we feel is important in science education. It means that science education should be attentive to the cultural milieu in which its practices and activities are being performed. As such, current contexts of different kinds of oppression, including the effects of colonialism and patriarchy, should be part of the conversation in science education. For them, the core of "taking a more sociopolitical focus is that the systemic transformation of the social world in the name of equality become the very stakes of research and practice" (p. 326). The authors conclude by proposing some questions with practical implications—asking for microscale actions—for daily life in schools and the very science education community if we want to embrace a sociopolitical turn. Noteworthy in this paper is that they depart from Gutiérrez's (2013) definition of a sociopolitical turn in mathematics education. She describes such a turn in a detailed way:

> I use the term sociopolitical turn to reference a growing body of researchers and practitioners who seek to foreground the political and to engage in the tensions that surround that work. The sociopolitical turn signals the shift in theoretical perspectives that see knowledge, power, and identity as interwoven and arising from (and constituted within) social discourses. Adopting such a stance means uncovering the taken-for-granted rules and ways of operating that privilege some individuals and exclude others. Those who have taken the sociopolitical turn seek not just to better understand mathematics education in all of its social forms but to transform mathematics education in ways that privilege more socially just practices. (p. 40)

Among the movements inside mathematics education that would cling to the sociopolitical turn, Gutierrez mentions streams of critical mathematics education, critical race theory and post-structuralist perspectives for mathematics education. According to the author, those movements foreground questions of identity, power, subjectivity and agency in ways that challenge one's understanding of what mathematics is and the positioning of this field in social imageries. She notes that it is an emergent discussion (or at least it was 10 years ago) and analyses the challenges and a set of questions and ideas that are under-investigated, related to marginalized learners (beyond achievement gap framings), values and politics involved in knowledge making in mathematics, and a myriad of other questions grasping the role of identities and power both in mathematics and in mathematics education (Gutiérrez, 2013), which was a set of perspectives already emphasized in different ways by Valero (2004), almost 10 years earlier. It seems clear, then, that the sociopolitical perspectives have been on the table for a while now in mathematics education. According to those advocates, the questions raised by the sociopolitical turn should be considered "side by side" with the cognitive and psychological perspectives that would arguably be at the roots of mathematics education as a field (Valero, 2004; Pais & Valero, 2012). Another aspect that is considered by Valero (2004) and more prominently by Pais and Valero (2012) is that the current focus (or overemphasis) on teaching and learning processes and their contexts in the sociopolitical turn might be, paradoxically, de-politicizing research in mathematics education as they would not consider issues of equity, social justice, economy and politics in their broader complexity. Instead of politicizing mathematics education in the framing that seems to be foregrounded by the field, maybe considering mathematics education within the Political (with capital P) would be a path for not oversimplifying the larger struggles (Pais & Valero, 2012).

Even though Tolbert and Bazzul (2017) draw from the sociopolitical turn in mathematics education to defend a similar shift in science education, it is important to recognize earlier efforts to produce this movement inside our field. Roth and Desautels (2002) organized an edited volume that they described as an attempt to break with the reminiscences of scientism in science education, represented by the obsession for teaching the "science of scientists" instead of a context-sensitive version of science education that would take into account indigenous forms of knowing, common sense and vernacular knowledge. They observe that the alienation of school science from (political) problems of students' realities tends to conceal public participation in scientific and technological decisions. Their objective, then, is to reconnect those dimensions—and the featured chapters go in this direction. Those ideas are developed further by Roth and Calabrese Barton (2004) 2 years later, where the authors take on the perspectives of agency and identity as central to scientific literacy.

When tracing the road toward the sociopolitical turn, we can go even further back in time and find Hodson's call in 1994 to politicize science education (Hodson, 1994). Hodson rejects technological determinism and calls for the effective participation of the public in decisions around science and technology, defending that it is essential students understand the co-production of science and society, the link between scientific and technological decisions and the (unequal) distribution of wealth and power, then develop their own worldviews and prepare to take action. He reaffirms this in subsequent articles (Hodson, 1999, 2003, 2010), where he extends his theorization. Hodson's call was heard by many scholars who embraced activist (and action-oriented, more generally) pedagogies in science education. Alsop's and Bencze (2014) edited collection serves as an example of this orientation in science education, not to mention the Journal of Activist Science and Technology Education, launched in 2009, and edited by the same authors, and the STEPWISE (Science and Technology Education Promoting Wellbeing for Individuals, Societies and Environments) edited book (Bencze, 2017).

There are multiple initiatives in this regard, and, again, I am not reviewing all of them here. This task would entail, in the first place, the disentanglement of the multiple grammars and vocabularies that are used when referring to sociopolitical perspectives in science education, including "science education for social justice", "justice-centred science education", and its variants with ecojustice, anti-capitalist, and so on. Another aspect that would demand to be better unravelled would be the relationships with growing perspectives such as movements pro-decolonization and indigenization in (science) education, public pedagogies (Torres-Olave & Guerrero,

Chap. 5, this volume), pro-communitarian projects, and beyond. The importance of unpacking those movements is invaluable in that it can help in avoiding the encapsulation/jargonization of science education in sometimes hyper-coded subcommunities that serve to hide (or simply disregard) political interests and programs, as Bazzul (2014) cautioned before. Equally important is to identify also those uses of such jargons without a true political commitment or in a way that, even though it shows sympathy with the idea that science education would be intertwined with political issues, would not engage with the political in science education research and practice—that is, make it central the questions related to power and identity. This formulation is borrowed from Valero (2018), who makes the distinction between weak and strong political perspectives in mathematics education. Simonneaux (2014) also suggested a similar duopole for SSI, classified as "cold" or "hot" depending on some factors such as the engagement with problems of real world and politicization. Building on this viewpoint, one can talk about weak and strong perspectives in sociopolitical approaches to science education.

As for weak sociopolitical perspectives, a central aspect would be the adoption of an acritical stance towards scientific production. For instance, there has been a proliferation, in recent years, of a literature stream that advocates for widening the participation of marginalized groups in our society, including BIPOC (Black, Indigenous, People of Colour), women and so on, in the so-called STEM (Science, Technology, Engineering and Mathematics) careers. It is easily noticed several problems stemming from the exclusion of people from diverse backgrounds in the making of science: from examples of racial bias in technology—examples ranging from soap dispensers (Waight et al., 2022), and automated speech recognition (Koenecke et al., 2020)—, to distortions in the research on contraceptive methods (Oreskes, 2019), the cases abound in which science could be benefitted by the participation of a broader spectrum of the population. However, when left untouched, the discussions on knowledge production itself (and who benefits from it nowadays) may contribute to reproducing the same inequalities we witness. One can change the gender imbalance of engineers in a given company, but if the extractive practices in mines of this company in the global South stay the same, the marginalization based on gender, class and race will persist in those countries, as well as the aggression to the more-than-human critters that is meant by those activities. This is not to say that science education alone can change those practices—and here I recall Pais and Valero (2012) analysis on the depoliticization of politics when one takes this path –, but it can, at the very least, spark discussions that will create awareness in future scientists and non-scientists, which may, in turn, flip the current unbalanced power relationships.

A second example of weak sociopolitical perspectives in science education is the tendency to create amalgams of different teaching-learning strategies (or philosophical orientations) rendered as justice-centred ones by research, assembling diverse perspectives that sometimes are at odds with each other. For example, is it possible to criticize the systems of power that create hyper-surveillance over teachers' work and make the production of scientific teaching-learning sequences the core of one's scholarship? Or defending the need to contextualize science teaching in line with critical pedagogies, and at the same time, mostly disregard students' perspectives to favour canonical conceptual contents, bringing to classrooms ready-made "contextualizations" that diminish the role of a collective construction between students and teachers? All of those cases have in common sympathy for some aspects of sociopolitical perspectives in science education but fail to engage in a deeper analysis of the ways in which scientific knowledge production relates to power and identity, aiming at producing a different future.

On the other hand, strong sociopolitical perspectives in science education seek to analyze knowledge production historically and contemporarily, scrutinizing its relationships with other social institutions. It assumes that science is closely tied to the ways society organizes itself. For instance, the close relationship between science and capitalist global system of production was the object of diligent analyses by several historians of science (Rieppel et al., 2018) and also many STS scholars, both historically and in contemporaneity. Questions such as biopiracy, toxic effects of chemical industry, and the manipulation of public opinion by big corporations have brought to the fore several aspects of this relationship, including, for example, how capital might have shaped research priorities of scientists (idem). In science education, several scholars have successfully dissected those relationships and found ways of addressing them with students. One example is the case of toxic dust in Québec City, Canada. Bencze and Pouliot (2016), drawing from a broad range of sources and using actor-network theory, were able to map out a complex network of relationships, including several stakeholders and a number of human, social, institutional and non-human actors, to explain a controversy around toxic metal dust polluting the region and affecting the environment. From a historical perspective, Cardinot et al. (2023) explore with high school students how a funding agency created in the context of a rise in research of nuclear physics in Brazil showcased the complex relationships between science and institutional politics (both the government and army forces) in defining research priorities in that context.

Strong sociopolitical perspectives in science education also value engagement with real-world problems and contexts, as well as activist initiatives that foreground students' engagement in sociopolitical action in their communities. For example, Morales-Doyle (2017) reports a study in a chemistry class that supported students to succeed academically while taking up urgent issues of social and environmental justice identified by their communities, nurturing a deeper relationship with their communities and positioning the youth as public intellectuals. Marques and Reis (2017) developed a project with students 12–13 years old that entailed producing and spreading digital videos to denounce environmental pollution in a given community, experiencing an activist initiative, as defended by Hodson (2003, 2010). Strong perspectives in the sociopolitical turn of science education recognize the plurality of worldviews and ways of knowing and being, valuing non-hegemonic cosmogonies and respecting other ways of being in the world, as well as learning from them. Authors such as Aikenhead and Elliot (2010) emphasize the importance of not only recognizing different ways of knowing and being in the world but also integrating those worldviews into school science, benefitting indigenous and non-indigenous students. Examples of this successful integration in science education are explored in works such as Seehawer (2018), and Richards et al. (2008), among others. In the same vein, anti-racist approaches, and intersectional feminist approaches to science education are growing perspectives in the field, even though some of those do not challenge the status quo in some cases, when they do not question systems of scientific production and neoliberal features lurking in the scientific and educational systems in the most diverse ways, as Carter (2014) analyzes.

Of course, those two poles—weak and strong—are separated from each other by countless shades of gray, and the purpose of such classification is not to hierarchize and diminish efforts to embrace socio-political perspectives. Rather, it is a call for a more radical engagement with such a turn in science education, going, as much as possible, to the roots of our unequal societies and avoiding the adoption of it purely as a jargon, or as a way to "rebrand" perspectives that (intend to) leave power structures that gear inequalities untouched. For this, we should be attentive to the extent our (collective) practices are reflecting our words; as Freire (2000, p. 87) emphasizes, "there is no true word that is not at the same time a praxis. Thus, to speak a true word is to transform the world." In a certain sense, those considerations respond (at least partially) to the questions that open this section: betting on a sociopolitical turn in science education, would not mean replacing progressist movements in science education that are already taking place, such as the decolonial perspectives (dos Santos & Lima Junior, 2023), the surge of feminist and gender perspectives in science education (Hussénius, 2014), among others. Those are not competing paradigms. Avoiding lenses that highlight competitiveness and binarism of all sorts, I see the sociopolitical turn as a wave pushing and constructively interfering (to borrow a metaphor from physics) with those other perspectives which aim to promote a science education capable of articulating with several forces that seek to surmount the barbarism of a system that oppresses and destroys living and non-living entities in this world. If we want to achieve that aim, it is needed more than what the linguistic and socio-cultural turns were committed to when they were proposed.

1.4 Allowing Multiple Worlds in the Sociopolitical Turn: A Bridge Too Far?

It is clear that the pillars of a socio-political turn in science education would be around the issues related to power and identity related to knowledge production. However, our way of imagining politics is generally related to the social world we, humans of a particular place and time (or following those from a particular place and time), created. Institutions, collective spaces and practices, language, symbols and hierarchies. Those ways of making meaning of the world (and living with(in) the world) are not homogeneous for every earthly dweller. In line with this, Stengers (2005) proposed the idea of *cosmopolitics* as a way to acknowledge and work within fractured, multiple and irreducible worlds, both at the epistemological and ontological levels. As Law (2015) noted, this goes back to a very real problem of how we can live in a common world co-existing with different people and entities living in their own worlds. Or, in his own words, "[...] the differences are *not* simply matters of belief. They are also a *matter of reals*. What the world *is*, is also at stake. Our orderings (for instance, our propensity to distinguish between nature or the natural on the one hand, and culture or the social on the other) do not carry to other places." (p. 127, emphases in the original). This problem has occupied not only educators (and science educators), of course, but also (and mainly) philosophers, and anthropologists—and it is brought more evidently by some contributors to this volume, see, for instance, Nascimento et al. (Chap. 17, this volume) and Robles-Piñeros and Abella (Chap. 11, this volume)—however, it is fundamental for us engaged in building future worlds through education.

For this reason, I want to suggest here that a sociopolitical turn in science education would be incomplete if it does not engage with ontologies,[3] beyond the issues of power and identity related to knowledge production. That is to say that it is worth remembering that maybe the ultimate task of (science) education is building future worlds and the possibility of citizens to build their worlds/worldings.[4] In this vein, it becomes vital to understand that other possible worlds exist and that the key for us to find balance in our existence on this planet goes through the task of learning about this multiplicity and how science can be in service of perpetuating this multiplicity. If we take this multiplicity of worldings as seriously as Viveiros de Castro (2015) and his Amerindian perspectivism, every species that ceases to exist (in the sixth mass extinction we are living) is a world that falls. This has straightforward consequences, such as the need to protect biodiversity, not because we should be good stewards of the environment—as if nature were our warehouse, to borrow Krenak's (2022) words—, but because we lose the multiplicity of perspectives we could learn from when we fail to protect biodiversity. This would not be the only consequence, though.

One of the tasks that comes as a consequence of the engagement with ontologies is unlearning things. One has to unlearn lessons that are deeply entrenched into our worldviews when it comes to science; for example: nature is "out there" to be discovered and explained through models and theories; evolution (or science) is driven by competition; there are some fundamental kinds that compose matter; time is linear, etc. The point I want to make here is that we have to unlearn absolutist beliefs about ontological assumptions in science to be able to construct different worldings and be invitational about diversity. This has a major consequence in our daily work

[3] Ontology in the philosophy of science is a controversial and multifaceted theme, but those controversies are beyond the point I want to make here. I adopt a view of scientific ontology as interwoven with epistemology, as argued by Chakravartty (2017).

[4] Here I stay with the ambiguity between world and worldings. For neo-materialist theorists, the word worldings tries to overcome the duality between subject and environment. In this way, the world comes into being whenever a subject interacts with the several expressions of phenomena that are legible for the subject—to this unending process we call worlding(s). I will use worlding here when I feel this expression of enmeshment between subject and environment would be a better call to express what I am trying to convey.

in classrooms (and in informal settings). For example, we often affirm scientific knowledge (and the realities this entails) in opposition to other ways of knowing (and their worlds), as if only one ontological would be possible and plausible, like if rendering the Higgs boson one of the matter constituents would automatically disprove alternative views (with similar or dissimilar—does not matter—explanatory power) from, for instance, the Yanomami people, who inhabit Amazon region. This does not put into question the explanatory power and amazing things modern science can build. However, we should be conscious of the chain of relationships that are embedded in our conceptualizations and worldviews. Those conceptualizations and worldviews are tied to specific (mostly Eurocentric) ways of engaging with the world outside us, which are different from the ways several different peoples around the world engage with their worlds. From that, it follows that ontologics do not equate to worldviews when it comes to science education, but can start from that idea—the ways we see reality are pivotal to our worldings.

The theme of ontologies in a sociopolitical key has been something present in the works of many colleagues in the field, and I will mention some. Bazzul and Kayumova (2016), for example, use Deleuze and Guattari's theory of assemblages to map out current ontological spaces and envision new ones for science education research and practice. Sharma (2020) advocates science education should abandon "substantive ontologies" in favour of "post-human relational ontologies", also departing from the notion of assemblages, in particular, understanding the COVID-19 pandemic as an "epidemic assemblage." In a third example, joined by the aesthetics domain, Tolbert and Bazzul (2020), bringing diverse STS scholars to the conversation, propose that science education should make (aesthetically) visible with "whatever is necessary" to enlarge our political and ontological senses of kinship, pursuing multi-species justice. Many other examples of engagement with ontologies in science education abound in the literature (e.g. Kayumova & Dou, 2022; Verma et al., 2022; etc.), and my task here was not to review the different ways in which those scholars have engaged with the theme but suggest the importance of ontological debate in the sociopolitical turn in science education. Now, I turn to more general considerations about this turn.

1.5 A Program for a Sociopolitical Turn in Science Education: A Definition, Three Dimensions and Two Levels

After critically reviewing the meanings literature has attributed to the sociopolitical turn in science education and adding some insights to the review, it becomes productive to synthesize some ideas that could be fruitful to science education research and practice. First, as a way of a definition, the sociopolitical turn refers to a body of research in science education that has been engaging with political aspects of science teaching and learning and the discussion of science within the Political, which means exploring the different ways in which science can relate or be positioned in relation to larger political discussions of our time, like the limits of growth, social inequalities, and historical issues such as racism, sexism, and colonialism in all forms. For this, authors in the sociopolitical turn have been working on dissecting the relationships between knowledge production, power, identities and ontologies to overcome current imbalances and oppressions that plague our society and are driving the sixth mass extinction on Earth. The work done by those authors entails rethinking pedagogical practices and research directions in science education, as well as cornerstone concepts and theoretical foundations in which this field lies, with the intent of contributing to the transformation of our society through science education, sometimes reimagining its possibilities.

Thinking of this tentative definition as a departure point that tries to depict what has been done by this field rather than a definitive and monolithic picture of it, I imagine here three dimensions that would be key for the work already done and also to be pursued by those engaged in this venture. The three dimensions would be those aspects to be unveiled and/or reimagined in science education research and practice, as briefly stated below:

- *Power*: understanding and challenging how power creates regimes of truth, norms, hierarchization and oppression—how power plays out in science education research and practices.
- *Identities*: understanding and challenging which identities and "kinds of people" are forged by science education research and practices. Which identities are (historically) welcomed in scientific spaces and why (which are not, and why)?
- *Ontologies*: understanding and challenging the worldings that are created/imagined as possibilities through science education research and practices, and those competing worldings that are marginalized or not even considered by totalizing and absolutist views inside the field.

Those three dimensions overlap and could be shaped in different ways should they depart from diverse points of view and theoretical framings—and those are aspects I did not explore in this chapter. However, following previous work from colleagues in this field and elsewhere (Tolbert & Bazzul, 2017; Gutiérrez, 2013; Valero, 2004), those seemed a set of productive (and clear) ideas to work with in order to foreground a sociopolitical perspective in science education. As for the levels, I keep referring to science education practices and research, which denote precisely the two levels I can see those three dimensions working out, in practice.

- *Level of pedagogical practices (reflexive)*: In this level, one would analyze questions related to power, identities and ontologies in current practices in schools, higher education and informal education, proposing their transformation aiming at justice-centred practices and goals. How the curriculum, school actors and mediators' actions and practices, discourse in its various forms, environments and policies are enmeshed with questions of power and identity as well as are committed to certain ontological views over others.

- *Level of research (meta-reflexive)*: In this level, one would investigate how research in science education has engaged issues related to power, identity and ontologies, as well as how research communities themselves produce power and identity relations and have ontologies driving their research programs. What are the preferred themes of study, the groups that are formed, and their influence over the field?

The first level is something that Gutiérrez (2013) emphasized in mathematics education, with a focus on teaching and learning processes and how they could be studied and transformed through the lenses of a sociopolitical turn. It is worth remembering that some of those issues were already addressed by the sociocultural perspectives in science education (Lemke, 2001), in a way that science classrooms are hardly depicted by researchers nowadays as places "purely for science instruction" without the involvement of many of the factors the sociopolitical perspectives pay close attention (gender relations; discourse and power; learning as a process of becoming that involves identity work, and so on). However, sociopolitical perspectives are not satisfied with analyzing those relations but engage vividly in overcoming oppressive structures, representing a major shift compared to sociocultural research in science education. The second level attends to the call from Pais and Valero (2012) about the need for more *research on research* to grasp the directions of the research field and not to "depoliticize the politics," in their words. This is of utmost importance if we consider that science and science education sometimes are not at the center of decisive debates capable of changing our life conditions and overcoming the huge contradictions of our society. At the same time, science education sometimes seems to take a divergent path in relation to educational research when it comes to addressing inequalities of all sorts. Methodological choices derived from positivism, the straightforward (and sometimes neophile) adherence to policies, the focus on building "optimal" resources and many other aspects are evidence of that. I see Pais and Valero's (2012) concerns, then, in two ways: avoiding the overemphasis of research on teaching and learning processes as paths to the politicization of science education and seeking connections with the world of educational research outside of our tiny field of science education. In my view, individual studies are not able to weigh those emphases, but the research on research may grasp current foci of the field, and then we could be able to balance identified tendencies collectively.

This brief introduction has no purpose of fixing a definitive image of what a sociopolitical turn would entail. It is, rather, an invitation for those who find relevant the matters I outlined here to join this movement from whatever point of view they find more productive—or even oppose the points briefly sketched here should one judge them too narrow or shallow. In a recent debate at a conference, when dialoguing about the "shape" a certain research program in science education should have, a colleague remarked that maybe the more fundamental question instead of the "final" shape, as a fixed image of a research program, the important thing would be where we are moving towards, or what moves us in the pursuit of justice. May we recognize, in difference, that a concerted movement towards a sociopolitical turn in science education is long overdue. May we embrace the turbulent image of the eye of a storm, in which we recognize it is time to move together with and beyond the world around us, that is fallen over our heads (Kopenawa & Albert, 2013). We are now just a few, inhabiting the margins of a tiny field, but we are brave, ready to create new worlds in science education capable of forging eco-just futures.

1.6 Presenting the Chapters

This book is organized into three sections featuring authors from 13 countries and multiple perspectives, but one common commitment: building better collective futures through science education. Even though the three sections have some coherence in themselves, they are better understood holistically, all advancing the core perspectives of this book. This means that even though the sections have each one a specific purpose (deepening central concepts of the field, analyzing local realities and imagining the future of the field), all the chapters attempt to do a bit of each of those tasks. By analyzing local realities, chapters in the second section are also creatively imagining the future of the field. By examining central concepts of science education, chapters in the first section are also bringing local examples to uphold their theorization, and in the third section, chapters are examining central concepts and sometimes delving into particular realities to propose the future of the field. In that way, there is no preferential way for reading the book, nor the interpretations of each chapter are limited to the ones I gave while organizing this volume. That said, I shall present the chapters in the following paragraphs in the order they appear in the book.

The first section features essays addressing some of the central concepts and ideas of science education, such as scientific literacy, socioscientific issues and nature of science, and introducing new ones, standing as examples the ideas of slow violence and critical imagination in science education. Opening the first, Jesper Sjöstrom, drawing from the northern European/Scandinavian tradition *didaktik*, based on the German concept of *Bildung* develops an integrative *didaktik* model for complex and socio-political socioscientific issues in education. After a fulsome review of current perspectives in science education, mostly those oriented towards the so-called Vision-III of scientific literacy, and a discussion on the imperatives of teaching sciences in the Anthropocene, the author develops a model aimed at guiding educators in addressing wicked problems in the current scenario of rampant wicked problems with direct consequence for science teaching at various levels. In the third chapter, Ajay Sharma and Ramnarayan Kalyanaraman contrast the COVID-19 pandemic, classified as an episode of "quick violence," to another class of violence, the slow violence. They argue that persistent degradation of socioecological systems are a threat to communities around the world and remain still understudied and overlooked in science education scholarship and in school science more broadly. Hence, shedding light on slow violence would be a way of preparing students and communities to build more just post-pandemic worlds.

The Chap. 4 presents an essay by Haira Gandolfi that advocates for a more politicized perspective on Nature of Science (NOS), considering the history of colonization and over-exploitation between currently known global North and global South. Drawing from decolonial studies and critical pedagogies, and building on examples from the history of science, she develops both a theoretical approach for a politicized and decolonial NOS, as well as provide readers with practical aspects for implementing such an approach in school science teaching. In the last chapter of the first section, Chap. 5, Betzabé Torres-Olave and Gonzalo Guerrero depart from the analysis of an environmental conflict in Chile to develop two central ideas they consider key for post-pandemic worlds in science education: critical imagination and conscientization. With a final aim of making scientific socio-political issues more pedagogical and palatable for teaching contexts, they explore voices from park rangers and glaciers and what could be learnt from those voices, understanding science education as a collective task in which public science educators (humans and non-humans) play an important role.

The second section reunite chapters addressing local perspectives, case studies, and particular experiences providing science education with a rich collection of perspectives for post-pandemic worlds. The section starts with the study by Liliana Valladares, who, departing from a comprehensive review of the nature of scientific literacy and the sociopolitical turn, analyzes the Mexican educational policy called "New Mexican School" in what it aligns and misaligns with the premises of a sociopolitical turn in science education, crafting a thoughtful critique of the document and great reflection on the topics of this book. Matthew Weinstein and Alysa Schafer in their Chap. 7 explore an overlooked aspect in science teaching and learning during/post-COVID-19, which is the impact of the pandemic on LGBTQIA+ youth and their community. Through a series of narratives depicting transphobia and hate speech, they analyze the role of science lessons in reinforcing normative assumptions on gender and sexuality and propose curriculum work that would defy those narrow and harmful assumptions for post-pandemic futures. Ana Maria Navas Iannini and Erminia Pedretti tackle non-formal educational settings' responses to the pandemic. Through the analysis of online initiatives of three institutions in Brazil, the US and Canada during the pandemic, trying to understand how those institutions enacted several facets of the so-called fourth generation science museums and how those emerging roles and ways of doing might reinvent museums' practices gazing forward.

Flavia Rezende, Andreia Guerra and Fernanda Ostermann depart from observations of current political scenario in Brazil in how it weaves to the pandemic context and beyond to analyze the trends of science education scholarship in Brazil and how it tackles the idea of politics and politicization in a discursive analysis. The authors split the analysis into three main periods and make a detailed analysis finding "low concern of Brazilian science education scholarship with political and social aspects of reality" (p. 162), even in a very convoluted political scenario in recent years—which is also found elsewhere as pointed by research coming from other countries, in their analysis. In the sequence, Ana Dumrauf, Silvina Cordero, Verónica Becerro, Luciano Iribarren and Raúl Ithuralde share a journey of actions, experiences and reflections during and after the COVID-19 as they offered two editions of a Continuous Teacher Training program for primary teachers in Argentina. Grounded in a decolonial approach with a strong critique of Eurocentrism from a Latin-American angle, the authors present reflections based on the course that involved science and health education, questioning ways of knowing, feeling and living. From that conversation, they point out a set of recommendations for those seeking to implement decolonizing practices in science and health education. In the eleventh chapter, Jairo Robles-Piñeros and Susana Abella go in a similar direction to the Argentinian team, but focusing on theoretical considerations for intercultural science education in Latin America. Based on philosophical and theoretical-methodological foundations, the authors discuss the role of Intercultural Science Education (ISE) in the initial and in-service training of science teachers, from a Latin American perspective and aiming at furthering a socio-political agenda for science education in Latin America.

Now to Central America, Joanne Nazir explores how the government and citizens of Trinidad and Tobago reacted to the COVID-19 pandemic and how this relates to ongoing discussions around scientific literacy. From that discussion, Nazir emphasizes the need for explicit teaching of nature of science and science-in-context approaches to counter naïve/uninformed positioning in public debate, illustrating the discussion with concrete examples of the Caribbean country. Closing this section, Sara Tolbert and Mahdis Azarmandi bring voices from New Zealand Aotearoa to question, from a feminist standpoint, disembodying and desensitizing practices of science. Departing from students' narratives and embedded with pluriversal ontologies and worldmaking practices, they reflect on the role of the sacred in science education to post-pandemic worlds.

In the last part of the book, titled "Imagining justice-centred futures for post-pandemic science education", there are more experimental chapters, trying to analyze but also reinvent science education with diverse lenses and perspectives. Opening this section, Larry Bencze, Lyn Carter, and Isabel Martins bring their voices to a conversation around some pivotal aspects—namely, political and economical ones—of the pandemic. The chapter starts with a review by Lyn Carter on the polycrises associated with global geopolitical struggles in and beyond the pandemic; then, Isabel Martins goes local, analyzing social injustice mechanisms playing out in Brazilian context, which were in the spotlight during the pandemic due to poor management by the far-right government; finally, Larry Bencze reports on ongoing work to prepare secondary school students for critical and activist civic engagements in contexts of 'progressive' neoliberalism, wrapping up in a narrative that emphasizes the need for a sociopolitical turn in science education. In the fifteenth chapter, Jesse Bazzul analyzes the interplay between power, knowledge and subjectivity during the COVID-19 pandemic, extrapolating this analysis to other realities, that existed before the pandemic and persist after it. In order to be relevant to current contexts of precarity and threats to human and non-human life, Bazzul renews previous calls for teachers to engage with questions of collective existence, along with the questions of politics and subjectivity associated with the former.

Marc Higgins and Anastasia Sanchez turn to the engagement with affect in science education to put forth their analysis of pandemic and post-pandemic science education in a socio-political key. For them, our current catastrophic times need to be *felt* otherwise: feeling like a scientist upholds dominant *knowings* and marginalizes potential alternatives. Proposing an alternative to this science-education-as-usual, the authors present heart-centred orientations to science education, with six figurations of the heart: arrested, angry, wavering, grieving, hopeful, and dreaming. With those, they imagine futures yet-to-come, in which may be a world of many worlds. Matheus Nascimento, Bruno Ferreira, and I elaborate on what a multinatural science education would look like. Departing from a recap of some perspectives on the debate about multiculturalism in science education and grounded on the Amerindian perspectivism as well as indigenous ways of knowing in South America, we describe initial ideas on a multinatural perspective in science education, approximating it to some works on the field. Thiago Ranniery, in the eighteenth chapter, proposes a provocative thought experiment focused on the compared analysis of the work of Brazilian indigenous artist Jaider Esbell and science visualization researchers and visual artists Evan Ingersoll and Gael McGill to suggest science education as science *poethics*. With that, he intends to defy current Eurocentric pillars that sustain narrow understandings of science education, based on a radical division between representation and reality and between nature and culture.

The nineteenth chapter, written by Nathan Willig Lima and Isabel Martins introduces elements from afro-diasporic cosmovisions to introduce a radical cosmopolitical and decolonial perspective in science education. They justify this innovation based on the dilemmas, ruptures, and contradictions faced by society and science education. The authors propose a metaphoric encounter between Exu, an orixá (deity) from afrodiasporic mythologies and Einstein, as the representation of a science expert. Exu, in the tradition evoked by Lima and Martins, is the herald orixá, a forerunner, and his story is full of contradictions that always point to the impossible or absurd, but also to the creativity, relationality and plurality. In a way, the principles represented by Exu are instantiated by this book: we created an *encruzilhada* (see details in Lima and Martins's Chap. 19) where different worldviews come together not to create a bland synthesis, but to celebrate the difference, to make the worlds clash, to create a version of what was suggested by the title: multiple worlds in a world. In Umbanda religion (in which Exu is worshiped), any ritual starts with an offering to Exu, who is the master of communication and guarantees the connection between both worlds that encounter during the ritual. Exu opens the paths. Choosing this chapter as the last (or the first, operating the other way around, as Exu proposes) one in the book is not by chance: I hope Exu opens the paths of this book and that the message carefully crafted by authors, reviewers and everyone else involved in the production of this can reach scholars around the world and create new possibilities for science education—more just, politically engaged and that respects difference as a principle of co-existence. A world where genocides and ethnical cleansing are absolutely unacceptable and in which we are in a more sustainable planet for humans, non-humans, sentient and non-sentient beings. A world where many worlds/worldings can fit together.

References

Ahmed, N., Marriott, A., Dabi, N., Lowthers, M., Lawson, M., & Mugehera, L. (2022). *Inequality kills: The unparalleled action needed to combat unprecedented inequality in the wake of COVID-19*. Oxfam GB.

Aikenhead, G., & Elliot, D. (2010). An emerging decolonizing science education in Canada. *Canadian Journal of Science, Mathematics and Technology Education, 10*(4), 321–338.

Alsop, S., & Bencze, L. (2014). *Activist science and technology education*. Springer.

Alsop, S., & Bencze, L. (2020). Reimagining science and technology education in the COVID-19 portal. *Journal for Activist Science and Technology Education, 11*(2), i–vii.

Babinčáková, M., & Bernard, P. (2020). Online experimentation during COVID-19 secondary school closures: Teaching methods and student perceptions. *Journal of Chemical Education, 97*(9), 3295–3300.

Bazzul, J. (2014). Science education as a site for biopolitical engagement and the reworking of subjectivities: Theoretical considerations and possibilities for research. In L. Bencze & S. Alsop (Eds.), *Activist science and technology education* (pp. 37–53). Springer.

Bazzul, J. (2015). Towards a politicized notion of citizenship for science education: Engaging the social through dissensus. *Canadian Journal of Science, Mathematics and Technology Education, 15*, 221–233.

Bazzul, J., & Kayumova, S. (2016). Toward a social ontology for science education: Introducing Deleuze and Guattari's assemblages. *Educational Philosophy and Theory, 48*(3), 284–299.

Bencze, L. (Ed.). (2017). *Science and technology education promoting wellbeing for individuals, societies and environments: STEPWISE*. Springer.

Bencze, L. (2020). Post-pandemic Science & Technology Education: An opportune and necessary ideological battle. *Journal for Activist Science and Technology Education, 11*(2), 33–54.

Bencze, L., & Pouliot, C. (2016). Battle of the bands: Toxic dust, active citizenship and science education. *Journal for Activist Science and Technology Education, 7*(1), 1–21.

Bencze, L., Pouliot, C., Pedretti, E., Simonneaux, L., Simonneaux, J., & Zeidler, D. (2020). SAQ, SSI and STSE education: Defending and extending "science-in-context". *Cultural Studies of Science Education, 15*, 825–851.

Bichara, D. B., Dagher, Z. R., & Fang, H. (2021). What do COVID-19 tweets reveal about public engagement with nature of science? *Science & Education, 31*, 1–31.

Braund, M. (2021). Critical STEM literacy and the COVID-19 pandemic. *Canadian Journal of Science, Mathematics and Technology Education, 21*(2), 339–356.

Çalık, M. (2021). Using the Covid-19 pandemic as a Socioscientific issue to support the scientific habits of mind. *Problems of Education in the 21st Century, 79*(5), 694–699.

Cardinot, D., Moura, C., & Guerra, A. (2023). Challenging the "science from nowhere" perspective in the classroom: Action research about a historical case of Brazilian science. *Science & Education, 32*(2), 327–359.

Carter, L. (2014). The elephant in the room: Science education, neoliberalism and resistance. In J. L. Bencze & S. Alsop (Eds.), *Socio-political activism and science & technology education* (pp. 23–36). Springer.

Carter, L. (2020). What's in a name: Post-liberalism, COVID-19 and science (education). *Journal for Activist Science and Technology Education, 11*(2), 16–32.

Chakravartty, A. (2017). *Scientific ontology: Integrating naturalized metaphysics and voluntarist epistemology*. Oxford University Press.

Dillon, J., & Avraamidou, L. (2020). Towards a viable response to COVID-19 from the science education community. *Journal for Activist Science and Technology Education, 11*(2), 1–6.

dos Santos, R. L., & Lima Junior, P. (2023). Are we experiencing a Decolonial turn on Brazilian Scientific Education? *Revista Brasileira De Pesquisa Em Educação Em Ciências, e47934*, 1–34.

Erduran, S. (2020). Science education in the era of a pandemic: How can history, philosophy and sociology of science contribute to education for understanding and solving the Covid-19 crisis? *Science & Education, 29*, 233–235.

Freire, P. (2000). *Pedagogy of the oppressed: 30th anniversary* (M. Ramos, Trans.). Continuum.
Gamage, K. A., Wijesuriya, D. I., Ekanayake, S. Y., Rennie, A. E., Lambert, C. G., & Gunawardhana, N. (2020). Online delivery of teaching and laboratory practices: Continuity of university programmes during COVID-19 pandemic. *Education Sciences, 10*(10), 291.
Gandolfi, H. (2023). Special issue "reflecting on Freire: A praxis of radical love and critical hope for science education"—Theme: Transnational collaborations and solidarities: Transnational conversations about science education: Paulo Freire, interculturality and socio-political transformation. *Cultural Studies of Science Education, 18*(1), 159–173.
García-Carmona, A. (2021). Learning about the nature of science through the critical and reflective reading of news on the COVID-19 pandemic. *Cultural Studies of Science Education, 16*(4), 1015–1028.
Gerard, L., Wiley, K., Debarger, A. H., Bichler, S., Bradford, A., & Linn, M. C. (2022). Self-directed science learning during COVID-19 and beyond. *Journal of Science Education and Technology, 31*, 1–14.
Gutiérrez, R. (2013). The sociopolitical turn in mathematics education. *Journal for Research in Mathematics Education, 44*(1), 37–68.
Hodson, D. (1994). Seeking directions for change: The personalisation and politicisation of science education. *Curriculum Studies, 2*(1), 71–98.
Hodson, D. (1999). Going beyond cultural pluralism: Science education for sociopolitical action. *Science Education, 83*(6), 775–796.
Hodson, D. (2003). Time for action: Science education for an alternative future. *International Journal of Science Education, 25*(6), 645–670.
Hodson, D. (2010). Science education as a call to action. *Canadian Journal of Science, Mathematics and Technology Education, 10*, 197–206.
Hussénius, A. (2014). Science education for all, some or just a few? Feminist and gender perspectives on science education: A special issue. *Cultural Studies of Science Education, 9*, 255–262.
Johns Hopkins University. (n.d.). Johns Hopkins University. Coronavirus Resource Center. https://coronavirus.jhu.edu/
Jones, K. E., Patel, N. G., Levy, M. A., Storeygard, A., Balk, D., Gittleman, J. L., & Daszak, P. (2008). Global trends in emerging infectious diseases. *Nature, 451*(7181), 990–993.
Kayumova, S., & Dou, R. (2022). Equity and justice in science education: Toward a pluriverse of multiple identities and onto-epistemologies. *Science Education, 106*(5), 1097–1117.
Koenecke, A., Nam, A., Lake, E., Nudell, J., Quartey, M., Mengesha, Z., et al. (2020). Racial disparities in automated speech recognition. *Proceedings of the National Academy of Sciences, 117*(14), 7684–7689.
Kopenawa, D., & Albert, B. (2013). *The falling sky*: Words of a Yanomami shaman. Harvard University Press.
Krenak, A. (2022). *Futuro ancestral* (Edição do Kindle). Companhia das Letras.
Law, J. (2015). What's wrong with a one-world world? *Distinktion: Scandinavian Journal of Social Theory, 16*(1), 126–139.
Lemke, J. L. (2001). Articulating communities: Sociocultural perspectives on science education. *Journal of Research in Science Teaching, 38*(3), 296–316.
Magesh, S., John, D., Li, W. T., Li, Y., Mattingly-App, A., Jain, S., Chang, E., & Ongkeko, W. M. (2021). Disparities in COVID-19 outcomes by race, ethnicity, and socioeconomic status: A systematic-review and meta-analysis. *JAMA Network Open, 4*(11), e2134147–e2134147.
Marques, A. R., & Reis, P. (2017). Producción y difusión de vídeos digitales sobre contaminación ambiental. Estudio de caso: Activismo colectivo basado en la investigación. *Revista Eureka sobre enseñanza y divulgación de las ciencias, 14*(1), 215–226.
Means, A. J., & Slater, G. B. (2021). Collective disorientation in the pandemic conjuncture. *Cultural Studies, 35*(2–3), 514–522.
Morales-Doyle, D. (2017). Justice-centered science pedagogy: A catalyst for academic achievement and social transformation. *Science Education, 101*(6), 1034–1060.

Moura, C. B., Nascimento, M. M., & Lima, N. W. (2021). Epistemic and political confrontations around the public policies to fight COVID-19 pandemic. *Science & Education, 30*(3), 501–525.

Oreskes, N. (2019). *Why trust science?* Princeton University Press.

Pais, A., & Valero, P. (2012). Researching research: Mathematics education in the political. *Educational Studies in Mathematics, 80*, 9–24.

Platto, S., Xue, T., & Carafoli, E. (2020). COVID19: An announced pandemic. *Cell Death & Disease, 11*(9), 1–13.

Raveendran, A., & Bazzul, J. (2021). Socialized medicine has always been political: COVID-19, science and biopower in India. *Cultural Studies of Science Education, 16*(4), 995–1013.

Reiss, M. J. (2020). Science education in the light of COVID-19: The contribution of history, philosophy and sociology of science. *Science & Education, 29*(4), 1079–1092.

Rezende, F., Ostermann, F., & Guerra, A. (2021). South epistemologies to invent post-pandemic science education. *Cultural Studies of Science Education, 16*(4), 981–993.

Richards, J., Hove, J., & Afolabi, K. (2008). *Understanding the aboriginal/non-aboriginal gap in student performance: Lessons from British Columbia* (Commentary No. 276). C.D. Howe Institute.

Rieppel, L., Lean, E., & Deringer, W. (2018). Introduction: The entangled histories of science and capitalism. *Osiris, 33*(1), 1–24.

Rincon-Flores, E. G., & Santos-Guevara, B. N. (2021). Gamification during Covid-19: Promoting active learning and motivation in higher education. *Australasian Journal of Educational Technology, 37*(5), 43–60.

Rocha, R., Atun, R., Massuda, A., Rache, B., Spinola, P., Nunes, L., et al. (2021). Effect of socioeconomic inequalities and vulnerabilities on health-system preparedness and response to COVID-19 in Brazil: A comprehensive analysis. *The Lancet Global Health, 9*(6), e782–e792.

Roth, W.-M., & Calabrese Barton, A. (2004). *Rethinking scientific literacy*. Routledge Falmer.

Roth, W.-M., & Desautels, J. (2002). Science education as/for socio-political action: Charting the landscape. In W.-M. Roth & J. Desautels (Eds.), *Science education as/for socio-political action* (pp. 1–16). Peter Lang.

Saribas, D., & Çetinkaya, E. (2021). Pre-service teachers' analysis of claims about COVID-19 in an online course. *Science & Education, 30*(2), 235–266.

Seehawer, M. (2018). South African science teachers' strategies for integrating indigenous and Western knowledges in their classes: Practical lessons in decolonisation. *Educational Research for Social Change, 7*(SPE), 91–110.

Sharma, A. (2020). Phronetic science for wicked times. *Journal for Activist Science and Technology Education, 11*(2), 7–15.

Simonneaux, L. (2014). From promoting the techno-sciences to activism—A variety of objectives involved in the teaching of SSIs. In L. Bencze & S. Alsop (Eds.), *Activist Science and Technology Education* (pp. 99–111). Springer.

Sjöström, J., & Eilks, I. (2018). Reconsidering different visions of scientific literacy and science education based on the concept of *Bildung*. In Y. Dori, Z. Mevarech, & D. Baker (Eds.), *Cognition, metacognition, and culture in STEM education* (pp. 65–88). Springer.

Stengers, I. (2005). The cosmopolitical proposal. In B. Latour & P. Weibel (Eds.), *Making things public: Atmospheres of democracy* (pp. 994–1003). MIT Press.

Tolbert, S., & Bazzul, J. (2017). Toward the sociopolitical in science education. *Cultural Studies of Science Education, 12*(2), 321–330.

Tolbert, S., & Bazzul, J. (2020). Aesthetics, string figures, and the politics of the visible in science and education. *Journal of Curriculum and Pedagogy, 17*(1), 82–98.

Valero, P. (2004). Socio-political perspectives on mathematics education. In P. Valero & R. Zevenbergen (Eds.), *Researching the socio-political dimensions of mathematics education* (Vol. 35, pp. 5–23). Springer.

Valero, P. (2018). Political perspectives in mathematics education. In S. Lerman (Ed.), *Encyclopedia of mathematics education*. Springer.

Valladares, L. (2021). Scientific literacy and social transformation: Critical perspectives about science participation and emancipation. *Science & Education, 30*(3), 557–587.

Verma, G., Campbell, T., Melville, W., & Park, B. Y. (2022). Science teacher education and a sociopolitical turn: The implications for democratic citizenship, and environmental and social justice. *Journal of Science Teacher Education, 33*(5), 459–465.

Viveiros de Castro, E. (2015). *Metafísicas Canibais* [Cannibal metaphysics]. Cosac Naify.

Waight, N., Kayumova, S., Tripp, J., & Achilova, F. (2022). Towards equitable, social justice criticality: Re-constructing the "black" box and making it transparent for the future of science and Technology in Science Education. *Science & Education, 31*(6), 1493–1515.

WHO. (2020, March 11). *WHO Director-General's opening remarks at the media briefing on COVID-19*—11 March 2020. https://www.who.int/director-general/speeches/detail/who-director-general-s-opening-remarks-at-the-media-briefing-on-covid-19%2D%2D-11-march-2020. Accessed 18 Mar 2023.

Cristiano B. Moura is an assistant professor at the Faculty of Education at Simon Fraser University, Canada. He has a PhD in Science, Technology and Education from CEFET-RJ, Brazil. His main research interest is fostering justice-centred science education through historical approaches, drawing mainly from Cultural History of Science, Science Studies and Decolonial theories. He was recently appointed editor-in-chief for the Science & Education journal.

Part II
Examining and Proposing Key Concepts and Ideas for Science Education in Post-pandemic Worlds

Chapter 2
Complex and Political Socioscientific Issues Education in the Anthropocene: Towards an Integrative Didaktik Model Driven by Transdisciplinarity, Relationality and Responsibility

Jesper Sjöström

2.1 Introduction

We are living in the Anthropocene epoch, a human-influenced age where humans affect the development of the whole globe in a major way (e.g. Crutzen, 2006; Steffen et al., 2011, 2015; Lewis & Maslin, 2015). This has consequences for all education (e.g. Blenkinsop et al., 2022), not at least science education (Gray & Colucci-Gray, 2014; Gilbert, 2016; Gray, 2018; Jeong et al., 2021; Wallace et al., 2022; Fazio, 2023) and broader also STEAM education (Guyotte, 2020). Mitchell and Stones (2022, p. 4) write: "the Anthropocene calls for school education to enable a critical ethical disposition that can respond to the existential, socio-economic and political nature of this epoch". There is consequently a need for a sociopolitical turn in (science) education, because "knowledge and politics are intertwined all the time" (Moura et al., 2021, p. 518). The so-called planetary health education framework, based on tensions between local and global priorities, emphasizes interconnection within nature, complexity, socio-ecojustice, and systems thinking and change (Redvers et al., 2023, p. e78). According to Mitchell and Stones (2022, p. 10), the "Anthropocene is super-complex" and highlights people-nature relationships.

Blenkinsop et al. (2022) suggest the term "wild pedagogies" as an umbrella term for meeting educational demands of the Anthropocene. Somewhat similarly, but more connected to "students' life-world becoming" or *Bildung*, Jakubik (2023) suggested the term "wisdom pedagogy". Here in this paper, the mixed term "wisdom

J. Sjöström (✉)
Faculty of Education and Society, Malmö University, Malmö, Sweden
e-mail: jesper.sjostrom@mau.se

C. B. Moura (ed.), *A Sociopolitical Turn in Science Education*, Contemporary Trends and Issues in Science Education 62,
https://doi.org/10.1007/978-3-031-78586-3_2

pedagogies" will be used for eco-reflexive *Bildung*-oriented teaching methods in the Anthropocene.

The paper starts with a Sect. (2.2) about science education in the Anthropocene. The concepts of *Bildung* and didaktik are also introduced in this section in relation to the Anthropocene. The following Sect. (2.3) is about knowledge bases of school science in the Anthropocene. For instance, it is argued that the interdisciplinary field of Environmental Humanities has an important role as one of several disciplinary bases for a "school science discipline". So-called didaktik models are introduced in the next Sect. (2.4) connected to eco-reflexive *Bildung*, a *Bildung*-conceptualisation for the Anthropocene. The following Sect. (2.5), which is the main one, is about formulating/modelling an "integrative didaktik model" in response to the sociopolitical turn in science education. This integrative didaktik model is successively builded. The section consists of five subsections: (1) Vision III[1]-education; (2) Didaktik models for "science-in-context"-education; (3) Didaktik model for eco-reflexive science education; (4) Mixing with ideas from other similar frameworks; and (5) Adding a Vision III-atmosphere. The next Sect. (2.6) is about "wisdom pedagogies", which also can be described as Vision III-driven didaktik models at the praxis level. The paper finishes with concluding remarks (Sect. 2.7). The main contribution of the paper is the construction of the integrative didaktik model driven by transdisciplinarity, relationality and responsibility in line with a sociopolitical turn in science education. Other contributions include a discussion of related teaching methods for praxis (wisdom pedagogies), an introduction of some key concepts and ideas for the whole book, and a review of relevant literature.[2]

2.2 *Bildung*, Didaktik and Science Education in the Anthropocene

Bildung is a central concept in Central European/Scandinavian educational theory. Andersen (2022, p. 186) defines it as being "the combination of education and moral and emotional development; the cultivation and development of personal character, autonomy, and sense of responsibility." *Bildung* is a dynamic non-instrumental concept and stands for both knowledge products and processes in a broad sense (e.g. Andersen, 2022; Sjöström & Eilks, 2020). According to the *Global*

[1] The three visions of scientific literacy and science education will be briefly introduced in the middle of Sect. 2.2.

[2] This paper is the third of three complementing, but at the same time independent, publications: "Vision III of scientific literacy and science education: an alternative vision for science education emphasising the ethico-socio-political and relational-existential" (Sjöström, 2024); "Science education for transformation: theorizing the three visions as a basis for new didaktik models for world-view awareness and socio-political transformation" (Sjöström, 2025, accepted); and "Complex and political socioscientific issues education in the Anthropocene: towards an integrative didaktik model driven by transdisciplinarity, relationality and responsibility" (this paper).

Bildung Manifesto, referred by Andersen (2022, p. 188), it is "a force of peaceful societal transformation". *Bildung* is, according to Jakubik (2023, pp. 46–47), about "developing leaners' knowledge, skills, and competences" as well as "moral virtues and wisdom"; this is based "on connecting the self with the world, on attitudes and values, on integral human formation, and on cultivating the person as a whole".

Bildung has a long and multifaceted history of ideas including, for instance, humanistic values, transparency for spirituality and critical-democratic citizenship. It is a complex concept with at least six historical elements: biological-organic growth processes/humanism, connection to ancient cultures, spiritual elements, romanticism, enlightenment thoughts, and emancipation (e.g. Sjöström & Talanquer, 2018; Zidny et al., 2020). The latter-mentioned sociopolitical dimension is the latest addition, mainly added after WWII (e.g. Sjöström et al., 2017). With *Bildung* the self is connected to the world from ethical perspectives. However, a traditionally rather individual-focused view on *Bildung* needs to be complemented with critical and collective perspectives (e.g. Kvamme, 2021).

The notion of *Bildung* is central in didaktik (e.g. Jank, 2014; Sjöström & Eilks, 2020; Deng, 2022). Didaktik is the name of the education tradition in northern Europe and Scandinavia (e.g. Gundem, 2010; Krogh et al., 2021). It can be seen as both the art of teaching and as the discipline for teachers and teacher educators (e.g. Hopmann, 2007; Ingerman & Wickman, 2015). Seen as a discipline for the teaching profession, didaktik gives theoretical support for professional teachers' well-informed didaktik decisions on didaktik questions such as the why, the what, and the how to teach a certain content (including relevant knowledge, values, socioscientific issues, etc.) to a certain group of students.

Regarding education, didaktik and *Bildung* have a common history of ideas. According to Duit (2015, p. 325), didaktik is based on *Bildung* and "stands for a multifaceted view of planning and performing instruction". A main idea of didaktik is that teachers have agency to make responsible choices in relation to the development of their students' *Bildung*. In explicit *Bildung*-oriented didaktik, content (in a broad sense) is considered fundamental (e.g. Deng, 2022). Important in didaktik research is developing different types of frameworks and tools for teachers that can be of help in their didaktik choices. With one word, it is called "didaktik models" (see further Sect. 2.4).

For grasping complexity in society, Andersen (2022) has developed an educational framework/model called "The Bildung Rose". It is a relatively straightforward "tool for addressing, grasping, and taking responsibility for society" (p. 186) that "allows a richer understanding of the relationship between transferable/horizontal and non-transferible/vertical knowledge" (p. 210). It consists of seven connected domains: production; technology; aesthetics; power; science (in a rather classical and narrow sense); narrative; and ethics. In *Bildung* the content in the different domains collaborate for: deep education and understanding; meaningful and purposeful development; and sustainable prosperity (p. 198). When one domain dominates over the others, the consequence is—according to Andersen (2022, p. 197)—exploitation, disruption, confusion, abusiveness, arrogance, narrow mindedness or anxiety. More philosophically and without an explicit educational tool,

Rowson (2019) has discussed *Bildung* in relation to future education and sustainability issues. He portrayed *Bildung* as being a values-driven applied philosophy of education and connected it to, for instance, transdisciplinarity and transformative education.

According to Biesta (2012, p. 817), "the role of the individual in the process of Bildung, […] has to be understood as a reflexive process", that is, a process where the individual establishes both a relationship and a critical stance towards the existing culture and society. It is about "being in responsible relation with other human beings and, by extension, with the natural world more generally" (Biesta, 2013, p. 739). Biesta regards this as highly political.

To specify the meaning of *Bildung* today, we have to base it on the fact that we live in the Anthropocene (e.g. Kvamme, 2021; Krogh et al., 2022). When explicitly oriented towards ecological awareness, such an orientation can be called eco-reflexive *Bildung* (e.g. Sjöström et al., 2016; Sjöström, 2018). From a posthuman perspective, *Bildung* is bound "to the world's materiality with culture and society as a part" (Yavuzkaya et al., 2022, p. 4).[3]

Science education in the Anthropocene (e.g. Jeong et al., 2021) can be seen as an intersection of Socioscientific issues (SSI), Environmental [and Sustainability] Education (ESE) and Futures Education (Hervé & Panissal, 2022). All these three connected science education approaches have recently been reviewed/discussed. Dillon and Herman (2023) discussed environmental education in the light of science education and Laherto et al. (2023) future-oriented science education oriented towards building sustainability competences. Very recently, two systematic reviews about mainstream SSI-teaching were published (Högström et al., 2024; Viehmann et al., 2024). In theory, SSI is about applying science in society, focusing controversial issues. Högström et al. (2024) characterize SSI as: complex issues of clear relevance for the future society; the issues encompass different possible values positions and scientific knowledge has a crucial role in the decision-making process. However, in practice, teachers seem to have a rather instrumental view on SSI (teaching objectives, topics and methods) (Högström et al., 2024). Therefore the authors argue for a shift from Vision I to Vision II. In 2007, Roberts first suggested his two well-known visions of scientific literacy. Simplified, Vision I can be described as science without society (internal view), whereas Vision II is about contextual application of scientific knowledge in life and society (external view). However, a pervasive sociopolitical and ecological problematization is missed in Vision II (e.g. Viehmann et al., 2024; Sjöström, 2024).

Therefore, I think SSI education in the Anthropocene should be driven by a Vision III of scientific literacy and science education (e.g. Sjöström & Eilks, 2018; Sjöström, 2024; Yavuzkaya et al., 2022). The Vision III-version suggested by me and coauthors is sociopolitically oriented (e.g. Sjöström & Eilks, 2018). It includes critical aspects that aim at philosophical reflection and sociopolitical actions. In this knowledge, values and actions are intertwined. Scientific literacy

[3] See also Taylor's (2017) elaboration on the concept of *Bildung* from a posthuman perspective.

from a Vision III-perspective aims at: "Socio-political action for justice, democracy, [and] critical sustainability" (Valladares, 2021, p. 569). According to Fuchs and Tan (2022, p. 12) "Vision III aims at science education to have socio-political functions, enabling students to take full and active roles as engaged citizens". Similarly, Birdsall (2022, p. 236) encourages the importance of opportunities to explore the political dimension of socioscientific issues to develop "abilities to engage in socio-political action-taking as informed citizens". According to Ibrahim et al. (2022, p. 41) "Vision III-science incorporates critical theories, in which students may act as transformative activists".

Vision III-driven education is typically about complex and political SSI:s such as a pandemic or climate change, or more concrete cases like exploring plastic problems from different perspectives (Danckwardt-Lillieström et al., 2024) or the "negative health outcomes and environmental degradation from artisanal coltan mining in the Congo" (Ibrahim et al., 2022, p. 33). More generally, it is about so-called *wicked problems* (e.g. Agustian, 2023; Lönngren & van Poeck, 2021) that have a (trans) disciplinary knowledge base and need creative solutions. According to Gilbert (2016, p. 192), "[e]verything is now complex – that is, deeply entangled, interconnected, unpredictable and open. Any given problem has multiple dimensions".

Science education as well as STEAM education can catalyze the shaping of informed, responsible and active citizens (e.g., Elmose & Roth, 2005; Lotz-Sisitka et al., 2015; Jeong et al., 2021) with *Bildung*. Environmental citizenship (Hadjichambis et al., 2020; Ariza et al., 2021) may be central in this. A Vision III-oriented science education includes philosophical, ethical and sociopolitical perspectives. The focus is on problematization, understanding uncertainties and balancing benefits and risks of science (e.g. Sjöström & Eilks, 2018). In addition to the traditional three legs of science education—concepts, Nature of Science (NOS) and STSE (Science, Technology, Society, Environment)—Hodson (2003) also suggested *sociopolitical activism* (as a fourth leg). Such a critically oriented approach can be connected to eco-reflexive *Bildung* and Vision III (e.g. Sjöström et al., 2016). Hodson (2011) has written the following about traditional SSI-oriented teaching: "My inclination is towards a much more radical, politicized form of SSI-oriented teaching and learning in which students not only address complex and often controversial SSI, and formulate their own position concerning them, but also prepare for, and engage in, sociopolitical actions that they believe will 'make a difference'." (p. 31).

Bildung brings with it a broad view of knowledge and knowings, which, in addition to being disciplinary, is also, for instance, cultural, practical, and tacit (Deng, 2020). "Science knowings" is an important knowledge base for politized SSIs as well as for contemporary *Bildung* (e.g. Elmose & Roth, 2005; Sjöström et al., 2016; Yavuzkaya et al., 2022). In line with this, science and technology education must be designed with the purpose of educating critical-democratic citizens (e.g. Sjöström & Eilks, 2018) with knowledge *in* and *about* science as well as with action competence.[4] Hodson (2011) argues for a critical scientific, technological and environmental literacy—shortened to *critical scientific literacy*.[5] The aim of it is to have knowledge and abilities for being an autonomous, responsible-taking and action-knowledgeable citizen, working in the interest of socio-ecojustice and global sustainability. By referring to the German *Bildung*-scholar Wolfgang Klafki, Elmose and Roth (2005, p. 21) called this *Bildung*, and it "involves competences for self-determination, constructive participation in society, and solidarity towards persons limited in the competence of self-determination and participation". It is praxis-oriented, in addition to being oriented towards consciousness and critical literacy. In line with this, Mogensen and Schnack (2010, p. 60) argued that their concept of action competence is "closely linked to democratic, political education and to [...] the notion of 'Bildung'."

A basis for "action *Bildung*"—from an Anthropocene perspective—is *agential literacy* based on a "posthuman ethics of mattering" (Jeong et al., 2021, p. 810). The term agential literacy is borrowed from Karen Barad (2000). Jeong et al. (2021, pp. 811–812) write: "Acting on scientific literacy conceived from a posthumanist perspective [...] calls for ethical questions pertaining to humans' participation in science and doing so *responsibly*. Scientific literacy, therefore, is not merely a human endeavor or simply a way of knowing *about* the natural world; rather, it becomes a way of learning how to intra-act responsibly within the natural world (Barad, 2000)."

The rest of this paper will be oriented towards different aspects of didaktik for eco-reflexive action *Bildung*. Klafki framed his *Bildung* concept with epochal key problems, such as the climate change (e.g. Sjöström & Eilks, 2020; Deng, 2022). It pronounces the importance of both deep and powerful knowledge from disciplines as well as interdisciplinarity and values. The position is that knowledge and values are not possible to separate. Hopmann (2007, p. 117) has written: "Didaktik and Bildung require normativeness [...] they challenge the teacher to be aware of the unavoidable normativeness in every dealing with whatever subject matter." The teacher has an active role in local "curriculum making" (enacted classroom curriculum) (Hopmann, 2007). Questions about why and what is important for "curriculum making", both on an institutional level and at the classroom level (e.g. Deng, 2022; Mitchell & Stones, 2022). Deng (2022, p. 605) writes: "At the heart of *Bildung*-centred *Didaktik* are three essential ideas – (1) *Bildung*, (2) a theory of content, and (3) teaching as an encounter between content and students." Worldview perspectives and corresponding visions and ideologies affect local "curriculum making" and other didaktik choices (e.g. Lacey, 2009; Sjöström, 2018). An integrative worldview[6] can be described with terms like: holism, complexity, plurality, relational thinking, spiritual knowledge, place-based knowledge, and transformation.

---

[4] For a discussion of a contemporary conceptualization of action competence, see: Sass et al. (2020).

[5] For a recent review about critical scientific and environmental literacies, see: Guerrero and Sjöström (2024).

[6] For a recent discussion of worldviews beyond a sustainable development discourse, see: Fitzpatrick (2023). See also: Sjöström, 2025, accepted.

Recently, Redvers et al. (2023, p. e80) (for higher education) discussed the main didaktik questions based on three different views on education: behaviourist/informative, constructivst/formative and praxis/transformative. It is the latter—praxis/transformative view—that is in line with an eco-reflexive *Bildung* view. From such a perspective diversity and participation are valued (the who-question). Personal growth and change (knowledge, behaviours and attitudes) as well as social change, are the goals (the why-question). Factual knowledge and contextualization need to be complemented with critical analysis and compassion (the what-question). An example of a teaching method enhancing reflexivity is place-based education (the how-question) (see futher Sect. 2.6 about wisdom pedagogies).

2.3 Knowledge Bases of School Science in the Anthropocene

In this section, I want to make some comments regarding school science and corresponding academic disciplines, how they relate to each other and their knowledge bases. According to Deng (2022, p. 602) "school subjects and an academic discipline are fundamentally different", although of course related. Consequently, school science has another and much broader knowledge base than the (traditional) science disciplines in themselves.

With reference to Bruno Latour, Gilbert (2016) claims that "the Anthropocene challenges scientists to think very differently about science – about what it is, what it is for, and what and who it should engage with" (p. 196). Thus, "we have to accept that science education as we have known must be transformed. Substantial rethinking – of its *content*, its *purposes* and its *relationships* – is required." (p. 188). For this, there is need for "a broadening of the traditional knowledge base of school science" (p. 189). Similarly, Pedretti writes (in Bencze et al., 2020, p. 830): "What counts as knowledge must be expanded to include (and legitimize) knowledge that is relational, holistic, interdisciplinary, outward in its orientation and grounded in students' experiences, relationships and contexts".

Therefore, science education in the Anthropocene is not only education *in* science, but as much education *about* science (in society) and how to use science (in a broad sense), for instance for sociopolitical activism (in a broad sense). For this, there is need for a (trans)disciplinary knowledge basis that is much broader than only traditional and specialized science disciplines in their self. Important examples of academic and humanistic (sub)disciplines that can provide nourishment to "a school science discipline" is Science and Technology Studies (STS) (e.g. Feinstein, 2011; El Halwany et al., 2021; Moura et al., 2021; Valladares, 2022), Environmental Humanities (e.g. Hutchings, 2014; O'Gorman et al., 2019) and Futures Studies (e.g. Carter & Smith, 2003; Rasa et al., 2024). School science in the Anthropocene also needs curriculum integration (Salinas et al., 2023) and to relate to knowledge development in the field of Environmental and Sustainability Education (ESE) (e.g. Lotz-Sisitka et al., 2015; Herranen et al., 2021; Dillon & Herman, 2023).

At some universities, it will hopefully be possible to develop "an Anthropocene-oriented school science discipline" (can also be called "Anthropocene-oriented science didaktik") as a basis for science teacher education (see further discussion about this in e.g. Yavuzkaya et al., 2024), whereas at others it will be more about mixing and integrating knowledges from different disciplines, not least from social sciences and the humanities (see e.g. Kahn & Zeidler, 2016). Thus, there will in the Anthropocene be a (trans)disciplinary knowledge base for science education. Bernstein (2015) writes: "Transdisciplinarity today is characterized by its focus on 'wicked problems' that need creative solutions, its reliance on stakeholder involvement, and engaged, socially responsible science." Guyotte (2020) highlights three guiding concepts in her discussion of a philosophy of STEAM education in the Anthropocene: transdisciplinarity, relationality and responsibility. She argues for an "education that can foster spaces for transdisciplinary conversations surrounding critical issues of the Anthropocene – those of ethics, sustainability, and relationality" (p. 769).

Only some of the knowledges from the traditional and specialized science disciplines can be regarded as relevant for school science. It is such knowledge that can form a basis for students' development of *powerful knowings* (Carlgren, 2020; Yavuzkaya et al., 2022), or with other words: "knowledge-as-a-means-for-cultivation-of-human-powers" (Carlgren, 2020, p. 324) or *Bildung*. Discussions about relevant content for teaching are central in teachers' discipline: didaktik.

2.4 Didaktik Models for Eco-Reflexive *Bildung*

Didaktik models can be used, for instance, as tools for analysis and for planning and design of teaching (e.g. Sjöström et al., 2020; Deng, 2022; Wickman et al., 2020). They can be seen as a bridge between theoretical perspectives and teaching practice (see Fig. 2.1). The main aim of didaktik models is to support teachers in their complex didaktik choices (e.g. Wickman et al., 2020). They can be used in the design and analysis of teaching, but also, for instance, for critical meta-reflection about teaching traditions or dilemmas, and in-action.

There are three main types/levels of didaktik models, although many models are mixtures of these:

META-LEVEL: "Metamodels", which mainly support teachers answering the why-question (about intentions, aims and objectives of teaching). It is about overall ideas of teaching and critical meta-reflection.

ANALYSIS/PLANNING-LEVEL: Content models, which support teachers answering the what-question (about content for instruction in a broad sense). It is about planning and analysis of teaching.

PRAXIS-LEVEL: Models for improved practice, which support teachers answering how-questions (about methods of instruction and media used in instruction). It is about teaching methods in a broad sense.

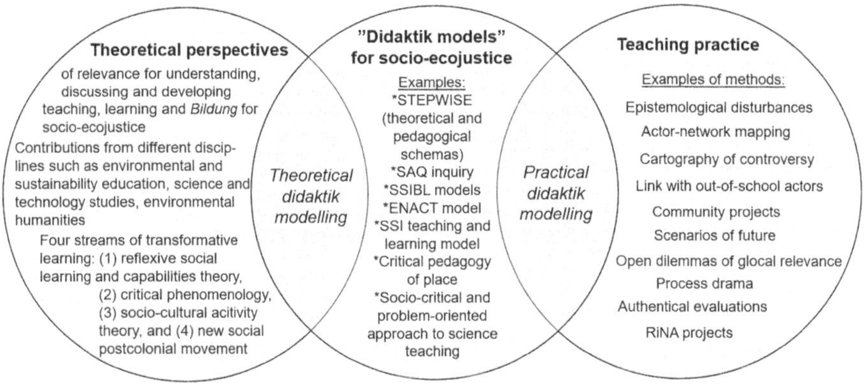

Fig. 2.1 Didaktik models (in a broad sense) as a bridge between theoretical perspectives and teaching practice/methods. They support teachers in their didaktik choices and reflection in relation to didaktik questions (who, why, what, how...). This figure is based on a similar, but more general one, published in e.g. Sjöström (2022, p. 253). This figure focuses on didaktik models for socio-ecojustice. The four streams of transformative learning (in the left circle) are from Lotz-Sisitka et al. (2015). (References to other parts of the figure (including explanations of abbreviations) can be found elsewhere in the text)

In the didaktik tradition the why- and what-questions are important to consider before coming to the how-question. Klafki developed didaktik models for *Bildung*-oriented didaktik (e.g. Sjöström & Eilks, 2020). Actually, didaktik models as an idea can be traced back to Klafki's didaktik analyses in the late 1950s, with five areas of questions. For the science subjects, Ingo Eilks together with coworkers have formulated and empirically evaluated a didaktik model for socio-critical and *Bildung*-oriented science teaching (e.g. Marks et al., 2014). I have, together with Eilks in particular, further elaborated on the *Bildung*-concept in relation to the science subjects (e.g. Sjöström, 2013, 2018; Sjöström et al., 2017; Sjöström & Eilks, 2018, 2020; Sjöström & Talanquer, 2018; Zidny et al., 2020; Yavuzkaya et al., 2022).

Didaktik, as teachers' discipline, should be based on critical-democratic values and theories and research approaches from the humanities and social sciences—mainly from different educational sciences—in addition to knowledge bases in and about different disciplinary knowledge areas, such as the natural sciences. Furthermore, didaktik is also a design science and as in other design sciences modelling is central. When didaktik models are formulated, used and/or developed, this may be referred to as "didaktik modelling" (e.g. Ingerman & Wickman, 2015; Sjöström, 2019, 2022; Wickman et al., 2020). Figure 2.1 differs between "theoretical didaktik modelling" and "practical didaktik modelling".

The illustration in Fig. 2.1 focuses on "science-in-context". Theoretical perspectives (left circle) are taken from, for instance, the fields of Environmental Humanities and Science and Technology Studies (STS). Other examples of theories that can be placed in the left circle are Young and Mullers' theory of powerful knowledge (Deng, 2022) and Callon and Latours' actor-network theory (Moura et al., 2021). Regarding the middle circle, examples of "science-in-context"-didaktik models/

frameworks mentioned are STEPWISE, SAQ, SSIBL, ENACT, the SSI teaching and learning model, and critical pedagogy of place. Explanations/references to these models as well as the examples of teaching methods (right circle) will be given in the next sections. See especially Sects. 2.5.2 and 2.5.4 for explanations/references to the "science-in-context"- teaching models. Wisdom pedagogies are discussed in Sect. 2.6.

In the following and main section of the paper, an integrative didaktik model in response to the sociopolitical turn in science education will be suggested/modelled.

2.5 Towards an Integrative Didaktik Model in Response to the Sociopolitical Turn in Science Education

An integrative didaktik model is successively builded in this section, which consists of the following five subsections: (1) Vision III-education; (2) Didaktik models for "science-in-context"-education; (3) Didaktik model for eco-reflexive science education; (4) Mixing with ideas from other similar frameworks; and (5) Adding a Vision III-atmosphere.

2.5.1 Vision III-Education

Vision III-teaching is most in line with the *Bildung*-oriented didaktik tradition. Teaching in a Vision I-context can be described as traditional: preparing for tests (Vogelzang et al., 2020) and closed questions and answers (Valladares, 2021). In a Vision II-context relevance (from a Western modern science perspective) is pronounced (e.g. Vogelzang et al., 2020; Zidny et al., 2020) as well as collaborative work and authentic evaluations (Valladares, 2021). If Vision I is pronouncing conceptual knowledge and science procedures, Vision II is pronouncing applications and social context (Sjöström & Eilks, 2018). However, according to Ibrahim et al. (2022, p. 41) there are "tendencies of most mainstream science education to fall under Vision-I/II's umbrella". Vision III adds critical perspectives to both Vision I and II (e.g. Sjöström & Eilks, 2018; Valladares, 2021; Sjöström, 2024) as well as indigenous science perspectives (Zidny et al., 2020). Furthermore, it adds emphases on complexity, the political, environmental engagement and responsible knowing-in-action (Sjöström, 2024).

In Vision II-science teaching the goal is to develop "science advocates", whereas it in Vision III-science teaching it is to develop "honest science broker" (Valladares, 2021). The latter is emphasizing critical-reflexive thinking; it is about learning to develop a critical-reflexive attitude grounded in substantive understanding of relevant concepts (Vogelzang et al., 2020). Knowledge in and about science contributing to responsible citizenship and engagement for transforming actions are in focus (e.g. Sjöström & Eilks, 2018; Valladares, 2021). Both Vision II and III are interested in SSI, but they have different focus. Important individual decisions are in focus from a Vision II-perspective, whereas collective discussions and the sociopolitical context are in focus from a Vision III-perspective, as well as uncertainty, risks and global relevance (e.g. Valladares, 2021).

Vision III-education can be seen as synonymous to a critical-eco-reflexive *Bildung*-oriented didaktik (Sjöström, 2024). It includes, for instance, a critical-reflexive attitude, broad and critical interdisciplinary understanding (including NOS, STSE and environmental SSIs), focus on meaningfulness and global relevance, actor-network mapping (e.g. Bencze, 2017; Ibrahim et al., 2022, pp. 42–43), and teachers first putting important issues (wicked problems) on the table and then making the content (in a broad sense) free (e.g. Van Poeck & Östman, 2020; Yavuzkaya et al., 2022) (see further Sect. 2.6 about wisdom pedagogies).

2.5.2 *Didaktik Models for "Science-in-Context"-Education*

An umbrella term for working with complex sociopolitical issues in science education is "science-in-context" (Bencze et al., 2020). Under this heading Larry Bencze et al. some years ago discussed four didaktik programs/frameworks/models: STEPWISE (Science & Technology Education Promoting Wellbeing for Individuals, Society & Environments) (see also e.g.: Bencze et al., 2012; Bencze, 2017, 2022), Socially Acute Questions (SAQ) education (section written by Laurence Simonneaux and Jean Simonneaux), (mainstream) Socioscientific Issues (SSI) education (section written by Dana Zeidler) and Science-Technology-Society-Environment (STSE) education (section written by Erminia Pedretti) (Bencze et al., 2020). Some of these authors have, in other contexts, instead of "science-in-context", used STSE (Pedretti & Nazir, 2011) or SSI (Simonneaux, 2014) as broad umbrella terms for socio-oriented science education (see further below).

Although the teaching has the same origin, there are some differences in ontological, epistemological and axiological positions (Bencze et al., 2020). Included in this, there are differences in worldview perspectives (e.g. Fitzpatrick, 2023; Sjöström, 2018; Sjöström, 2025, accepted). One difference in teaching practice is teachers' view of how political the science teaching might be. Gilbert (2016, p. 190) writes: "possibly through a commitment to science's objectivity, many teachers aim to treat SSIs in a 'balanced' way, *avoiding* discussions of economic or political interests and/or socio-cultural values". In a sociopolitical Vision III-driven science education such discussions are rather encouraged.

Pedretti and Nazir (2011) captured a range of different goals and pedagogies through six "STSE currents": Application/design; Historical; Logical reasoning; Value centered; Sociocultural; Socio-ecojustice. The three latter, and especially socio-ecojustice, are in line with the focus of this paper. Dominant approaches in the "socio-ecojustice STSE education current" are described as creative, affective, reflexive, critical, experiential, and placed-based.

Similar to Pedretti and Nazir, Simonneaux (2014) discussed different "SSI-orientations" using a continuum from "cold" (mainly emphasizing, for instance, monodisciplinarity, scientific learning and epistemic values) to "hot" (also emphasizing transdisciplinarity, political citizenship and philosophical values). In the middle of Simonneaux's model we find mainstream SSI, for example, characterized by knowledge *about* science, critical thinking, and scientific citizenship. Although contextualization is then emphasized, the focus is on cognition and evidence-based argumentation. This is problematized at the "hot" end, which also contains philosophical reflection and sociopolitical activism. Vision I can be placed at the cold end, and Vision III at the hot end of Simonneaux's continuum.

Similarly, Levinson (2017) compared mainstream SSI (e.g. Zeidler et al., 2005)—placed in the middle of Simonneaux's continuum—with other science-society education approaches, such as STEM (Science-Technology-Engineering-Mathematics), SAQ, and STEPWISE. He described the educational purpose of mainstream STEM, which is at the cold end of Simonneaux's continuum, as providing human capital. The education purpose of mainstream SSI is the development of scientific knowledge needed for socioscientific reasoning, that of SAQ to develop a critical discourse, and that of STEPWISE to develop knowledge for activism and socio-ecojustice. Both SAQ and STEPWISE are at the hot end of Simonneaux's continuum.

The interdisciplinary (project) area "Environmental Citizenship in the 21st Century Education" (Hadjichambis et al., 2020) has ideas and models that can be used in and transformed into science teaching. Some of the ideas and models are for stimulating youth activism (Reis, 2020). There are many pedagogies for sustainability and transformative learning based on head, hands and heart (Sipos et al., 2008). These include action learning, critical emancipatory pedagogy, pedagogy for eco-justice, and traditional ecological knowledge, among others. Many of these and other similar approaches and models are presented in a meta-model called the "pedagogical landscape of Education for Environmental Citizenship" (Hadjichambis & Paraskeva-Hadjichambi, 2020, p. 247) (see further the next section about wisdom pedagogies).

Four concrete examples of didaktik programs/models that take the global challenges into consideration are as follows: (1) the STEPWISE theoretical framework and pedagogical approach for activist Science and Technology education developed by Larry Bencze (2017, e.g. p. 20 and 27; see also Bencze, 2022) and his coworkers, (2) the so-called SSIBL-model (SSI + IBSE) developed by Ralph Levinson and the PARRISE consortium (2017, p. 480) (see also: Ariza et al., 2021; Georgiou & Kyza, 2023), (3) the ENACT model developed by Hyunju Lee and her coworkers (e.g. Hwang et al., 2023), and (4) the model/framework for socio-critical and *Bildung*-oriented science teaching ("Framework outlining the socio-critical and problem-oriented approach to science teaching") developed by Ingo Eilks and his coworkers (e.g. Marks & Eilks, 2009; Marks et al., 2014). The latter mentioned framework and a developed version of it will be described in the next subsection.

Before elaborating on Eilks' didaktik model and a developed version of it, I will mention another more theoretically oriented—or vision-oriented—didaktik model that I have developed together with two PhD students, Merve Yavuzkaya and Paul Clucas (Yavuzkaya et al., 2022). It is a reflection model for teachers, promoting powerful knowings and agency in science/chemistry teaching. It also includes the teacher's consideration of student transformation of content (i.e., *Bildung* processes). The model can further be connected to the development of reflective judgment (Zeidler et al., 2009) and a risk understanding (e.g. Schenk et al., 2019; Develaki, 2024) based on critical realism as an ethico-onto-epistemology (e.g. Zembylas, 2006; Levinson, 2018) in line with Barad's agential literacy. Furthermore, it can be discussed in terms of sociopolitically oriented Vision III (e.g. Sjöström & Eilks, 2018), emphasizing "engagement with social participation and emancipation" (Valladares, 2021, p. 557) as well as sociopolitical actions.

2.5.3 *Didaktik Model for Eco-Reflexive Science Education*

Eilks and coworkers' framework (Marks et al., 2014) (called didaktik model in Sjöström et al., 2020) outlining a socio-critical and problem-oriented approach to science teaching consists of the following four columns: objectives (this first column corresponds to column two in the framework shown in Fig. 2.2); criteria for selecting issues and approaches (third column); methods (fourth column); and structure of the lesson plan (fifth column).

Some years ago, I extended and developed Eilks' model to a model for eco-reflexive didaktik (Sjöström, 2019). The developed model is shown in Fig. 2.2. The new model connects even more explicitly than its precursor to the didaktik-tradition

Philosophical ideas	Objectives (WHY?)	Criteria for selecting issues and approaches (WHAT?)	Teaching methods (HOW?)	Structure of lessons and projects
Holism and Eco-reflexivity	Reflexive *Bildung* and Transformative learning	Sustainability issues	Integrating cognitive and affective domains (incl. dialogic processes)	1. Problem analysis
Critical realism	Problematization, Emancipation, Subjectification, Praxis	Relevant, controversial and complex socio-scientific issues	Inquiry-based learning (incl. practical work)	2. Clarifying the science background and context
Egalitarianism and Altruism	Vision III of scientific literacy	Authenticity	Relevant media (incl. authentic)	3. Resuming the socio-political-scientific controversy
Humanization (plural knowledge and mutifacetted problematization)	Learning *in* and *about* science (incl. NOS and STSE)	Includes scientific knowledge and processes (incl. uncertainty)	Methods structuring scientific reasoning and controversial debating	4. Alternatives and examples of actions for socio-ecojustice
Reconstructionism and Critical pedagogy	Promotion of critical and responsible citizenship	Allows for open discussion (incl. different discourses)	Methods provoking the explication of individual opinions and actions	5. Meta-reflection (incl. analysis of knowledge emphases and possible activism)

Fig. 2.2 Didaktik model for eco-reflexive science teaching (from Sjöström, 2019, p. 50). It was developed based on a framework (first published in: Marks & Eilks, 2009) outlining a socio-critical and problem-oriented approach to chemistry/science teaching

and the main didaktik questions (why, what, how). Compared to the model by Eilks, an extra column (column one in the figure) was added. Column one concerns philosophical ideas behind the developed model. Many of the ideas in this column (holism, realism, egalitarianism, altruism) were taken from Bencze and Carter (2011). Column two relates to objectives (the why-question). Ideas were taken from e.g. Sjöström and Eilks (2018). Column three deals with the content (the what-question), and ideas were taken from e.g. Sjöström et al. (2016). Column four concerns the practice (how-question) and ideas were taken from e.g. Bencze (2017), and Eilks and his coworkers (e.g. Marks et al., 2014).

Column five in the model shown in Fig. 2.2 deals with the concrete structuring of lessons and projects. This fifth column concerns: (1) first making a problem analysis, and (2) then clarifying the science background and context. One way to get an overview of the context from a critical perspective is to construct—or at least to study an already existing—actor-network map. By constructing actor-network maps, students are enlightened about problematic actants and become motivated to address sociopolitical issues (Bencze, 2017) (see also: Ibrahim et al., 2022). The teaching is then followed by: (3) resuming the sociopolitical-scientific controversy, (4) identifying alternatives and examples of actions for socio-ecojustice, and, finally, (5) meta-reflection, including analysis of knowledge emphases and possible activism.

2.5.4 Mixing with Ideas from Other Similar Frameworks

In Fig. 2.3, the five steps for structuring socio-critical and problem-oriented science teaching, that is the steps in the fifth column in Fig. 2.2, are mixed with similar teaching steps in some other similar frameworks/models like SSIBL and STEPWISE. Compared to the fifth column in Fig. 2.2, an extra step—the fourth step—have been added in Fig. 2.3. The background to this is explained below.

The sequence illustrated in Fig. 2.3 (see below in the text for a detailed description) is an integrated structure/sequence for "science-in-context"-projects. It can be described as a "synthesis didaktik model" based on several previously published SSI-oriented didaktik models. It is especially based on the five boxes in the fifth column in the model for eco-reflexive didaktik shown in Fig. 2.2 (from Sjöström,

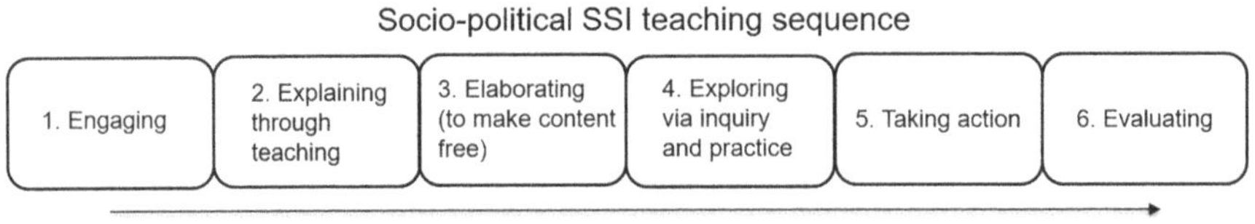

Fig. 2.3 Sociopolitical socioscientific issues teaching sequence (mainly based on Eilks' framework, the ENACT model, SSIBL models and the 5E instruction model—see references in the text). What is in the six boxes are described in detail in the text

2019). In addition to this fifth column, the following "didaktik models"/sources were used in the "didaktik modelling":

- Eilks and his coworkers' "Framework outlining the socio-critical and problem-oriented approach to science teaching" (e.g. Marks & Eilks, 2009; Marks et al., 2014)
- Bencze and his coworkers' STEPWISE models (e.g. Bencze et al., 2012; Bencze, 2017, 2022)
- The 5E instructional model—engagement, explanation, elaboration, exploration, evaluation (Bybee, 2014; see also Friedrichsen et al., 2016 where the model is used in an SSI context)
- Sadler and his coworkers' model for socioscientific teaching and learning—focal issue, socioscientific ideas/practices, culminating experience (e.g. Sadler et al., 2017)
- Lee and her coworkers' ENACT model—Engage, Navigate, Anticipate, Conduct, Take action (e.g. Hwang et al., 2023)
- Levinson and his coworkers' SSIBL-models—ask, find out, act, reflect (e.g. Levinson & the Parrise Consortium, 2017; Amos et al., 2020; Ariza et al., 2021; Georgiou & Kyza, 2023)
- Model for SAQ inquiry and the S3R framework (see Simonneaux and Simonneaux in Bencze et al., 2020)

Ideas/models in the following eight publications were also used/integrated, to a larger or lesser extent:

- Chang Rundgren and Rundgren (2010) (the SEE-SEP model—highlights Science, but also Ethics/Morality, Policy, Sociology/Culture, Environment and Economy; Knowledge, Values and Personal experience are all important) (see also Chang Rundgren, 2023, where the SEE-SEP model is integrated with several other didaktik models, such as SSIBL)
- Hodson (2011) (eight steps for sociopolitical action in science education)
- Kahn and Zeidler (2019) (socioscientific reasoning—input from complex socioscientific issues, inquiry, skepticism, and reflective moral judgement, which is based on e.g. rational thinking, perspective taking, and empathy) (see also e.g. Zeidler et al., 2019; Romine et al., 2020)
- Birdsall (2022) (ideas behind a socioscientific issues/Vision III teaching approach)
- Mitchell (2022) (didaktik models in geography education in the Anthropocene for powerful knowledge and social justice)
- Yavuzkaya et al. (2022) (vision-oriented didaktik model promoting powerful knowings and agency—including, for instance, powerful knowings-oriented didaktik question areas)
- Van Poeck et al. (2024) (philosophically oriented teaching theory—staging and performing)
- Hadjichambis and Paraskeva-Hadjichambi (2020) (individual and collective actions)

Below a detailed description of the six steps in the sociopolitical SSI teaching sequence illustrated in Fig. 2.3 are presented (the texts are mainly taken from the main references described above):

1. (Engagement) **E**ngage[7] in SSI:s; ASK: Problem analysis; Students ask and reflect (with teacher stimuli—"pre-putting" on the table); Report on the inquiry; Identifying
2. (Explanation) **N**avigate SSI:s; FIND OUT 1: Clarifying the science background and (personally relevant) context; Collection/Analysis of information; Teachers prepare by working with powerful knowings-oriented didaktik question areas and arranging the classroom environment (in a broad sense); Teachers teaches—putting on the table; the dramaturgy of teaching: *staging* the scene of objects/phenomena that are brought to the students' attention and *performing* interventions within and upon the staged setting; disciplinary key ideas; science practices; thinking scientifically; crosscutting concepts—Evolution of the question
3. (Elaboration) **A**nticipate consequences; FIND OUT 2: Beliefs and values of different stakeholders; Resuming the sociopolitical-scientific controversy (perceptions of risk; deliberative discussions; socioscientific reasoning practices, including perspective taking); Reflexivity/subjectivity of investigators; highlight Science, but also Ethics/Morality, Policy, Sociology/Culture, Environment and Economy; Knowledge, Values and Personal experience are all important; six dimensions of reasoning: problematization, interactions, knowledges, uncertainties and risks, values, and governance; Make content free (by for instance critical thinking, highlighting ethical aspects and considering values positions)—Deliberation
4. (Exploration) **C**onduct scientific and engineering practices; FIND OUT 3: Students practice (with teacher support) (small-group works, students design and perform small-scale RiNA-projects "to address self-identified harms in STSE relationships"); Explicitness/construction of reasoning; Searching for solutions
5. **T**ake action; ACT: Which alternatives are there?; individual and/or collective actions for socio-ecojustice; student-led actions (taking sociopolitical action); Possible answers & actions
6. (Evaluation) REFLECT: Meta-reflection (including epistemic analysis); synthesis of key ideas; evaluating consequences

2.5.5 Adding a Vision III-Atmosphere

The six steps illustrated in Fig. 2.3 should be performed in a "Vision III-atmosphere". Support for teachers' thinking about the values basis and their answers on especially why- and what-questions, but also on how-questions, in this atmosphere are given by the first four columns in the didaktik model presented in Fig. 2.2. How a Vision

[7] The bolded letters are forming "ENACT".

III-atmosphere is surrounding the sociopolitical SSI-teaching sequence is illustrated in Fig. 2.4. Ideas for this atmosphere-illustration are taken from models in Yavuzkaya et al. (2022, p. 11) and Birdsall (2022, p. 248). Inspiration to the keywords added in the "atmosphere" were also gained from some other sources, such as an environmental education model by Cantell et al. (2019, p. 719) emphasizing future orientation, hope, knowledge, values, worldview, identity, thinking skills, emotions, motivation, participation, and action.

Vision III-driven (critical-eco-reflexive *Bildung*-oriented) didaktik includes for instance critical-reflexive attitude, broad and critical interdisciplinary understanding (including NOS, STSE and environmental socioscientific issues), focus on meaningfulness and global relevance, actor-network mapping, and teachers first putting important issues on the table and then make the content (in a broad sense) free (e.g. Van Poeck & Östman, 2020; Yavuzkaya et al., 2022). Additional more simple keywords in the new integrative didaktik model (Fig. 2.4) could have been: ecological worldview, future orientation, contextualization, and participation. However, these perspectives can be seen as integrated in the more complex keywords and phrases used in the model.

The new integrative didaktik model in Fig. 2.4 should, like all didaktik models, not be seen as finalized. In this paper it has theoretically been formulated/modelled based on several other models and publications, some of them empirical. After the new model has been mangled and tested in practice (practical didaktik modelling) (e.g. Wickman et al., 2020; Sjöström, 2022) in different contexts and together with

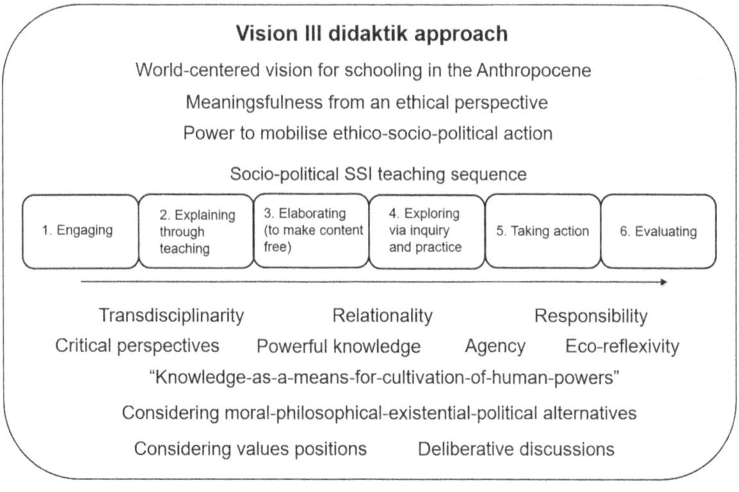

Fig. 2.4 Sociopolitical socioscientific issues teaching sequence driven by Vision III. The framing is inspired by models in Yavuzkaya et al. (2022, p. 11) and Birdsall (2022, p. 248). The three terms transdisciplinarity, relationality and responsibility are taken from Guyotte (2020). The quote "knowledge-as-a-means-for-cultivation-of-human-powers" is from Carlgren (2020, p. 324). Moral-philosophical-existential-political alternatives are connected to Vision III (e.g. Sjöström, 2018, 2024)

different teaching methods (mainly wisdom pedagogies), it will be adjusted. Most probable there will also be need for special versions of the model focusing on specific school levels, school subjects and/or sociopolitical issues.

2.6 Wisdom Pedagogies

This final section before the concluding remarks focuses on methods for teaching praxis in line with Vision III. In relation to the model in Fig. 2.1, we are now mainly in the right circle. It can also be described as didaktik models on the praxis-level—or in this case—as wisdom pedagogies (e.g. Jakubik, 2023). There have been suggested many different pedagogies for sustainability and transformative learning based on head, hands and heart (e.g. Sipos et al., 2008). These include, for instance, action learning, critical emancipatory pedagogy, pedagogy for eco-justice, and pedagogies based on indigenous knowledge. Related to such teaching methods is the so-called transdisciplinary pedagogy with its connections to STEAM and posthumanism (e.g. Burnard et al., 2021) as well as to ecopedagogy (e.g. Gadotti, 2010; Norat et al., 2016; Misiaszek, 2021) and so-called wild pedagogies (e.g. Blenkinsop et al., 2022).

In their paper about science education in the Anthropocence, Jeong et al. (2021) claim that a re-imagined science education requires new pedagogies. These may be undisciplined and unconventional, such as place-based pedagogies (read more about place-based education in, for instance: Dillon & Herman, 2023, pp. 721–725). Such Vision III-oriented teaching methods open up for posthumanism.

Recently, Jakubik (2023) discussed the need for wisdom pedagogies for enhancing students' *Bildung*, self-cultivation and life-world becoming. Such pedagogies are about "applying different knowledge wisely, with values, attitudes, authenticity, responsibility, [and] practical wisdom (phronesis)" (p. 56). They can, according to this author, facilitate emancipatory competence and critical knowing. It is about "Learning by life-world becoming, reflecting, questioning, criticizing feelings, passion, compassion, holistic understanding, [and] wise actions" (p. 56), driven by sustainability mindset principles, such as interconnectedness, oneness with nature, ecoliteracy, long term-thinking, reflection, and mindfulness (Rimanoczy, 2020).

Fuchs and Tan (2022, p. 14) argue that a Socially Responsible Science Education (SRSE) "that aims to achieve Vision III of scientific literacy" can be supported by the four frameworks "STSE, Socio-Scientic Issues-based, Critical Pedagogy of Place and STEPWISE" (all of them mentioned earlier in this paper). Similarly, Mang et al. (2021, p. 555) write: "Using an SSI-based STEAM approach can help to achieve Vision III by making science learning more relatable through holistic experiences that place equal emphasis on emotional learning, values, systematic thinking, and multiple perspectives (Sjöström et al., 2017; Sjöström & Eilks, 2018)."

Ibrahim et al. (2022) discuss the following five examples of STEPWISE pedagogies related to Vision III: (1) actor-network mapping as art experience; (2) art as a teaching medium of teaching ideologies influencing science; (3) art provocations of socioscientific issues as examples of student engagement in Vision III scientific literacy; (4) art provocations supporting student emancipation in Vision III scientific literacy; and (5) artistic designs of place that operate to uphold power. Bencze (2017) and his coworkers have studied many so-called RiNA-projects (=Research-informed & Negotiated Actions) as part of their STEPWISE framework. One of their findings is that by constructing actor-network maps, students are enlightened about problematic actors and become motivated to address socio-eco challenges.

However, many other teaching methods have also been suggested. The methods (or teaching strategies) mentioned in the circle to the right in Fig. 2.1 are, in addition to RiNA projects and actor-network mapping, epistemological disturbances (e.g. Bencze et al., 2020, p. 839), cartography of controversy (e.g. Bencze et al., 2020, p. 840), link with out-of-school actors (e.g. Bencze et al., 2020, p. 840), community projects (e.g. Pedretti & Nazir, 2011), scenarios of future (e.g. Bencze et al., 2020, p. 840), open dilemmas of glocal relevance (e.g. Bencze et al., 2020, p. 840; Pedretti & Nazir, 2011), drama (e.g. García-Puchades & Martos-García, 2022; Ødegaard, 2023; Danckwardt-Lillieström et al., 2024), and authentical evaluations (e.g. Valladares, 2021, p. 569). Other examples are, for instance, debates and role plays (e.g. Bencze et al., 2020, p. 839; Pedretti & Nazir, 2011; Avsar Erumit et al., 2024), deliberative discussions involving values, ethics and risk (e.g. Birdsall, 2022), inclusion of alternative knowledge systems (e.g. Pedretti & Nazir, 2011; Zidny et al., 2020), and engaging in sociopolitical actions (e.g. Bencze, 2017; Birdsall, 2022, pp. 246–247). Examples of critical pedagogy of place-methods are local park restoration projects and initiatives which study and clean up local watersheds (see references in Fuchs & Tan, 2022). Other examples of "nontraditional teaching features" are integrative contents, multiple relationships between science and society, and the expansion of classroom space (Marcelino et al., 2023).

As mentioned above didaktik—as teachers' discipline—contributes with theoretical support for teachers' well-informed decisions on didaktik questions such as the why, the what, and the how to teach a certain content to a certain group of students. Not at least when teaching sociopolitical issues, it is important that the science teacher is a "critical scholar and not a rationalist technician" (Moura et al., 2021, p. 519). It is teachers' agency, grounded in their critical-eco-reflexive *Bildung*, that will guide their didaktik choices (e.g. Mitchell & Stones, 2022; Yavuzkaya et al., 2022).

2.7 Concluding Remarks

This paper has discussed didaktik models[8] for socio-ecojustice-oriented teaching praxis. The main disciplinary subject area has been the sciences, but the discussion is hopefully useful for all teaching analysis and planning related to sociopolitical

[8] Including, as described in the paper, both frameworks for local curriculum making and teaching planning and practice.

issues. The discussed models and pedagogies are driven by environmental citizenship and a sociopolitically oriented Vision III of scientific literacy and science education. More concretely, the paper has discussed examples of educational frameworks and practices for socio-ecojustice and eco-reflexive *Bildung*. Vision III emphasizes moral-philosophical-existential-political alternatives and sociopolitical actions. A point of departure is an awareness that we are living in a post-pandemic world in the Anthropocene era. *Bildung* is a central concept in central European/ Scandinavian educational theory. It has a long and multifaceted history of ideas including, for instance, humanistic values and critical-democratic citizenship. Didaktik, which is closely connected to *Bildung*, can be seen as the "profession discipline" of teachers and therefore the academic field in the core of teacher education and teacher professional practice and development. A main idea of didaktik is that teachers have the agency to make autonomous and responsible choices in relation to curricula interpretations, pedagogies and the development of their students' *Bildung*. Important in didaktik research is to develop different types of arrangements, frameworks and tools for teachers, which can be supportive in their didaktik choices. With one word it is called didaktik models. Their role is to bridge theory and practice. There are three levels of didaktik models: meta-level; analysis/ planning-level; and praxis-level. From the perspective of critical-eco-reflexive *Bildung* the interest in this paper is didaktik models enhancing eco-reflexivity and "action *Bildung*". An integrative didaktik model driven by Vision III has been developed/modelled. It is based on the following six steps in a teaching sequence: (1) engaging in a sociopolitical SSI, (2) explaining as a result of teaching, (3) elaborating, (4) exploring, (5) taking action, and (6) evaluating. All this is performed in a "Vision III-atmosphere". Wisdom pedagogies is used as a term for corresponding praxis-oriented teaching methods.

Acknowledgments I am grateful for valuable comments from Editor Cristiano Moura, Professor Erminia Pedretti as well as two anonymous reviewers on earlier versions of the paper.

References

Agustian, H. Y. (2023). The critical role of understanding epistemic practices in science teaching using wicked problems. *Science & Education*. Online first: 22 September 2023. https://doi.org/10.1007/s11191-023-00471-2

Amos, R., Knippels, M. C., & Levinson, R. (2020). Socioscientific inquiry-based learning: Possibilities and challenges for teacher education. In M. Evagorou, J. A. Nielsen, & J. Dillon (Eds.), *Science teacher education for responsible citizenship: Towards a pedagogy for relevance through socioscientific issues* (pp. 41–61). Springer.

Andersen, L. R. (2022). The Bildung rose: An educational framework for grasping complexity. *Futures of Education, Culture and Nature – Learning to Become, 1*, 186–211.

Ariza, M. R., Christodoulou, A., van Harskamp, M., Knippels, M. C. P., Kyza, E. A., Levinson, R., & Agesilaou, A. (2021). Socioscientific inquiry-based learning as a means toward environmental citizenship. *Sustainability, 13*, 11509.

Avsar Erumit, B., Namdar, B., & Oğuz Namdar, A. (2024). Promoting preservice teachers' global citizenship and contextualised NOS views through role-play activities integrated into place-

based SSI instruction on climate issues. *International Journal of Science Education, 46*, 590–619.
Barad, K. (2000). Reconceiving scientific literacy as agential literacy. In R. Reid & S. Traweek (Eds.), *Doing science + culture* (pp. 221–258). Routledge.
Bencze, L. (Ed.). (2017). *Science and technology education promoting wellbeing for individuals, societies and environments—STEPWISE*. Springer.
Bencze, L. (2022). Politicized socioscientific issues education promoting ecojustice. In Y.-S. Hsu et al. (Eds.), *Innovative approaches to socioscientific issues and sustainability education: Linking research to practice* (pp. 351–365). Springer Nature.
Bencze, L., & Carter, L. (2011). Globalizing students acting for the common good. *Journal of Research in Science Teaching, 48*, 648–669.
Bencze, L., Sperling, E., & Carter, L. (2012). Students' research-informed socioscientific activism: Re/visions for a sustainable future. *Research in Science Education, 42*, 129–148.
Bencze, L., Pouliot, C., Pedretti, E., Simonneaux, L., Simonneaux, J., & Zeidler, D. (2020). SAQ, SSI and STSE education: Defending and extending "science-in-context". *Cultural Studies of Science Education, 15*, 825–851.
Bernstein, J. H. (2015). Transdisciplinarity: A review of its origins, development, and current issues. *Journal of Research Practice, 11*, article R1.
Biesta, G. (2012). Becoming world-wise: An educational perspective on rhetorical curriculum. *Journal of Curriculum Studies, 44*, 815–826.
Biesta, G. (2013). Responsive or responsible? Democratic education for the global networked society. *Policy Futures in Education, 11*, 733–744.
Birdsall, S. (2022). Socioscientific issues, scientific literacy, and citizenship: Assembling the puzzle pieces. In Y.-S. Hsu, R. Tytler, & P. J. White (Eds.), *Innovative approaches to socio-scientific issues and sustainability education: Linking research to practice* (pp. 235–250). Springer Nature.
Blenkinsop, S., Morse, M., & Jickling, B. (2022). Wild pedagogies: Opportunities and challenges for practice. In M. Paulsen, J. Jagodzinski, & S. M. Hawke (Eds.), *Pedagogy in the Anthropocene: Re-wilding education for a new earth* (pp. 33–51). Palgrave Macmillan.
Burnard, P., Colucci-Gray, L., & Sinha, P. (2021). Transdisciplinarity: Letting arts and science teach together. *Curriculum Perspectives, 41*, 113–118.
Bybee, R. W. (2014). The BSCS 5E instructional model: Personal reflections and contemporary implications. *Science and Children, 51*, 10–13.
Cantell, H., Tolppanen, S., Aarnio-Linnanvuori, E., & Lehtonen, A. (2019). Bicycle model on climate change education: Presenting and evaluating a model. *Environmental Education Research, 25*(5), 717–731.
Carlgren, I. (2020). Powerful knowns and powerful knowings. *Journal of Curriculum Studies, 52*, 323–336.
Carter, L., & Smith, C. (2003). Re-visioning science education from a science studies and futures perspective. *Journal of Future Studies, 7*, 45–54.
Chang Rundgren, S. N. (2023). Demonstrating didactic models for ESD and Bildung in school education. *Multidisciplinary Journal of School Education, 12*(2(24)), 15–31.
Chang Rundgren, S. N., & Rundgren, C.-J. (2010). SEE-SEP: From a separate to a holistic view on socioscientific issues. *Asia-Pacific Forum on Science Learning and Teaching, 11*, article 2.
Crutzen, P. J. (2006). The "anthropocene". In E. Ehlers & T. Krafft (Eds.), *Earth system science in the Anthropocene* (pp. 13–18). Springer.
Danckwardt-Lillieström, K., Andrée, M., & Rundgren, C. J. (2024). Process drama as a tool for participation in explorations of 'wicked problems' in upper secondary chemistry education. *LUMAT: International Journal on Math, Science and Technology Education, 12*, 50–79.
Deng, Z. (2020). *Knowledge, content, curriculum and Didaktik: Beyond social realism*. Routledge.
Deng, Z. (2022). Powerful knowledge, educational potential and knowledge-rich curriculum: Pushing the boundaries. *Journal of Curriculum Studies, 54*, 599–617.
Develaki, M. (2024). Uncertainty, risk, and decision-making: Concepts, guidelines, and educational implications. *Science & Education*. Online first: 15 July 2024. https://doi.org/10.1007/s11191-024-00544-w

Dillon, J., & Herman, B. (2023). Environmental education. In N. G. Lederman, D. L. Zeidler, & J. S. Lederman (Eds.), *Handbook of research on science education: Volume III* (pp. 717–748). Routledge.

Duit, R. (2015). Didaktik. In R. Gunstone (Ed.), *Encyclopedia of science education* (pp. 325–327). Springer.

El Halwany, S., Zouda, M., & Bencze, J. L. (2021). Stepping into STS literature: Some implications for promoting socioecological justice through science education. *Cultural Studies of Science Education, 16*, 1083–1096.

Elmose, S., & Roth, W.-M. (2005). *Allgemeinbildung*: Readiness for living in risk society. *Journal of Curriculum Studies, 37*, 11–34.

Fazio, X. (Ed.). (2023). *Science curriculum for the Anthropocene, Volume 2: Curriculum models for our collective future*. Springer International Publishing.

Feinstein, N. (2011). Salvaging science literacy. *Science Education, 95*, 168–185.

Fitzpatrick, H. (2023). A review of worldviews beyond sustainability: Potential avenues for human-nature connectedness. *Visions for Sustainability, 19*, 9–57.

Friedrichsen, P., Sadler, T., Graham, K., & Brown, P. (2016). Design of a socioscientific issue curriculum unit: Antibiotic resistance, natural selection, and modeling. *International Journal of Designs for Learning, 7*, 1–18.

Fuchs, T. T., & Tan, Y. S. M. (2022). Frameworks supporting socially responsible science education: Opportunities, challenges, and implementation. *Canadian Journal of Science, Mathematics and Technology Education, 22*, 9–27.

Gadotti, M. (2010). Reorienting education practices towards sustainability. *Journal of Education for Sustainable Development, 4*, 203–211.

García-Puchades, W., & Martos-García, D. (2022). Politicizing the learning of sustainable development goals through drama performances and agonistic debates: A teacher training experience from the perspective of radical democracy. *Environmental Education Research, 28*, 1656–1671.

Georgiou, Y., & Kyza, E. A. (2023). Fostering chemistry students' scientific literacy for responsible citizenship through socioscientific inquiry-based learning (SSIBL). *Sustainability, 15*, 6442.

Gilbert, J. (2016). Transforming science education for the Anthropocene—Is it possible? *Research in Science Education, 46*, 187–201.

Gray, D. (2018). Science education futures. Science education as if the whole earth mattered. *Visions for Sustainability, 9*, 60–76.

Gray, D., & Colucci-Gray, L. (2014). Globalisation and the Anthropocene: The reconfiguration of science education for a sustainable future. *Sisyphus-Journal of Education, 2*, 14–31.

Guerrero, G., & Sjöström, J. (2024). Critical scientific and environmental literacies: A systematic and critical review. *Studies in Science Education*. Online first: 13 May 2024. https://doi.org/10.1080/03057267.2024.2344988

Gundem, B. (2010). Didactics–Didaktik–Didactique. In C. Kridel (Ed.), *Encyclopedia of curriculum studies* (pp. 293–294). SAGE.

Guyotte, K. W. (2020). Toward a philosophy of STEAM in the Anthropocene. *Educational Philosophy and Theory, 52*, 769–779.

Hadjichambis, A. C., & Paraskeva-Hadjichambi, D. (2020). Education for environmental citizenship: The pedagogical approach. In A. C. Hadjichambis et al. (Eds.), *Conceptualizing environmental citizenship for 21st century education* (pp. 237–261). Springer.

Hadjichambis, A. C., Reis, P., Paraskeva-Hadjichambi, D., Činčera, J., Boeve-de Pauw, J., Gericke, N., & Knippels, M. C. (Eds.). (2020). *Conceptualizing environmental citizenship for 21st century education*. Springer.

Herranen, J., Yavuzkaya, M., & Sjöström, J. (2021). Embedding chemistry education into environmental and sustainability education: Development of a didaktik model based on an eco-reflexive approach. *Sustainability, 13*, 1746.

Hervé, N., & Panissal, N. (2022). Writing fictional short stories about the Anthropocene: Effects on students' futures thinking. *Frontiers in Education, 7*, 842252.

Hodson, D. (2003). Time for action: Science education for an alternative future. *International Journal of Science Education, 25*, 645–670.
Hodson, D. (2011). *Looking to the future: Building a curriculum for social activism*. Springer.
Högström, P., Gericke, N., Wallin, J., & Bergman, E. (2024). Teaching socioscientific issues: A systematic review. *Science & Education*. Online first: 03 July 2024. https://doi.org/10.1007/s11191-024-00542-y
Hopmann, S. (2007). Restrained teaching: The common core of *Didaktik*. *European Educational Research Journal, 6*, 109–124.
Hutchings, R. (2014). Understanding of and vision for the environmental humanities. *Environmental Humanities, 4*, 213–220.
Hwang, Y., Ko, Y., Shim, S. S., Ok, S. Y., & Lee, H. (2023). Promoting engineering students' social responsibility and willingness to act on socioscientific issues. *International Journal of STEM Education, 10*, 1–16.
Ibrahim, S., Del Gobbo, D., El Halwany, S., Zouda, M., Milanovic, M., Hassan, N., Krstovic, M., Kofman, N., & Bencze, L. (2022). STEAM not STEaM: Revisioned pedagogies prioritizing social justice and ecological sustainability in STEAM education. *Journal for Activist Science and Technology Education, 12*, 33–52.
Ingerman, Å., & Wickman, P.-O. (2015). Towards a teachers' professional discipline. In P. Burnard, B.-M. Apelgren, & N. Cabaroglu (Eds.), *Transformative teacher research: In theory and practice for the C21st* (pp. 167–179). Sense Publishing.
Jakubik, M. (2023). Cultivating the future in higher education: Fostering students' life-world becoming with wisdom pedagogy. *Trends in Higher Education, 2*, 45–61.
Jank, W. (2014). Didaktik, Bildung, content—On the writings of Frede V. Nielsen. *Philosophy of Music Education Review, 22*, 113–131.
Jeong, S., Sherman, B., & Tippins, D. J. (2021). The Anthropocene as we know it: Posthumanism, science education and scientific literacy as a path to sustainability. *Cultural Studies of Science Education, 16*, 805–820.
Kahn, S., & Zeidler, D. L. (2016). Using our heads and HARTSS*: Developing perspective-taking skills for socioscientific reasoning (*Humanities, ARTs, and Social Sciences). *Journal of Science Teacher Education, 27*, 261–281.
Kahn, S., & Zeidler, D. L. (2019). A conceptual analysis of perspective taking in support of socioscientific reasoning. *Science & Education, 28*, 605–638.
Krogh, E., Qvortrup, A., & Graf, S. T. (Eds.). (2021). *Didaktik and curriculum in ongoing dialogue*. Routledge.
Krogh, E., Qvortrup, A., & Graf, S. T. (Eds.). (2022). *Bildung, knowledge, and global challenges in education: Didaktik and curriculum in the Anthropocene era*. Routledge.
Kvamme, O. A. (2021). Rethinking *Bildung* in the Anthropocene: The case of Wolfgang Klafki. *HTS Theological Studies, 77*, 1–9.
Lacey, H. (2009). The interplay of scientific activity, worldviews and value outlooks. *Science & Education, 18*, 839–860.
Laherto, A., Rasa, T., Miani, L., Levrini, O., & Erduran, S. (2023). Future-oriented science education building sustainability competences: An approach to the European *GreenComp* framework. In X. Fazio (Ed.), *Science curriculum for the Anthropocene, Volume 2: Curriculum models for our collective future* (pp. 83–105). Springer International Publishing. https://doi.org/10.1007/978-3-031-37391-6_5
Levinson, R. (2017). SAQs as a socio-political programme: Some challenges and opportunities. *Sisyphus-Journal of Education, 5*, 25–39.
Levinson, R. (2018). Realising the school science curriculum. *The Curriculum Journal, 29*, 522–537.
Levinson, R., & the PARRISE consortium. (2017). Socioscientific inquiry-based learning: Taking off from STEPWISE. In L. Bencze (Ed.), *Science and technology education promoting wellbeing for individuals, societies and environments—STEPWISE* (pp. 477–502). Springer.
Lewis, S., & Maslin, M. (2015). Defining the Anthropocene. *Nature, 519*, 171–180.

Lönngren, J., & Van Poeck, K. (2021). Wicked problems: A mapping review of the literature. *International Journal of Sustainable Development & World Ecology, 28*, 481–502.

Lotz-Sisitka, H., Wals, A. E., Kronlid, D., & McGarry, D. (2015). Transformative, transgressive social learning: Rethinking higher education pedagogy in times of systemic global dysfunction. *Current Opinion in Environmental Sustainability, 16*, 73–80.

Mang, H. M. A., Chu, H. E., Martin, S. N., & Kim, C. J. (2021). An SSI-based STEAM approach to developing science programs. *Asia-Pacific Science Education, 7*, 549–585.

Marcelino, L. V., Dias, E. D., Rüntzel, P. L., Milli, J. C., Santos, J. S., Souza, L. C., & Marques, C. A. (2023). Didactic features specific to green chemistry teaching in the Journal of Chemical Education. *Journal of Chemical Education, 100*, 2529–2538.

Marks, R., & Eilks, I. (2009). Promoting scientific literacy using a socio-critical and problem-oriented approach to chemistry teaching: Concept, examples, experiences. *International Journal of Environmental and Science Education, 4*, 131–145.

Marks, R., Stuckey, M., Belova, N., & Eilks, I. (2014). The societal dimension in German science education – From tradition towards selected cases and recent developments. *Eurasia Journal of Mathematics, Science & Technology Education, 10*, 285–296.

Misiaszek, G. W. (2021). Ecopedagogical literacy of a pandemic: Teaching to critically read the politics of COVID-19 with environmental issues. *The Journal of Environmental Education, 52*, 358–369.

Mitchell, D. (2022). GeoCapabilities 3—Knowledge and values in education for the Anthropocene. *International Research in Geographical and Environmental Education, 31*, 265–281.

Mitchell, D., & Stones, A. (2022). Disciplinary knowledge for what ends? The values dimension of curriculum research in the Anthropocene. *London Review of Education, 20*, 23.

Mogensen, F., & Schnack, K. (2010). The action competence approach and the 'new' discourses of education for sustainable development, competence and quality criteria. *Environmental Education Research, 16*, 59–74.

Moura, C. B., Nascimento, M. M., & Lima, N. W. (2021). Epistemic and political confrontations around the public policies to fight COVID-19 pandemic: What can science education learn from this episode? *Science & Education, 30*, 501–525.

Norat, M. D. L. A. V., Herreria, A. F., & Rodriguez, F. M. M. (2016). Ecopedagogy: A movement between critical dialogue and complexity: Proposal for a categories system. *Journal of Education for Sustainable Development, 10*, 178–195.

O'Gorman, E., Van Dooren, T., Münster, U., Adamson, J., Mauch, C., Sörlin, S., et al. (2019). Teaching the environmental humanities: International perspectives and practices. *Environmental Humanities, 11*, 427–460.

Ødegaard, M. (2023). Using drama in science education and for sustainability issues. In D. McGregor & D. Anderson (Eds.), *Learning science through drama: Exploring international perspectives* (pp. 69–86). Springer.

Pedretti, E., & Nazir, J. (2011). Currents in STSE education: Mapping a complex field, 40 years on. *Science Education, 95*, 601–626.

Rasa, T., Lavonen, J., & Laherto, A. (2024). Agency and transformative potential of technology in students' images of the future: Futures thinking as critical scientific literacy. *Science & Education, 33*(5), 1145–1169.

Redvers, N., Guzmán, C. A. F., & Parkes, M. W. (2023). Towards an educational praxis for planetary health: A call for transformative, inclusive, and integrative approaches for learning and relearning in the Anthropocene. *The Lancet Planetary Health, 7*, e77–e85.

Reis, P. (2020). Environmental citizenship and youth activism. In A. C. Hadjichambis et al. (Eds.), *Conceptualizing environmental citizenship for 21st century education* (pp. 139–148). Springer.

Rimanoczy, I. (2020). *The sustainability mindset principles: A guide to developing a mindset for a better world.* Routledge.

Roberts, D. A. (2007). Scientific literacy/science literacy. In S. K. Abell & N. G. Lederman (Eds.), *Handbook of research on science education* (pp. 729–780). Lawrence Erlbaum.

Romine, W. L., Sadler, T. D., Dauer, J. M., & Kinslow, A. (2020). Measurement of socioscientific reasoning (SSR) and exploration of SSR as a progression of competencies. *International Journal of Science Education, 42*, 2981–3002.

Rowson, J. (2019). *Bildung in the 21st century—Why sustainability prosperity depends upon reimagining education* (CUSP essay in series on the Morality of Sustainable Prosperity, No 9). Retrieved from the World Wide Web August 09, 2024 at cusp.ac.uk/essay/m1-9

Sadler, T. D., Foulk, J. A., & Friedrichsen, P. J. (2017). Evolution of a model for socioscientific issue teaching and learning. *International Journal of Education in Mathematics, Science and Technology, 5*, 75–87.

Salinas, I., Fernández, M. B., Johnson, D., & Bastías, N. (2023). Freire's hope in radically changing times: A dialogue for curriculum integration from science education to face the climate crisis. *Cultural Studies of Science Education, 18*, 21–39.

Sass, W., Boeve-de Pauw, J., Olsson, D., Gericke, N., De Maeyer, S., & Van Petegem, P. (2020). Redefining action competence: The case of sustainable development. *The Journal of Environmental Education, 51*, 292–305.

Schenk, L., Hamza, K. M., Enghag, M., Lundegård, I., Arvanitis, L., Haglund, K., & Wojcik, A. (2019). Teaching and discussing about risk: Seven elements of potential significance for science education. *International Journal of Science Education, 41*, 1271–1286.

Simonneaux, L. (2014). From promoting the techno-sciences to activism—A variety of objectives involved in the teaching of SSIs. In L. Bencze & S. Alsop (Eds.), *Activist science and technology education* (pp. 99–111). Springer.

Sipos, Y., Battisti, B., & Grimm, K. (2008). Achieving transformative sustainability learning: Engaging head, hands and heart. *International Journal of Sustainability in Higher Education, 9*, 68–86.

Sjöström, J. (2013). Towards *Bildung*-oriented chemistry education. *Science & Education, 22*, 1873–1890.

Sjöström, J. (2018). Science teacher identity and eco-transformation of science education: Comparing Western modernism with Confucianism and reflexive *Bildung*. *Cultural Studies of Science Education, 13*(1), 147–161.

Sjöström, J. (2019). Didactic modelling for socio-ecojustice. *Journal for Activist Science and Technology Education, 10*, 45–56.

Sjöström, J. (2022). Didaktik modelling illustrated by sustainability teaching arrangements in preschool. *Educare, 5*, 249–280.

Sjöström, J. (2024). Vision III of scientific literacy and science education: an alternative vision for science education emphasising the ethico-socio-political and relational-existential. *Studies in Science Education*, Online first: 02 November 2024. https://doi.org/10.1080/03057267.2024.2405229

Sjöström, J. (2025, accepted). Science education for transformation: theorizing the three visions as a basis for new didaktik models for worldview awareness and sociopolitical transformation. In J. L. Bencze (Ed.), *Building Networks for Critical and Altruistic Science Education*. Springer Nature.

Sjöström, J., & Eilks, I. (2018). Reconsidering different visions of scientific literacy and science education based on the concept of *Bildung*. In Y. Dori, Z. Mevarech, & D. Baker (Eds.), *Cognition, metacognition, and culture in STEM education—Learning, teaching and assessment* (pp. 65–88). Springer.

Sjöström, J., & Eilks, I. (2020). The *Bildung* theory—From von Humboldt to Klafki and beyond. Chapter 5. In B. Akpan & T. J. Kennedy (Eds.), *Science education in theory and practice* (Springer Texts in Education) (pp. 55–67). Springer.

Sjöström, J., & Talanquer, V. (2018). Eco-reflexive chemical thinking and action. *Current Opinion in Green and Sustainable Chemistry, 13*, 16–20.

Sjöström, J., Eilks, I., & Zuin, V. G. (2016). Towards eco-reflexive science education: A critical reflection about educational implications of green chemistry. *Science & Education, 25*, 321–341.

Sjöström, J., Frerichs, N., Zuin, V. G., & Eilks, I. (2017). Use of the concept of *Bildung* in the international science education literature, its potential, and implications for teaching and learning. *Studies in Science Education, 53*, 165–192.

Sjöström, J., Eilks, I., & Talanquer, V. (2020). Didaktik models in chemistry education. *Journal of Chemical Education, 97*, 910–915.

Steffen, W., Persson, Å., Deutsch, L., Zalasiewicz, J., Williams, M., Richardson, K., et al. (2011). The Anthropocene: From global change to planetary stewardship. *Ambio, 40*, 739–761.

Steffen, W., Richardson, K., Rockström, J., Cornell, S. E., Fetzer, I., Bennett, E. M., et al. (2015). Planetary boundaries: Guiding human development on a changing planet. *Science, 347*, 1259855.

Taylor, C. A. (2017). Is a posthumanist *Bildung* possible? Reclaiming the promise of *Bildung* for contemporary higher education. *Higher Education, 74*, 419–435.

Valladares, L. (2021). Scientific literacy and social transformation: Critical perspectives about science participation and emancipation. *Science & Education, 30*, 557–587.

Valladares, L. (2022). Post-truth and education: STS vaccines to re-establish science in the public sphere. *Science & Education, 31*, 1311–1337.

Van Poeck, K., & Östman, L. (2020). The risk and potentiality of engaging with sustainability problems in education—A pragmatist teaching approach. *Journal of Philosophy of Education, 54*, 1003–1018.

Van Poeck, K., Vandenplas, E., & Östman, L. (2024). Teaching action-oriented knowledge on sustainability issues. *Environmental Education Research, 30*, 334–360.

Viehmann, C., Fernández Cárdenas, J. M., & Reynaga Peña, C. G. (2024). The use of socioscientific issues in science lessons: A scoping review. *Sustainability, 16*, 5827.

Vogelzang, J., Admiraal, W. F., & van Driel, J. H. (2020). Effects of Scrum methodology on students' critical scientific literacy: The case of green chemistry. *Chemistry Education Research and Practice, 21*, 940–952.

Wallace, M. F., Bazzul, J., Higgins, M., & Tolbert, S. (Eds.) (2022). *Reimagining science education in the Anthropocene*. Springer Nature.

Wickman, P.-O., Hamza, K., & Lundegård, I. (2020). Didactics and didactic models in science education. In P. J. White, R. Russell, J. Ferguson, & J. Cripps Clark (Eds.), *Methodological approaches to STEM education research. Volume 1* (pp. 34–49). Cambridge Scholar Publishing.

Yavuzkaya, M., Clucas, P., & Sjöström, J. (2022). ChemoKnowings as part of 21st century *Bildung* and subject didaktik. *Frontiers in Education, 7*, 869156.

Yavuzkaya, M., Hasslöf, H., & Sjöström, J. (2024). Becoming a science teacher in the Anthropocene: Scientific knowledge as 'matters of concern'. *International Journal of Science Education*, Online first: 25 June 2024. https://doi.org/10.1080/09500693.2024.2367174

Zeidler, D. L., Sadler, T. D., Simmons, M. L., & Howes, E. V. (2005). Beyond STS: A research-based framework for socioscientific issues education. *Science Education, 89*, 357–377.

Zeidler, D. L., Sadler, T. D., Applebaum, S., & Callahan, B. E. (2009). Advancing reflective judgment through socioscientific issues. *Journal of Research in Science Teaching, 46*, 74–101.

Zeidler, D. L., Herman, B. C., & Sadler, T. D. (2019). New directions in socioscientific issues research. *Disciplinary and Interdisciplinary Science Education Research, 1*, 1–9.

Zembylas, M. (2006). Science education as emancipatory: The case of Roy Bhaskar's philosophy of meta-reality. *Educational Philosophy and Theory, 38*, 665–676.

Zidny, R., Sjöström, J., & Eilks, I. (2020). A multi-perspective reflection on how indigenous knowledge and related ideas can improve science education for sustainability. *Science & Education, 29*, 145–185.

Jesper Sjöström is a full Professor of Science Education at Malmö University, Sweden. He is also a full trained upper-secondary school teacher oriented toward chemistry and general science/science studies. Two important concepts in his research are eco-reflexive Bildung and didaktik. Related to this he is interested in Bildung-oriented didaktik models and modeling as well as in the philosophy of science education. An important framing for him is the Anthropocene era and related global environmental and sociopolitical issues. Based on this he is asking how an eco-reflexive Bildung-oriented science education may look like?

Chapter 3
The (Un)intelligibility of Slow Violence in Science Education

Ajay Sharma and Ramnarayan Kalyanaraman

3.1 Introduction

COVID-19 pandemic has proved to be nothing but an unprecedented disaster in the way it left a ghastly trail of death and destruction since 2019 in all parts of the world. According to the World Health Organization, nearly 7 million worldwide deaths can be attributed to the pandemic (https://covid19.who.int/). By all measures, the violence unleashed by the pandemic has not only been fast, but also so spectacular and direct in an in-your-face kind of way. Even though we now appear to be in post-pandemic times, episodes of fast violence, such as the 2023 floods in Libya that killed more than 11,000 people, continue to sear our collective memoryscapes on a continual basis. But not all forms of violence catch our attention or are so easily recognizable. There are forms of violence that are neither spectacular nor swift and flow beneath the radar undetected and unrecognized by all except the ones who are directly afflicted. Even the ones who bear the brunt of such forms of violence may at times struggle to understand their nature.

This chapter is about such forms of slow violence. We focus specifically on the slow violence that is brought on by capitalism-induced disruptions of the socioecological systems of our world (Nixon, 2011). This is not just because socioecological slow violence disproportionately affects the poor and marginalized communities wherever it rears its head. We also wrote this chapter because, unfortunately, as we make the case ahead, slow violence is rendered largely unintelligible by the science education paradigm dominant in many countries around the world. This matters

A. Sharma (✉)
University of Georgia, Athens, GA, USA
e-mail: ajay@uga.edu

R. Kalyanaraman
Independent Scholar, Uttarakhand, India

C. B. Moura (ed.), *A Sociopolitical Turn in Science Education*, Contemporary Trends and Issues in Science Education 62,
https://doi.org/10.1007/978-3-031-78586-3_3

because we believe that a science education that makes this form of violence visible and intelligible will better prepare students to act against it with justice and equity. It has been long recognized that science education should help to prepare students for performing their diverse responsibilities as adult citizens. One of these roles is positively influencing public policy at all levels in ways that reflect current scientific understanding, and a respect for evidence and democratic decision-making (Anderson et al., 2006).

It is also well-known that despite decades worth of well-intentioned efforts, our progress towards meeting this goal has been woefully inadequate. Research indicates that this shortcoming is especially evident on socioecological issues where, for instance, despite possessing a favorable orientation towards environmentalism most adults in the United States are ill-prepared by science education to understand and act on such issues as well-informed, agentive citizens (Anderson, 2017). Some inadequacies in science education, such as nature-social dualism inherent in curricular representations of the natural world, that contribute to adult citizens' woeful understanding of the world around them are well known, and efforts to make improvements have been long underway (Sharma & Buxton, 2018). For instance, over the last few decades, a loosely aligned collective of efforts to reform science education has emerged to bring society within the ambit of science education, especially in the context of issues related to justice and equity in the science-society interface. Here I am referring to efforts like Socioscientific Issues in science education (SSI), Science, Technology, Society and Environment education (STSE), Citizen-science, and Socio-scientific Issue-based learning (SSIBL). Unfortunately, while some science teachers and schools have adopted pedagogical practices and orientations suggested by these reform efforts, the hegemony of the dominant science education paradigm has maintained.

In wider education research one does find some sporadic references to or focus on slow violence, such as Cairns' (2022) insightful paper on pedagogies of slow violence, and Higham's (2022) entry in the Oxford Research Encyclopedia of Education on the everyday forms of slow violence that gets inflicted on students in schools. Unfortunately, though, our review of research literature suggests that aforementioned reform efforts have not sufficiently recognized the importance of addressing the problem of slow violence through science education. There is, therefore, an urgent need for science educators to address this lacuna to better serve their goals of justice and equity-oriented science education in our schools. We begin by first introducing to the readers the nature of slow violence in the context of socioecological issues facing our planet. Then we segue into a detailed explanation of how slow violence becomes unintelligible in the dominant science education paradigm. Finally, we outline a vision of science education that might make slow violence intelligible and actionable for the students.

3.2 Understanding Slow Violence

Scientific consensus has been emphatically clear for quite some time that the current pace and scale of anthropogenic socioecological changes occurring on the planet pose an unprecedented challenge for sustainability of life and human societies (Rockström et al., 2009). Further, it has also been evident that globalized hyper-capitalism is the overwhelmingly dominant cause for most of these anthropogenic changes (Moore, 2017). Sometimes these anthropogenic changes to socioecological systems manifest in quick episodes of explosive, spectacular violence, such as extra-strong hurricanes, flash floods and landslides and causing tremendous loss of life and property in a short span. But, as Nixon (2011) showed, some kinds of violence caused by capitalism induced socioecological changes also happen slowly and almost imperceptibly. Through case studies from around the world, Nixon presented this *slow violence* as a "violence of delayed destruction that is dispersed across time and space, an attritional violence that is typically not viewed as violence at all ... a violence that is neither spectacular nor instantaneous, but rather incremental and accretive, its calamitous repercussions playing out across a range of temporal scales" (p. 2).

Nixon (2011) cites climate change, toxic drift, acidification of oceans, and thawing of cryosphere as some examples of socioecological crises that spawn slow violence on communities around the world in ways that "present formidable representational obstacles that can hinder our efforts to mobilize and act decisively" (p. 2), and thus are "underrepresented in strategic planning as well as in human memory" (p. 3). Key characteristics of slow violence are:

(a) Extended temporality: As mentioned, slow violence occurs unhurriedly over time. It is not one event. Rather, it is a gradual process that keeps developing, often hidden from its victims. For instance, the steady loss of habitable land for the small island nations of Maldives, Kiribati or Tuvalu has been a very slow process occurring over several decades. This loss is caused by slowly rising sea levels because of global climate crisis that in turn lead to increased flooding, intrusion of saltwater in freshwater supplies, and destruction of coral reefs that protect the islands against storms. Research suggests that the rate of sea level rise in and around Maldives is only about 0.8–1.6 millimeters per year (Bindoff et al., 2007). To an untrained human eye this would appear as an almost imperceptibly slow rise. But, over time the small changes in sea level add up to yield disastrous consequences as research suggests that much of the land in these island nations will become uninhabitable by 2050s (Storlazzi et al., 2018).

(b) Spatial dispersion: Slow violence is often dispersed and can causally connect regions that may not be contiguous with each other. For instance, greenhouse gases produced in one region contribute to global climate crisis and thus influence communities around the world. Similarly, pesticides used on farms in one place have been found to adversely impact communities far from those lands (Harrison, 2011).

(c) Invisibility: Because of its slow and dispersed nature, often slow violence cannot be easily discerned. An apropos example would be the recent drinking water crisis in Flint, Michigan, USA (Pauli, 2019). The presence of lead in the water piped to homes in Flint was invisible to those consuming it as its presence cannot be smelled, tasted, or seen in water. Further, because lead was present in low quantities it took years of exposure to show itself in the form of delayed health effects, such as poor cognitive development and learning disabilities in children—issues that could easily be misdiagnosed, especially as it was poor, Black families that were disproportionately exposed to lead in water (Clark, 2018). Of course, as Davies (2022) pointed out in his study of the people living with toxic pollution in the 'cancer alley' region of Louisiana, many a times the slow violence may be quite visible to the people suffering from it, but their stories "do not *count*, thus rendering certain populations and geographies vulnerable to sacrifice" (p. 409). That is, lacking the spectacle and directness of more direct forms of 'fast' violence, such cases often get delegated to the margins in media as well as policy discourses. Davies (2022) implicates the "politics of indifference" as a key factor that invisiblizes slow violence suffered by the marginalized communities for the rest of the world. Referring to this representational hurdle, Nixon (2011) poses an important question that deserves to be answered:

> In an age when the media venerate the spectacular, when public policy is shaped primarily around perceived immediate need, a central question is strategic and representational: how can we convert into image and narrative the disasters that are slow-moving and long in the making, disasters that are anonymous and that star nobody, disasters that are attritional and of indifferent interest to the sensation-driven technologies of our world image? (p. 3).

(d) Uncertain attribution: It can often be exceedingly difficult to unravel the causes of slow violence as the issues that cause slow violence also often happen to be 'wicked' problems from a social policy perspective (Fladvad, 2021). Wicked problems show many characteristics that make attribution of causes particularly problematic. They are difficult to define, unique, and lend themselves to multiple competing explanations (Peters, 2017). For instance, it is not clear how to apportion responsibility in cases of a wicked problem like climate crisis. There is no consensus who deserves the most blame—fossil fuel corporations, governments, consumers, citizens, or some combination of all of these actors?

(e) Disproportionate impact on the poor and marginalized: Nixon's (2011) study as well as the vast corpus of research on slow violence that his study led to have amply shown that the impact of slow violence in most cases is disproportionately borne by the poor and marginalized. Unfortunately, neither do they usually have the resources to cope with the slow violence nor the political voice and heft to prevent or mitigate its effects. This can be clearly seen in the case of pesticide poisoning among agricultural workers in most parts of the world (Perfecto, 2019). Agricultural workers, especially in poor or developing countries, like India, are disproportionately exposed to pesticides and other chemicals used in farming. For instance, the rise in cancer cases because of long-term exposure to pesticides has ravaged poor farming communities in much of the rural Punjab region of India (Ahluwalia & Kaur, 2021).

(f) Delayed Impacts: As we saw above in the cases of pesticide poisoning among agricultural workers, the impact of slow violence is often felt after a long-time lag. Another example would be the delayed impacts of exposure to asbestos in workers employed in construction, shipbuilding and other related industries. Research has shown that malignant mesothelioma, a particularly lethal form of cancer, is strongly correlated with exposure to asbestos (Alpert et al., 2020). Even though asbestos has been banned in most countries since the 1970s, the disease shows a long latency and the cases of asbestos exposure linked malignant mesothelioma are still being reported in countries that have long banned it (Petrof et al., 2020; Huh et al., 2022).

Case studies documented by Nixon (2011) show that global hyper-capitalism is directly implicated in the rise of slow violence in different parts of the world. Following Nixon, other scholarship on slow violence has also amply documented how through boundless commodification of nature as well as labor, turbo-capitalism has served to enable a regressive redistribution of wealth and power from the poor to the rich while externalizing environmental and social costs and inflicting slow violence upon the poor and marginalized (El, 2020). This form of virulent, runaway capitalism is sustained and legitimized by the hegemonic discourse of neoliberalism. After this introduction to slow violence, we now shift our attention to science education to show how slow violence is unintelligible in the science education paradigm dominant in many parts of the world.

3.3 The Unintelligibility of Slow Violence in Science Education

It is true that many of the socioecological crises that inflict slow violence are now considered important topics to be taught in science classrooms around the world. Efforts to educate children about climate crisis are good examples (Henderson & Drewes, 2020). However, we contend that even on such topics, science education often renders slow violence unintelligible to students. Our argument highlights discursive, curricular as well as pedagogic dimensions of science education. Of course, because we are speaking of science education across many societies and diverse sociocultural and political contexts, our thesis perforce makes broad generalizations. Nonetheless, our hope is that readers will see our core arguments as plausible hypotheses that highlight an important issue and thus deserve to be further explored through multiple empirical investigations in diverse educational contexts.

We begin by bringing our readers' attention to the profoundly significant reform in science education that has been sweeping schools in the United States as well as many other capitalist societies. One of us has written about this reform effort as the neoliberal STEM-ification of science education (Sharma, 2016). This effort entails instrumentalization of science education for the social efficiency purposes of preparing workers for the economy. As a result, technical problem-solving and engineering-type tasks seem to have increasingly replaced open-ended, curiosity-driven scientific investigations as the primary focus of scientific inquiry in science classrooms (Rudolph, 2005). This trend to instrumentalize science learning for problem solving is also evident in the recent changes in the school science curricular standards in the United States. For instance, while engineering always had an implicit presence in the earlier National Science Education Standards (National Research Council, 1996) in the United States, these standards were supplanted by the Next Generation Science Standards in 2013. The new standards positioned engineering in a more central role in science education both as a pedagogical approach and as an important content area in itself (Purzer et al., 2014). Expectedly, corporations have tended to support such efforts. For instance, British Petroleum has sought to promote their own interests through a neoliberal model of STEM education in primary and secondary schools in the United Kingdom (Tannock, 2007). We see similar critiques against STEM-ification of science education emerging from other capitalist societies (Mendick et al., 2017). One can safely surmise that this trend coupled with the neoliberal accountability measures and standardization of curricula has further diminished whatever little curricular space science educators may have had for enacting place-based Science-Technology-Society-Environment (STSE) or Socioscientific Issues (SSI) focused science pedagogy where issues of slow violence could be naturally contextualized for the students (Carter, 2016).

Second, as we well know, school science is not the same as scientists' science. Science taught to students is recontextualized from scientific research communities so as to fit the systemic constraints of mass, public education (Sharma & Anderson, 2009). However, under the hegemonic influence of neoliberalism, the recontextualization process transmogrifies scientists' science in ways that make teaching about slow violence almost impossible for even the most enterprising science educators. Quite like the fast food served in fast-food restaurants, the current system of "fast schooling" is designed to offer pre-packaged and pre-processed, bite-sized units of science content to the students. These units are first conveniently broken into measurable scientific knowledge and skill units and then transmitted to the students at a fast pace optimized to produce maximum student performance in terms of quantifiable measures of learning. As a result, units and lesson plans rarely last beyond a few weeks, and the topics and issues taught once are either never taught again or are re-visited (often by some other teacher) after arbitrarily determined and rather long gaps. It is not surprising then that a recent systematic review of climate change education indicates that though they are generally recognized as ineffectual still top-down, didactic approaches continue to dominate teaching of climate change in science classrooms (Rousell & Cutter-Mackenzie-Knowles, 2020).

Naturally, this would make teaching about slow violence in school settings exceedingly difficult. Of course, a science teacher could always talk about slow violence in an abstract way or present condensed case studies of distant communities ravaged by slow violence. But it is difficult to see how the slow temporality, spatially dispersed, and complex nature of slow violence could be reduced to bite-sized curriculum content without obscuring both the mundane, unspectacular everyday nature of violence and the oppressive logics of capitalism that sustains it. It is no surprise then that as Cairns (2022) suggests when environmental justice issues are taught in science classrooms, educators feel compelled to engage in pedagogies of immediacy that "mobilize feelings of personal accomplishment through immediate action in the local environment, with an eye to creating good environmental citizens of the future" (p. 522). Such pedagogies also carry risks that, without a sustained investigation of underlying structural causes of socioecological crises, a classroom based, bite-sized representation of slow violence may reduce such topics to ""ruin porn", which trades in sensationalized imagery and obscures the historical and structural roots of crisis" and "feed narratives of "damage" that depict communities as "defeated and broken"" (Cairns, 2022, p. 524).

Third, socio-ecological crises that inflict slow violence are typically categorized as *wicked problems* within the field of social policy (Camillus, 2008). Rittel and Webber (1973) developed the concept of wicked problems to explain how certain social policy problems cannot be solved through science alone. They listed ten key characteristics of social policy problems that make them wicked, such as there is often no clear, definitive way to define such problems, they never get completely solved, solutions are not right/wrong, but better/worse, and such problems straddle organizational and disciplinary boundaries. Socio-ecological crises have been seen as typically wicked because they are "defined by high complexity, uncertainty, and contested social values'' (Miller, 2013, p. 279). Further, these issues are usually born from "the functioning and evolution of interconnected and complexly interacting socio-ecological systems'' and defy solutions because "they are multicausal, intertwined with other problems, and value-laden" (Metzger & Curren, 2017; p. 94). One also wonders if, with the ever-increasing excess of nature-social dualism busting "quasi-objects" or "naturecultures" (Latour, 2012) on the planet, such as the climate crisis and genetically modified animals, socioecological problems may be becoming even more wicked over time (Sharma & Buxton, 2018). Since scientists' science alone is not enough to understand socioecological wicked problems, it is no surprise that school science too is ill-equipped to help students comprehend how socioecological problems could perpetrate slow violence on the poor and marginalized.

Fourth, there exists a serious paradigm lag between scientists' science and school science. In today's world, where the majority of the biosphere can be located in agricultural and settled anthropogenic biomes, ecologists no longer believe in the nature-social dualism that characterized much of ecological sciences in the twentieth century (Schmitz, 2016). Unfortunately, however, school science is still stuck with the old ecology paradigm. It continues to reify the nature-social dualism by excluding humans from the 'natural world'. In such a 'natural world' socioecological issues typically point to harm occurring mostly to the nonhuman world. As a result, students get the idea that if polar ice caps are melting then it is the polar bears that are threatened but not the humans. And in a few situations where humans are represented as getting harmed, these unfortunate individuals are invariably shown to be situated in lands far away from most students in capitalist societies (Sharma &

Buxton, 2018). This distance may not only de-temporalize and de-individualize the threatened communities and thus reducing them to as caricatures of unlucky victims of unstoppable tragedies, we would argue that it also encourages a cult of futility that may engender "denial, paralysis, apathy or even perverse reactive behavior" (Hulme, 2007, p. 818).

This paradigm lag permits continuation of other outdated ideas in the curriculum as well that are not very helpful in understanding socioecological crises. These include the use of the balance of nature metaphor to understand homeostasis in natural systems and the "black boxed" cybernetic system approach of the ecosystem ecology (Sharma & Buxton, 2018). For instance, following the traditional ecosystem ecology perspective, students learn to trace matter cycles and energy flows in closed ecosystems. However, in the Anthropocene it would be unrealistic to expect closed ecosystems. That is, matter and energy flows across ecosystem boundaries are not the exceptions but rather the norm in the world we live in. Further, it is not just matter and energy flows but also organism and information flows (in the form of flow of genes as well as transfer of visual, auditory, and chemical signals) that have been found to be critical for understanding socioecological systems in the Anthropocene (Jianguo, 2013). For instance, current polar ecology can hardly be understood without considering the historic ongoing polewards movement of many temperate zone species (Thomas, 2020).

Fifth, at their core, slow violence issues are ethical issues as they deal with understanding the nature of harm done to human (and nonhuman) communities, apportioning responsibility and identifying ways to render justice to the involved entities. However, an explicit instruction in ethics would be an anomaly in most science classrooms. In its bid to come across as objective as a method of inquiry as well as a body of knowledge about the world, the dominant school science paradigm tends to ignore the ethical dimensions of socioecological issues that manage to find a place in the science curricula (Sharma, 2020). Further, while ethical literacy finds its place as an important component of the null curriculum, neoliberal ethical norms get passed on to the students as part of the hidden curriculum in science classrooms. These ethical norms reify human exceptionalism, normalize instrumental economic rationality, and present environmental action as a desirable personal virtue to the students (Sharma & Alvey, 2021).

Finally, while science educators have been advocating for the use of controversial issues in science learning, controversial issues also happen to arouse strong emotions (Reiss, 2022). In recognition, science educators and researchers have long emphasized the importance of emotions in science teaching and learning (Zembylas, 2005). However, research indicates that the affective dimension continues to be ignored, and scientific knowledge and cognitive understanding remain the dominant pedagogic goals of science learning (Rousell & Cutter-Mackenzie-Knowles, 2020). Further, most teacher preparation programs fail to adequately prepare teachers to understand and respond to students' emotions in ways that contribute to their learning (Waajid et al., 2013). It is no wonder then that most teachers feel unprepared and thus tend to avoid controversial topics that can be expected to evoke strong emotions amongst students (Pace, 2021). As a result, the dominant science education paradigm has little space for socioecological topics involving slow violence as they can definitely be expected to arouse strong emotions amongst students on account of the unjustifiable suffering endured by marginalized communities ravaged by slow violence.

While we certainly hope to see more research on this issue, the arguments outlined above give us reasonable basis to suspect that slow violence is effectively unintelligible in the currently dominant science education paradigm. Thus, if the goal is to help students understand slow violence as an important component of socio ecological issues then we may have to commit to working towards something akin to a Kuhnian paradigm shift in science education. We do not expect this task to be simple, quick, or even imminent. Currently, under the hegemony of progressive neoliberalism, the science education policy environment exists in a kind of equilibrium that, while open to small, incremental changes, tends to be hostile to any attempt for transformational change (Sharma, 2023). As has been seen often in social policy research, policy environment tends to become receptive to efforts for bringing about a paradigm shift only when external socio-political conditions themselves undergo a significant change that in turn makes 'business as usual' science education policy untenable for the policy and political elite (Cairney, 2019). Nevertheless, it would be a mistake to rest on our oars waiting for an opportune policy climate to arrive. Instead, what may help is to keep working on the conceptual scaffolding that will be needed for the paradigm shift when the right time arrives. Towards that end, the following section presents some characteristics that we hope to see in a new and transformed science education paradigm, one that should make slow violence intelligible and actionable for most science students.

3.4 Making Slow Violence Intelligible in Science Education

As long as science education remains STEM-ified, and thus geared towards producing STEM-qualified workers for the economy, it is difficult to imagine a science education that makes slow violence intelligible. So, the first step would be to continue ongoing efforts to denormalize and delegitimize the neoliberal hegemony over science education through sustained critique and building of alternative visions for a just and equity focused science education. Thanks to neoliberalism, social efficiency goals are dominant not just in science education, but in all disciplinary areas. Thus, following the strategy suggested by Laclau and Mouffe (2000) for fighting against the hegemony of oppressive discourses, science educators might also need to link up with similar efforts in other disciplinary areas through creation of "chains of equivalences" that establish commonalities between different groups of educators contesting the neoliberal hegemony and unite them in opposition as a contingently allied common front.

As mentioned earlier, in the current science education paradigm, students get to learn about socioecological crises that are mostly situated in communities that are spatially and/or temporally distant from theirs. At best, this can offer only a sanitized, superficial and problematic understanding of slow violence to students. Unfortunately, despite sustained advocacy by votaries of socioscientific issues (SSI), science, technology, society and environment education (STSE), and other similar justice and equity-oriented place-based pedagogical approaches, students rarely get to learn about their socioecological world through local, community situated student-led investigations. This will need to change if we wish to make slow violence intelligible for the students. Slow violence's slowness, mundane, unspectacular and spatially dispersed nature works to obscure itself. But if students investigate, they will find that slow violence exists in all communities, including their own.

Now, 'place' is often construed as 'local' in place-based education. However, as Ulrich Beck (2009) has argued, we live in a global, hyper-connected "risk society" where potentially catastrophic dangers are increasingly global rather than local. As a result, according to Beck, the risks facing any community have become increasingly delocalized at three levels: (a) spatial—risks no longer stay within any geographical, political boundary; (b) temporal—as discussed earlier many socioecological issues have a long latency period and thus effects cannot be readily anticipated and restricted within a short time span; and (c) social—through long and complex chains of effects the social effects of socioecological issues are often felt in faraway communities. This calls for place-based pedagogical approaches that situate the learning of science in local community contexts. But, equally importantly, these approaches should also lead students to interrogate the nature of 'locality' in learning about slow violence with an 'eco-cosmopolitan' spirit that encourages a sense of shared responsibility for the global environment, transcending local and national boundaries. Further, because exploitative and extractive capitalist systems are integral to the complex causal webs implicated in slow violence, these place-based approaches (as some of these approaches already are) will also need to be critical in intent so as to enable students to understand the socioeconomic dimension of slow violence and better appreciate its underlying issues of justice and equity (Gruenewald, 2001).

Further, as mentioned earlier, the slow, dispersed nature of slow violence cannot be taught in the standardized, fragmented, bit-sized curriculum units through pedagogies of immediacy (Cairns, 2022). Not only is slow violence difficult to see and measure, attempts to teach slow violence in the 'fast schooling' mode run the risk of reinscribing the psychologically debilitating narratives of damage in students whose communities are suffering from slow violence. Slow violence entails gradual, incremental changes in a "zone of ordinariness" (Berlant, 2007) that would be invisible to *fast observation* that pedagogies of immediacy (Cairns, 2022) depend upon. Thus, some environmental educators have been advocating adoption of *slow pedagogies* for environmental education in general (Payne & Wattchow, 2008) as well as in the specific case of teaching about slow violence (Cairns, 2022).

As O'Lear (2021) comments, "In communities affected by slow violence, such as places where toxic pollution has harmful impacts over time, much can be gained by paying attention to the spatial and temporal scales of lived experience" (p. 7). Thus, scholars who study geographies of slow violence emphasize the importance of *slow observation* that allows one to notice the slow degradation of the environment as a lived experience. Indeed, as Davies (2018) avers *slow observation* is "a critical means through which polluted communities understand the lived reality of persistent environmental threats" (p. 1538). Following these critical geographers, we expect slow pedagogies to adopt the methodology of slow observation to "make visible and accessible forms of harm and depleted realities that are perceived only through spending time" in a community ravaged by slow violence (O'Lear, 2021, p. 12). Educators will need to work with students to help them explore their own community's past to better understand the incremental nature of slow violence. For instance, Cairns (2022) showed how an environmental justice focused organization engaged in *pedagogies of excavation* with local students that led them to "excavate histories of slow violence to develop a structural analysis of contemporary conditions" that helped them understand slow violence plaguing their community (p. 20).

Scientists alone cannot investigate and document slow violence occurring in communities all over the world. However, in a welcome development, we also see that in recent years communities themselves have stepped up to seek collaboration with local and academic experts to monitor and document incremental, dispersed destruction of their local socio-ecological regimes through citizen-science based participatory ecological monitoring projects (Farquhar, 2012). Further, we are also much encouraged to note that in many parts of the world, the younger generation, particularly the indigenous youth, has begun to play a leading role in initiating and sustaining efforts to mitigate the effects of climate change and other socioecological crises (Ritchie, 2021). Thus, we see significant potential in the civic science based pedagogies to emerge as critically oriented slow pedagogies that are well-suited for understanding slow violence (Dillon et al., 2016). Participation in long-term, slow observation based citizen science and citizen social science projects should make visible for students slow violence "that is neither spectacular nor instantaneous, but rather incremental and accretive" (Nixon, 2011; p. 2) and afflicts their communities. These projects paired with critical oriented exposure to the capitalist, structural underpinnings of slow violence should also empower students to pursue environmental justice actions that redress harms done to their communities.

Of course, we will need to make systemic and curricular reforms to create conditions that make it easier and feasible for educators to implement slow pedagogies in their science classrooms. We imagine such reforms to include abandonment of standardized science curricula and high stakes teacher accountability regimes in favor of teacher professional empowerment, curricular freedom, and democratic accountability (Cochran-Smith et al., 2017). Further, given that socioecological issues that spawn slow violence are quintessential wicked problems, we cannot expect students to understand slow violence with the help of science alone. In recent years, sustainability science has become the preferred approach for understanding wicked socio-ecological issues because of its transdisciplinary focus on understanding "the interactions between natural and social systems, and with how those interactions affect the challenge of sustainability: meeting the needs of present and future generations while substantially reducing poverty and conserving the planet's life support systems'' ("Sustainability Science", n.d.). Thus, we expect that in the new science education paradigm science educators will be encouraged and enabled to adopt a sustainability science perspective on slow violence and the socioecological issue engendering it. Moving away from its current industry focused STEM moorings, sustainability science oriented school science will bring science closer to social studies for making slow violence intelligible for the students.

In contrast with the traditional ecological perspective, sustainability science is based on the new ecological paradigm that moves away from the nature-social dualism by seeing our environment as "both the product of and the setting for human interactions" (Scoones, 1999; p. 479), and assuming that "human political, cultural, religious, and economic institutions influence how nature works" and "feedbacks from nature can instigate institutional change in a co-dependent way" (Schmitz, 2016; p. 7). The new ecological paradigm stresses the importance of (a) nonlinear dynamics in matter cycles and energy flows, (b) cross-scale interactions and feedback relations that have both ecological and social dimensions, (c) self-organization in complex systems, (d) ecological (material, energy, organism and information) flows across ecosystem boundaries, (e) social-ecological resilience, and (f) environmental governance in understanding our human dominated world (Sharma & Buxton, 2018). As a result, a sustainability science oriented school will not only be more up-to-date with the current state of knowledge in ecological science, it should also be better equipped to help students understand disturbed and degraded socio-ecological regimes that produce slow violence.

Finally, as one of us have argued earlier, transdisciplinary and sustainability science oriented school science will also need to have ethical literacy embedded in it as an integral element (Sharma, 2020). Slow violence poses difficult ethical dilemmas. Deciding what is the right and fair course of action is rarely easy, especially when the causal webs associated with slow violence spread far and wide both spatially and temporally in all sorts of complicated ways. Thus, students will need not only a basic understanding of different ways to think ethically about a situation, but also repeated exposure to ethical dilemmas so that over time students can develop a habitual disposition of *phronesis* for dealing with ethically fraught situations. That is, "the kind of prudential judgment by which equivocal circumstances are negotiated and acted upon with a view for the common good" (Chia et al., 2013, p. xv).

3.5 Conclusion

Post-pandemic times are not normal times. Current era of neoliberalism endorsed globalized hyper-capitalism is causing increasingly frequent episodes of quick violence. But they get duly reported and attended to by the state and society. Unfortunately, the currently hegemonic economic system is also creating vast geographies of slow violence that often evade notice, get no mention in the media, and receive no ameliorating response from anyone. For now, it may be easy to pretend that slow violence exists only in a few faraway communities and as such may not deserve any curricular space in a science classroom. But that can only provide a short time, false comfort because the fact of the matter is that slow violence occurs everywhere, even in the most prosperous communities in the west. Thus, if the capitalist plunder of the world goes on unchecked and unregulated for a few more decades, the luxury of pretending that all is well and we are safe from slow violence may not be available to even the most privileged communities in the west.

Finally, while the importance and legitimacy of social efficiency goals for science education cannot be denied, science education cannot be normalized to exist for the sake of producing well-qualified workers for the economy alone. Further, science may have been used to commodify our world and enrich the already rich. But it has also been used and should continue being used to bring about a more equitable and just world—a world where slow violence gets readily recognized and attended to. That would also mean that we make social and ecological justice important goals of science education. Unfortunately, as we made the case in this chapter, the currently dominant school science paradigm is not quite compatible with social and ecological justice goals for science education as it faces a serious intelligibility problem with regards to slow violence. The wicked nature of slow violence is partly to be blamed. But, more importantly, as we explained it is also the current overall neoliberal orientation of science education as well as schooling in general, paradigm lag between scientists' science and school science, lack of transdisciplinarity and ethical literacy, and exclusion of the affective dimension that makes the current school science paradigm ill-equipped to prepare students to understand and act for justice and equity as regards slow violence.

The current neoliberal hegemony in science education may look formidable and enduring for now—enough to make visions of a paradigm shift towards a slow violence sensitive science education a distant and vague pipe dream. But if scholars like Fraser (2017) and Wainwright and Mann (2018) are to be believed, the ongoing global sociopolitical and ecological cataclysms may soon punctuate the neoliberal policy equilibrium, leading to destabilization that could upend neoliberal hegemony and usher in a range of possible dystopian and utopian futures. Critical scholars would be well advised to prepare for this day and create conditions for more hopeful futures by building the conceptual wherewithal needed for bringing about the paradigm shift in science education of the sort advocated in this chapter.

References

Ahluwalia, M., & Kaur, A. (2021). Assessment of chromosomal aberrations among agricultural workers exposed to pesticides in Punjab, India. *Journal of Biochemical and Molecular Toxicology, 35*(2), e22646.

Alpert, N., van Gerwen, M., & Taioli, E. (2020). Epidemiology of mesothelioma in the 21st century in Europe and the United States, 40 years after restricted/banned asbestos use. *Translational Lung Cancer Research, 9*(Suppl 1), S28.

Anderson, M. (2017). *For earth day, here's how Americans view environmental issues*. Pew Research Center. Retrieved from http://www.pewresearch.org/facttank/2017/04/20/for-earth-day-heres-how-americans-view-environmentalissues/

Anderson, C. W., et al. (2006, April). *Overview of NARST multiple paper set: Learning progressions toward environmental literacy*. Annual meeting of the National Association for Research in Science Teaching, San Francisco.
Beck, U. (2009). *World at risk*. Polity.
Berlant, L. (2007). Slow death (sovereignty, obesity, lateral agency). *Critical Inquiry, 33*(4), 754–780.
Bindoff, N. L., Willebrand, J., Artale, V., Cazenave, A., Gregory, J. M., Gulev, S., Hanawa, K., Le Quere, C., Levitus, S., & Nojiri, Y. (2007). Observations: Oceanic climate change and sea level. In S. Solomon, D. Qin, M. Manning, Z. Chen, M. Marquis, K. B. Averyt, M. Tignor, & H. L. Miller (Eds.), *Climate change 2007: The physical science basis. Contribution of working group I to the fourth assessment report of the intergovernmental panel on climate change*. Cambridge University Press.
Cairney, P. (2019). *Understanding public policy: Theories and issues*. Bloomsbury Publishing.
Cairns, K. (2022). Feeling environmental justice: Pedagogies of slow violence. *Curriculum Inquiry, 51*(5), 522–541.
Camillus, J. C. (2008). Strategy as a wicked problem. *Harvard Business Review, 86*(5), 98.
Carter, L. (2016). Neoliberalism and STEM education. *Journal for Activist Science and Technology Education, 7*(1), 31–41.
Chia, R., Holt, R., & Yuan, L. (2013). In praise of strategic indirection: Towards a non-instrumental understanding of phronésis as practical wisdom. In M. Thompson & D. Bevan (Eds.), *Wise management in organisational complexity* (pp. 53–67). Springer.
Clark, A. (2018). *The poisoned city: Flint's water and the American urban tragedy*. Metropolitan Books.
Cochran-Smith, M., Baker, M., Burton, S., Chang, W.-C., Cummings Carney, M., Fernández, M. B., Stringer Keefe, E., Miller, A. F., & Sánchez, J. G. (2017). The accountability era in US teacher education: Looking back, looking forward. *European Journal of Teacher Education, 40*(5), 572–588.
Davies, T. (2018). Toxic space and time: Slow violence, necropolitics, and petrochemical pollution. *Annals of the American Association of Geographers, 108*(6), 1537–1553.
Davies, T. (2022). Slow violence and toxic geographies: 'Out of sight' to whom? *Environment and Planning C: Politics and Space, 40*(2), 409–427.
Dillon, J., Stevenson, R. B., Wals, A. E. J., & Editors, G. (2016). Introduction. *Conservation Biology, 30*(3), 450–455. http://www.jstor.org/stable/24760973
El, E. (2020). *Slow violence in contemporary American environmental literature*. Cambridge Scholars Publishing. https://books.google.com/books?id=IjcTEAAAQBAJ
Farquhar, S. C. (2012). *Keeping track of nature: Interdisciplinary insights for participatory ecological monitoring*. University of Edinburgh.
Fladvad, B. (2021). Rethinking democracy in times of crises: Towards a pragmatist approach to the geographies of emerging publics. *Social Science Information, 60*(2), 230–252.
Fraser, N. (2017). From progressive neoliberalism to Trump—And beyond. *American Affairs, 1*(4), 46–64.
Gruenewald, D. A. (2001). *A critical pedagogy of place: A counter-discourse for sustainability*. The University of New Mexico.
Harrison, J. L. (2011). *Pesticide drift and the pursuit of environmental justice*. MIT Press.
Henderson, J., & Drewes, A. (2020). Teaching climate change in the United States. In *Teaching climate change in the United States* (pp. 1–10). Routledge.
Higham, L. (2022). Slow violence and schooling. In *Oxford Research Encyclopedia of Education*. Oxford University Press.
Huh, D.-A., Chae, W.-R., Choi, Y.-H., Kang, M.-S., Lee, Y.-J., & Moon, K.-W. (2022). Disease latency according to Asbestos exposure characteristics among malignant mesothelioma and Asbestos-related lung cancer cases in South Korea. *International Journal of Environmental Research and Public Health, 19*(23), 15934.
Hulme, M. (2007). Newspaper scare headlines can be counter-productive. *Nature, 445*(7130), 818–818.

Jianguo, W. (2013). Key concepts and research topics in landscape ecology revisited: 30 years after the Allerton Park workshop. *Landscape Ecology, 28*(1), 1–11.
Laclau, E., & Mouffe, C. (2000). *Hegemony and socialist strategy: Towards a radical democratic politics*. Verso.
Latour, B. (2012). *We have never been modern*. Harvard University Press.
Mendick, H., Berge, M., & Danielsson, A. (2017). A critique of the STEM pipeline: Young people's identities in Sweden and science education policy. *British Journal of Educational Studies, 65*(4), 481–497.
Metzger, E. P., & Curren, R. R. (2017). Sustainability: Why the language and ethics of sustainability matter in the geoscience classroom. *Journal of Geoscience Education, 65*(2), 93–100.
Miller, T. R. (2013). Constructing sustainability science: Emerging perspectives and research trajectories. *Sustainability Science, 8*(2), 279–293. https://doi.org/10.1007/s11625-012-0180-6
Moore, J. W. (2017). The Capitalocene, part I: On the nature and origins of our ecological crisis. *The Journal of Peasant Studies, 44*(3), 594–630.
National Research Council. (1996). *National Science Education Standards*. N. A. Press. http://www.nap.edu/readingroom/books/nses/html/index.html
Nixon, R. (2011). *Slow violence and the environmentalism of the poor*. Harvard University Press.
O'Lear, S. (2021). *A research agenda for geographies of slow violence: Making social and environmental injustice visible*. Edward Elgar Publishing.
Pace, J. L. (2021). *Hard questions: Learning to teach controversial issues*. Rowman & Littlefield Publishers.
Pauli, B. J. (2019). *Flint fights back: Environmental justice and democracy in the Flint water crisis*. MIT Press.
Payne, P. G., & Wattchow, B. (2008). Slow pedagogy and placing education in post-traditional outdoor education. *Journal of Outdoor and Environmental Education, 12*, 25–38.
Perfecto, I. (2019). Pesticide exposure of farm workers and the international connection. In *Race and the incidence of environmental hazards* (pp. 177–203). Routledge.
Peters, B. G. (2017). What is so wicked about wicked problems? A conceptual analysis and a research program. *Policy and Society, 36*(3), 385–396.
Petrof, O., Neyens, T., Nuyts, V., Nackaerts, K., Nemery, B., & Faes, C. (2020). On the impact of residential history in the spatial analysis of diseases with a long latency period: A study of mesothelioma in Belgium. *Statistics in Medicine, 39*(26), 3840–3866.
Purzer, S., Moore, T. J., Baker, D., & Berland, L. (2014). Supporting the implementation of NGSS through research: Engineering. *NARST Annual International Conference*.
Reiss, M. J. (2022). Learning to teach controversial topics. In *Handbook of research on science teacher education* (pp. 403–413). Routledge.
Ritchie, J. (2021). Movement from the margins to global recognition: Climate change activism by young people and in particular indigenous youth. *International Studies in Sociology of Education, 30*(1–2), 53–72.
Rittel, H. W., & Webber, M. M. (1973). Dilemmas in a general theory of planning. *Policy Sciences, 4*(2), 155–169.
Rockström, J., Steffen, W., Noone, K., et al. (2009). A safe operating space for humanity. *Nature, 461*, 472–475. https://doi.org/10.1038/461472a
Rockstrom, J., Steffen, W., Noone, K., Persson, A., Chapin, F. S., III, Lambin, E. F., Lenton, T. M., Scheffer, M., Folke, C., & Schellnhuber, H. J. (2009). A safe operating space for humanity: Identifying and quantifying planetary boundaries that must not be transgressed could help prevent human activities from causing unacceptable environmental change, argue Johan Rockstrom and colleagues. *Nature, 461*(7263), 472–476.
Rousell, D., & Cutter-Mackenzie-Knowles, A. (2020). A systematic review of climate change education: Giving children and young people a 'voice' and a 'hand' in redressing climate change. *Children's Geographies, 18*(2), 191–208. https://doi.org/10.1080/14733285.2019.1614532
Rudolph, J. L. (2005). Inquiry, instrumentalism, and the public understanding of science. *Science Education, 89*(5), 803–821.

Schmitz, O. J. (2016). *The new ecology: Rethinking a science for the Anthropocene*. Princeton University Press.

Scoones, I. (1999). New ecology and the social sciences: What prospects for a fruitful engagement? *Annual Review of Anthropology, 28*, 479–507. https://doi.org/10.2307/223403

Sharma, A. (2016). STEM-ification of education: The zombie reform strikes again. *Journal for Activist Science and Technology Education, 7*(1), 42–51.

Sharma, A. (2020). Phronetic science for wicked times. *Journal for Activist Science and Technology Education, 11*(2), 7–15.

Sharma, A. (2023). Can we be hopeful yet? In M. Weinstein et al. (Eds.), *Science educations for just and hopeful worlds*. Springer.

Sharma, A., & Alvey, E. M. (2021). The undercurrents of neoliberal ethics in science curricula: A critical appraisal. *Ethics and Education, 16*(1), 122–136.

Sharma, A., & Anderson, C. (2009). Recontextualization of science from lab to school: Implications for science literacy. *Science & Education, 18*(9), 1253–1275. https://doi.org/10.1007/s11191-007-9112-8

Sharma, A., & Buxton, C. (2018). *The natural world and science education in the United States*. Springer.

Storlazzi, C. D., Gingerich, S. B., Van Dongeren, A., Cheriton, O. M., Swarzenski, P. W., Quataert, E., Voss, C. I., Field, D. W., Annamalai, H., & Piniak, G. A. (2018). Most atolls will be uninhabitable by the mid-21st century because of sea-level rise exacerbating wave-driven flooding. *Science Advances, 4*(4), eaap9741.

Sustainability Science. Retrieved from http://sustainability.pnas.org/

Tannock, S. (2007). To keep America number 1: Confronting the deep nationalism of US higher education. *Globalisation, Societies & Education, 5*(2), 257–272.

Thomas, C. D. (2020). The development of Anthropocene biotas. *Philosophical Transactions of the Royal Society B, 375*(1794), 20190113.

Waajid, B., Garner, P. W., & Owen, J. E. (2013). Infusing social emotional learning into the teacher education curriculum. *International Journal of Emotional Education, 5*(2), 31–48.

Wainwright, J., & Mann, G. (2018). *Climate Leviathan: A political theory of our planetary future*. Verso Books.

Zembylas, M. (2005). Emotions and science teaching: Present research and future agendas. In *Beyond Cartesian dualism: Encountering affect in the teaching and learning of science* (pp. 123–132). Springer.

Ajay Sharma is an associate professor in the Department of Educational Theory and Practice, at the University of Georgia, Athens, USA. His current research centers on theoretical and ethnographic explorations of neoliberalism's impact on education. In his past research, Ajay Sharma has focused on representations of nature in science education, and classroom discourse in K-12 science classrooms from the perspectives of individual agency and equity.

Ramnarayan Kalyanaraman has worked, for the last 28 years, in the remote Greater Himalayan Gori River valley region located in Pithoragarh district in the Indian State of Uttarakhand. He is currently working in multiple Himalayan landscapes in the Kumaon region of Uttarakhand. His principal role has been that of a conservation facilitator and practitioner, rural livelihood specialist, community mobilizer, natural history and climate change educator and guide, and as a researcher in ecology and environment along with being an advocate of community rights and environmental justice.

Chapter 4
(Re)considering Science Education in the Face of Environmental Injustices in Pre- and Post-pandemic Worlds: Insights from a Critical-Decolonial Nature of Science

Haira Gandolfi

4.1 Environmental (In)justices and Education: An Overview

Scholarship in the field of environmental education (EE) has a longstanding history across the world, both in the global North and in the global South, and it has been particularly connected with issues around human engagement with and life in the natural world, including here occupation of land, exploitation of natural resources (e.g., water, plants, animals, minerals), and degradation of the natural world (e.g., air, water, soil). As exemplified by the COVID-19 pandemic, such interactions between humans and the natural world involve high levels of complexity around the deeply intertwined webs/systems of material nature that are constitutive of such interactions, as illustrated by terms that have become prominent in the public arena because of the pandemic, e.g., viruses, physical contact, airborne, masks, social distancing. In addition, the pandemic also brought to our attention the socio-political character grounding these human-nature interactions, such as the deep inequalities linked to possibilities for reducing contact with the coronavirus (e.g., social distancing) and to access to masks, health care (including vaccinations), etc. Thus, as we move to an apparent post-pandemic world, considering the kinds of relationships that we seek to rebuild (or perhaps maintain) with the natural world, including the associated inequalities that permeate such connections, becomes central to those of us interested in the connected fields of EE and science education.

In the specific case of EE, in the past two decades we have seen the emergence of debates about the role of education in what has been called *environmental (in)*

H. Gandolfi (✉)
University of Cambridge, Cambridge, UK
e-mail: heg38@cam.ac.uk

C. B. Moura (ed.), *A Sociopolitical Turn in Science Education*, Contemporary Trends and Issues in Science Education 62,
https://doi.org/10.1007/978-3-031-78586-3_4

justices[1] which, as is the case of the COVID-19 pandemic, are deeply intertwined with both the material and the socio-political aspects of human-nature interactions. However, unlike COVID-19, environmental injustices are not a simple case of *pre* and *post*: they have been deeply ingrained with our local and global centuries of histories of occupation, exploitation and degradation of the natural world. As noted by Acosta (2012) and Grosfoguel (2016), such kinds of engagement with the natural world and with-each-other-in-relation-to-the-natural-world have been central elements of many of our socio-political-economic systems, such as mercantilism, colonialism, and capitalism. In many ways, as argued by others, the COVID-19 pandemic can be seen as part of a wider crisis (Ahmed et al., 2022; Alsop & Bencze, 2020)—or part of a *polycrisis* (Morin & Kern, 1999)—around how such kinds of systems have been persistently shaping and impacting human-human and human-nature interactions. As such, environmental injustices have existed long before and through the pandemic and continue to exist as we move onto a post-pandemic world. Thus, across this chapter I seek to explore, from the perspective of such longstanding environmental injustices, the particular role(s) that science education can play as we reconsider our interactions with the natural world and with-each-other-in-relation-to-the-natural-world as part of post-pandemic praxis.

According to the *Rede Brasileira de Justiça Ambiental* (Brazilian Network of Environmental Justice) (RBJA, 2001, *my own translation*), the construct of environmental (in)justice involves:

(a) ensuring that no social group, based on ethnicity, race or class, bears a disproportionate share of the negative environmental consequences of economic operations, political decisions and federal, state, local programs, as well as the absence or omission of such policies;
(b) ensuring fair and equitable access, direct and indirect, to the country's environmental resources;
(c) ensuring broad access to relevant information on the use of environmental resources and the destination of waste and location of sources of environmental risks, as well as democratic and participatory processes in the definition of policies, plans, programs and projects that concern them;
(d) favouring the constitution of collective subjects of rights, social movements and popular organizations to be protagonists in the construction of alternative development models, which ensure the democratization of access to environmental resources and the sustainability of their use.

Overall, this term has been used to refer to the (un)equal access to the environment by and the (un)equal impact of environmental degradation to specific—and often marginalised—groups across the world (Schlosberg, 2013). This construct then asks us to approach issues to related to the environment, such as environmental

[1] Throughout this chapter, I will employ the joint term 'environmental (in)justice' to signpost that discussions in this area need to not only support a collective aim for 'environmental justice', but also explicitly recognise the socio-historical legacies and impacts of centuries of 'environmental injustices' faced by people and communities across the world.

disasters and the climate crisis, not only from a purely material and/or scientific perspective, but also from the socioeconomic, political and historical perspectives of communities that have been severely impacted by environmental issues for centuries (Acselrad, 2010). As such, across this scholarship, we often find three interconnected framings surrounding this term—(1) legal, (2) scientific, and (3) socio-political:

(1) Legal framing: grounded on equity in access to rights and legal procedures in cases of violation of environmental laws linked to people's lives in the natural world (Santos, 2018).
(2) Scientific framing: grounded on scientific knowledge and practices around preservation of the environment and the mitigation of environmental disasters and their impact, including technical planning and studies around, for instance, housing, energy generation, etc. (Mitchell, 2019).
(3) Socio-political framing: beyond the delimitation of what the legal and of scientific scopes above are, this framing places the notion of justice in dialogue with other fields of knowledge related to socio-political structures including, for instance, issues pertaining to culture and citizenship around access to clean water, secure food supply, etc. (Schlosberg, 2013).

Several scholars (e.g., Acselrad, 2010; Schlosberg, 2013) in the field have recently pointed out the limitations of working within the landscape of environmental (in) justice without considering the three framings outlined above interconnectedly. In particular, they criticise an apparent extensive focus on legal and scientific framings only, with less attention paid to socio-political voices and dynamics traditionally concerned with building wider social justice. An example of this kind of limitation can be found in the Brumadinho tragedy[2] in 2019 in Minas Gerais, Brazil: this event happened when a dam owned by the mining company Vale do Rio Doce collapsed causing the death of dozens of people and immense environmental impact in the area (and beyond). If approached from a legal framing only, environmental justice in such a case often ends up being simply related to applying fines and/or sanctions to the company from a financial point of view, which are usually of short-term and low-impact nature, as actually happened in the Brazilian case. Instead, meaningful consideration of socio-political framings in environmental (in)justices should go beyond legal and scientific delimitations, that is: the socioeconomic, political and historical perspectives of communities that have been severely impacted by environmental issues for centuries, should be central to considerations of environmental (in)justices.

The relevance of this socio-political framing is made even more evident when we delve deeper into the distribution of cases of environmental (in)justices across the world. Not unlike what we saw across the world during the COVID-19 pandemic (Ahmed et al., 2022), the *Most Affected People and Areas* (MAPA) in relation to

---

[2] See more here: https://www.bbc.co.uk/news/world-latin-america-51220373 and https://www.hrw.org/news/2019/01/30/trail-death-after-another-dam-collapses-brazil

measurements of environmental impact mostly coincide with groups facing socio-historical and political marginalisation, such as communities in the Global South, Black, Indigenous, people of colour (BIPOC) and LGBTQIA+ communities, women, etc. living anywhere in the world (Voskoboynik, 2018; Young, 2021). As outlined in Fig. 4.1 below, such communities and areas have been historically and contemporarily the most affected by environmental degradation, as illustrated by the overwhelming figures related to environmental disasters across places such as

Fig. 4.1 What is MAPA? (Source: Fridays for Future, retrieved from https://fridaysforfuture.org/newsletter/edition-no-1-what-is-mapa-and-why-should-we-pay-attention-to-it/)

Brazil (as exemplified above), Pakistan,[3] Black and Latine neighbourhoods in the USA (as exemplified in Fig. 4.2 below), the Congo Basin,[4] etc.

Of particular importance emerging from this socio-political framing is the recognition of the intertwined nature between environmental degradation and socio-historical processes related to issues of exploitation—of people, of land and of the

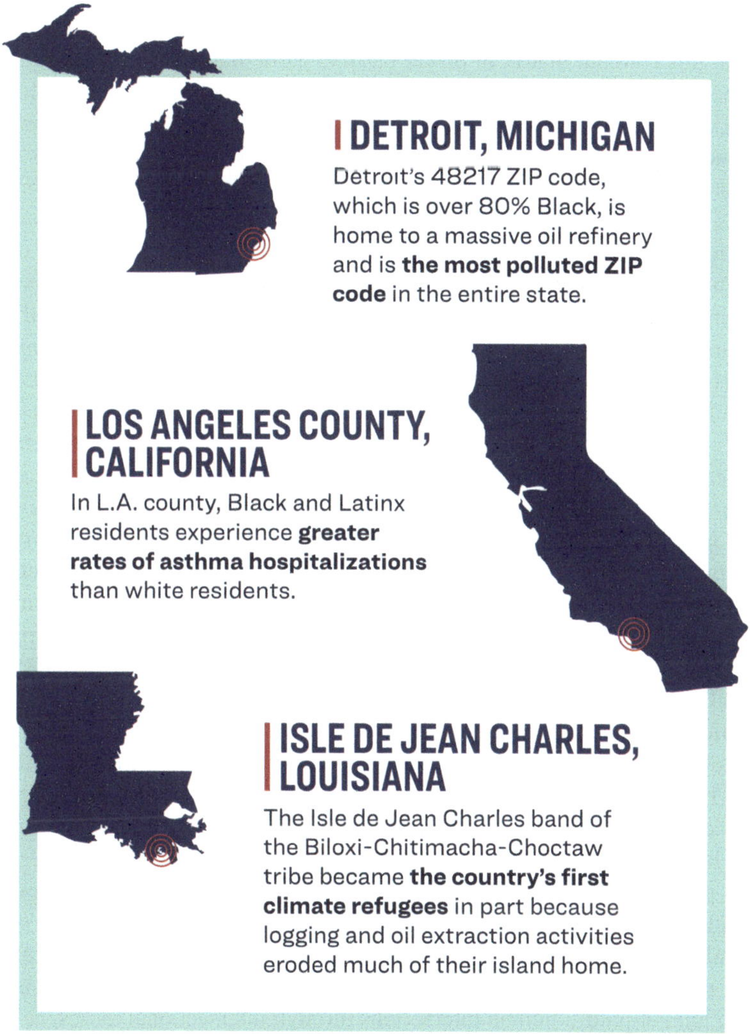

Fig. 4.2 Examples of cases of environmental injustice across the USA. (Source: https://elizabethwarren.com/plans/environmental-justice)

[3] https://www.theguardian.com/environment/2022/sep/15/pakistan-floods-made-up-to-50-worse-by-global-heating

[4] https://friendsofthecongo.org/climate-change/

environment more generally—across the centuries, especially connected to what Quijano (2007) calls the *colonial matrix of power*. According to this author, a colonial matrix of power comprises four (interconnected) domains: control of the economy, through land appropriation, exploitation of labour and natural resources; control of authority, through institutions such as army, law enforcement, schools, etc.; control of gender and sexuality, through moral teachings via family and education; and control of subjectivity and knowledge, through education, etc. Decolonial scholars like Quijano (2007), Mignolo and Walsh (2018), and Tuck and Yang (2012) argue that this colonial matrix of power has had particular relevance to issues relating to the human-nature interactions, especially through practices of exploitation colonialism: that is, grounded on biological and geopolitical management and appropriation of indigenous' lands, and on the control and exploitation of natural resources (such as minerals, plants, water) through displacement and genocidal strategies.

As a result of this colonial matrix of power, Le Grange (2023) argues that the killing and displacement of ecosystems—or *ecocide*—by colonial projects is still present in socio-scientific and technological approaches to environmental issues framed around purposes of, for instance, *progress, growth,* and *development* (Tuck et al., 2014). As highlighted by the historian of science Roy (2018), *"[modern] science was itself built upon a global repertoire of wisdom, information, and living and material specimens collected from various corners of the colonial world. Extracting raw materials from colonial mines and plantations went hand in hand with extracting scientific information and specimens from colonised people"*. Examples of such (neo)colonial socio-scientific/technological endeavours around human-nature interactions that are deeply connected with environmental (in)justices can be seen in the persistent positioning of formerly colonised communities—such as South American countries like Brazil (see Fig. 4.3) and Central and Southern African countries (see Fig. 4.4)—as global sources of minerals for manufacturing and electronics industries, while also bearing the brunt of environmental impacts from such kind of exploitation (Eckstein et al., 2021).

This chapter is then grounded on the position that environmental (in)justice needs to be approached beyond the legal and scientific framings outlined above to properly centre the socio-political (including historical) perspectives of communities that have been severely impacted by environmental issues for centuries, especially in places such as the Global South, where exploitation of natural resources has been both the livelihood (Roy, 2018; Voskoboynik, 2018) and the bane of environmental struggles (Eckstein et al., 2021). By engaging with this socio-political framing, environmental (in)justice allows us to approach environmental issues surrounding human-nature-human interactions through intersectional lenses, acknowledging the unequal exposition of already marginalised communities to the impact of environmental degradation resulting from land, knowledge, labour and monetary exploration by certain social groups driving the colonial matrix of power. In particular, when grounded on the kind of decolonial thinking outlined above, this framing prompts us to reflect on important socio-political questions as we reconsider human-nature-human interactions in our post-pandemic world around, for instance: what

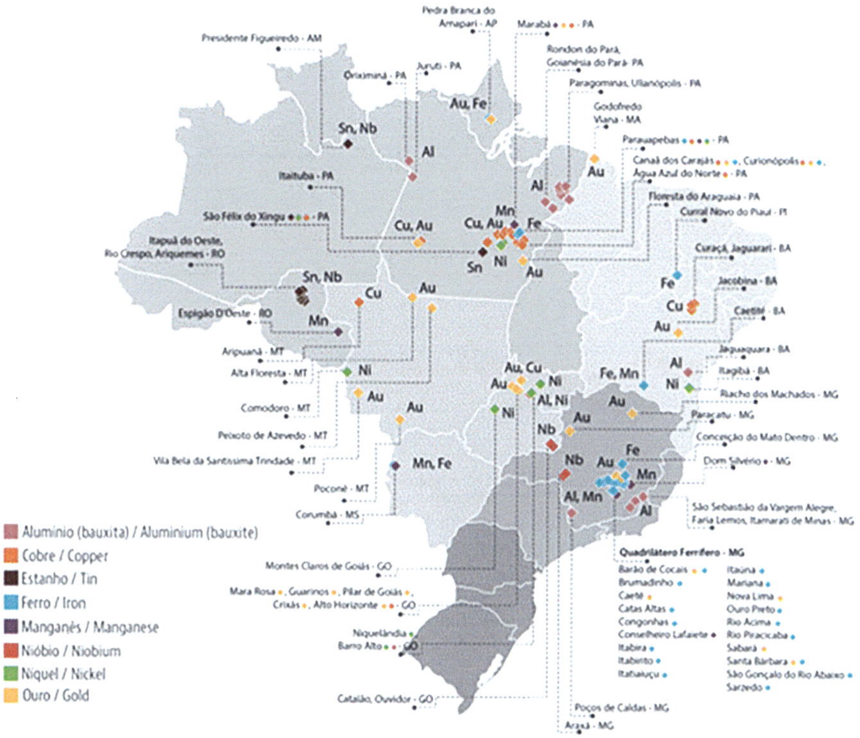

Fig. 4.3 Minerals and ores currently (2019) exploited across Brazil. (Source: Agência Nacional de Mineração/ANM, retrieved from: https://blog.jazida.com/mineracao-no-brasil/)

happens when human development in the natural world puts different communities and populations in an unequal position in relation to environmental impact, generating an uneven distribution of MAPA? How can calls for *sustainable development* consider both the environmental and the socioeconomic burden that a global over-reliance on natural resources has placed on certain communities? And what happens when our approaches to scientific and technological development, sometimes even the so-called *sustainable* ones, clash with quests for social justice?

This particular approach to environmental (in)justice has been, however, generally absent from the field of EE in the Global North, the context from where I write this chapter, even before the COVID-19 pandemic. While EE has taken many paths throughout its history and across the world,[5] in Global North countries like England, it has been closely connected with outdoor education and sustainable development goals/SDGs, with less attention paid to socio-political (and associated social justice) framings (Gandolfi, 2023a; Martin et al., 2013). Glackin and King (2020), for

[5] In Latin America, for example, it has been indeed historically tied to land and ecological justice movements (e.g., Ferguson et al., 2021; the Landless Workers Movement).

Fig. 4.4 Minerals and ores currently (2007) exploited across the African continent. (Source: https://www.mapsofworld.com/africa/minerals-and-ores.html)

instance, argue that whilst different political climates have influenced how environmental issues are addressed within the mainstream English education sector, there is a particular trend in how the *content* of EE is often approached by educational policies, which can be summarised as: *"a focus on teaching a) about the environment (i.e., content and skills about the environment), but not on teaching b) for (i.e., environmental advocacy and action) nor c) in (i.e., outside traditional classrooms) the environment"* (Gandolfi, 2023a, p. 3). This particular vision for EE means that environmental issues are often addressed in England through content-based approaches grounded on the learning of *facts* about the environment, ignoring the role of institutions, peoples and wider social structures within these issues (Tannock, 2020; Gandolfi, 2023a), and with a stark absence of any engagement with the complex socio-political, moral and ethical aspects that are centred by the notion of environmental (in)justice I outlined above.

In addition, as recently argued by Dunlop and Rushton (2022a) in England and by colleagues in other Global South and North contexts (e.g., Gandolfi et al., 2024; Eaton & Day, 2020; Layrargues, 2018; Misiaszek, 2019; Schilling Trein, 2018), such EE approaches often tend to position this area *for* the green market, that is, for the development of green technical knowledge, skills, etc., while maintaining the structures of global capitalism (e.g., neoliberalism) that drive such exploitative and unjust human-nature-human interactions unchallenged. In these cases, EE seems to have been occupied by neoliberal terms like *development, opportunities, net zero* and *green careers*[6] that also are central for maintaining the colonial matrix of power outlined by Quijano (2007): they do not seek to challenge the impact resulting from our current (capitalism) and old (mercantilism, colonialism) hegemonic socio-political-economic models in the appropriation of nature by certain parts of society, which also have also incurred in dispute over territories, conflicts of socio-environmental nature and, ultimately, intense and persistent structures of environmental injustices (Misiaszek, 2019).

This *occupation* of EE discourses and approaches by neoliberal and (neo)colonial ideas surrounding human-nature-human interactions (Eaton & Day, 2020; Tannock, 2020) then promotes specific views on environmental issues that are dissociated from global, socio-political and systemic perspectives, such as the role of corporate power and geopolitics in the climate crisis. For instance, findings from Misiaszek's (2019) research about EE with teachers in Brazil, Argentina and in the USA, and from Eaton and Day's (2020) study with teachers in Canada, show how very few discussions are actually found in EE practices around the unequal footing between our private spheres (i.e., individual actions) and public spheres (i.e., larger socio-political structures at both national and international stages) behind environmental challenges, and even less about those issues linked to the colonial matrix of power outlined by Quijano (2007). According to Maldonado-Torres (2007) and Quijano (2007), this kind of *obstruction of socio-political* discussions is a classic trope of (neo)colonial endeavours, in which these practices are used to justify exploitation of natural and human resources for the sake of wider scientific, technological, and social development.

It is then from this landscape of EE and environmental (in)justices that I continue to write this chapter, now moving onto exploring the particular role that science education, an area that has been traditionally linked to this landscape, can play within a socio-politically environmental (in)justice-informed perspective to EE.

---

[6] *Green careers* are those created within the existing capitalist job market, such as *environmental consultant*, with the aim of helping companies with strong links to environmental injustices, degradation and exploitation to rebrand/reposition themselves as *green*. Within certain EE landscapes, the relevance of this field has been framed exactly in relation to preparing young people for such careers.

4.2 Environmental Injustices, Science Education, and Nature of Science

Within contemporary scholarship around the intersection between science and EE, researchers in England (e.g., Dunlop et al., 2021; Glackin & King, 2020; Gandolfi, 2023a; Gandolfi et al., 2024), my context of work as a science educator, have identified a persisting trend in how the *content* of EE has been outlined policies and frameworks (e.g., national curriculum and examinations): a focus on teaching *the* scientific knowledge about the environment (e.g., rise in CO_2 levels in the atmosphere; temperature trends; life cycles of certain materials), but not on teaching *how* that knowledge relates to wider societal issues pertaining to the environment, including in relation to environmental (in)justices. Such trends are often grounded on discourses of *impartiality,*[7] positioning these discussions as belonging only to sociology, history and politics lessons, with science needing to *stick to the facts* about *the* environment. As recently argued by Rezende and Ostermann (2020), behind this scenario there is a prevalent view of science—both as discipline and as a human endeavour—as neutral and objective (or *technical rational*), devoid of any social, political and cultural aspects. And even though recent scholarship from the fields of History, Philosophy and Sociology of Science (e.g., Harding, 2008; Kelly et al., 1993; Roy, 2018)—alongside our recent unequal political and onto-epistemological experiences of the COVID-19 pandemic (see Moura et al., 2021)—have challenged such naïve technical rationality perspective of scientific development, socio-political discussions remain consistently absent from how science education engages with environmental and other socio-scientific issues like the pandemic, as I argued elsewhere (e.g., Gandolfi, 2019, 2021a, 2023b) alongside other colleagues working from diverse Global North contexts (e.g., Dunlop & Rushton, 2022b; Levinson, 2018; Tannock, 2020; Tolbert & Bazzul, 2017; Zeidler et al., 2019). According to this critical scholarship, the following practices tend to be the most common across within science education practices and policies:

- Science positioned as *objective* and *neutral*, with a focus on sticking to the *facts* about, for instance, *the* environment;
- Focus on individual actions and private lives in relation to socio-scientific issues;
- Absence of the *public sphere* when exploring socio-scientific issues (i.e., larger socio-political structures at both national and international stages);
- Backlash against socio-political discussions in relation to socio-scientific issues, including of social justice nature (e.g., neo/colonial legacies of environmental exploitation).

[7] See, for example, in the case of England: https://www.theguardian.com/education/2022/feb/17/guidance-on-political-impartiality-in-english-classrooms-confusing-say-teachers-unions & https://www.theguardian.com/education/2021/nov/05/do-not-encourage-pupils-to-join-climate-protests-says-draft-dfe-strategy

This scenario for science education then leads to an approach to the content of EE (and other socio-scientific issues such as the COVID-19 pandemic) within school science that is devoid of any engagement with complex socio-political and axiological aspects related to the nature of environmental issues and associated injustices outlined in the previous section (Gandolfi, 2023a), running in direct opposition to what international scholarship presents about the relevance of interdisciplinary, creative, ethical and socio-political thinking and practice to overcome these challenges (e.g., Latour, 2020). As such, it actually ends up diminishing the role that science education can occupy in helping address environmental—and other kinds of socio-scientific—(in)justices: instead of opening up schools', teachers' and students' engagement with the epistemological, social and axiological complexities of the links between environmental emergencies, injustices, scientific development and knowledge, this approach constrains science education experiences to solely technical aspects of scientific knowledge that are insufficient and not fit-for-purpose for dealing with challenging socio-scientific issues in a socially just manner (Bazzul & Tolbert, 2019; Gandolfi, 2024; Osborne et al., 2022; Tannock, 2020). In addition, as recently highlighted by Dunlop et al. (2022) in the particular case of education in England, such approaches shy away from what teachers and young people have recently asked for around more in-depth engagement with the complex nature of environmental issues, such as ideas around extractivism, consumption, and capitalism, which are central to environmental (in)justices and the colonial matrix of power (Quijano, 2007).

This is not to say, of course, that science educators have not previously highlighted the importance of such socio-political framing for science education, as illustrated by the wealth of diverse contributions to this book. Indeed, colleagues have been advocating this for decades now even before the COVID-19 pandemic: e.g., Bencze (2017), Bencze and Carter (2011), Levinson (2006), and Running-Grass (1995). But this concern still remains, even after the recent example of such socio-political groundings of scientific development made visible by the pandemic: if we want to indeed reposition science education as an important component of addressing all sorts of socio-scientific injustices, socio-political voices and framings need to be systematically embedded in it.

In what follows, I then introduce a critical-decolonial approach to science education (see also Gandolfi, 2024) to bring it in close alignment with the quest for environmental justice—and, more widely, with a social-political and social-justice informed turn in the field –, drawing on: existing critical and decolonial scholarship in science education (e.g., Carter, 2017; Gandolfi, 2021b, 2024; Ideland, 2018; Tolbert & Bazzul, 2017), my current experience as a science educator in England, and my previous experience as a chemistry teacher in a country with deep links to the (neo)colonial exploitation of nature and natural resources (Brazil). While I firmly believe that science education can actively contribute to discussions about and actions around environmental (in)justices, this needs to start from an in-depth reflection about how science teaching practices might instead perpetuate the discourses grounded on technical rationality that I outlined above, which insist in positioning science as neutral and objective and devoid of any social, political and cultural aspects. In particular, in the next section I will argue that we need to critically (re)think the approaches to the area known as *Nature of Science (NOS) teaching and learning* towards challenging a science education that is dissociated from global, socio-political and systemic perspectives, such as the role of corporate power, geopolitics and overexploitation of natural resources for technological development (Gandolfi, 2019, 2021b, 2024) in environmental disasters and injustices, and in many other socio-scientific issues, such as the COVID-19 pandemic.

4.2.1 Rethinking NOS for Challenging Environmental (In)justices

Broadly speaking, the area of NOS teaching and learning aims at fostering students' understanding of science—as a discipline and as a human endeavour—*as a process* and not simply as a *collection of facts and information* (Allchin, 2011; Erduran & Dagher, 2014), with the hope that such understanding would support people's deeper and more informed engagement with the complexities of a field that has great impact worldwide, including in civic decision-making and in socio-scientific issues (Gandolfi, 2021a, 2024; Ideland, 2018; Osborne et al., 2022). In general, NOS within science education is then argued to be an important vehicle for challenging the common view of science as neutral and objective (Allchin, 2020; Gandolfi, 2021a, 2024; Hodson, 2014; Park et al., 2023), that is, the naïve *technical rationality* perspective (Rezende & Ostermann, 2020) that I briefly outlined in the previous section.

Grounded on a wide variety of scholarship from correlated fields such as History, Philosophy and Sociology of Science, conceptualisations of NOS usually revolve around understanding how scientific knowledge and practices are developed through inquiry and interactions within society. There is, however, diversity to be found in this area in relation to what should constitute the *content* of NOS brought to science lessons across different age groups and curricular contexts (Park et al., 2023). While, in general, seminal science educators working on NOS tend to agree with the relevance of both epistemic (i.e., nature of its knowledge) and socio-institutional (i.e., its status as a social enterprise) dimensions of science as part of NOS teaching and learning (Allchin, 2011; Erduran & Dagher, 2014; Hodson, 2014; Lederman et al., 2002), recent literature has started to call attention to the tendency of most NOS-based practices to over-emphasise the epistemic dimension of NOS (e.g., the nature and processes of theory building, modelling, inferencing, observations and inquiring, etc.), while downplaying its socio-institutional dimension (e.g., Gandolfi, 2019; Park et al., 2023). Here we seem to go back to the challenges I raised earlier in this chapter around recognising the deep and complex connections of science with social, political and cultural axis; that is, of explicitly and meaningfully recognising science as a socio-political endeavour (Tolbert & Bazzul, 2017), as we are attempting to do across this book.

In particular, traditional scholarship in the field of NOS seems to remain grounded on a binary separation between epistemic and socio-institutional aspects of scientific development, instead of more fully embracing what recent scholarship from the fields of History, Philosophy and Sociology of Science (e.g., Ashby & Machin, 2021; Harding, 2008; Latour, 2020; Mitova, 2023; Raja et al., 2022; Roberts, 2009; Roy, 2018; Seth, 2009) has been consistently arguing for at least two decades: since scientific development is inherently socio-political, its epistemic elements are intertwined (or inseparable) from the social-institutional nature of science (Moura et al., 2021). Seth (2009, p. 374), for instance, in their editorial to a Special Issue on *Science, colonialism, postcolonialism*, summarised such argument from the specific perspective of scientific development and colonialisms: *"in the wake of the social turn in the history of science, scholars of science and colonialism have emphasised not only the discursive or ideological power of science, but also the ways in which Western scientific knowledge has been co-constituted with colonialism"*. That is, the socio-political nature of science [e.g., its "discursive or ideological power"] is not a particular, separate element of scientific knowledge development, but it is "co-constitutive"/intertwined with scientific knowledge (its *epistemological* elements).

An example for such non-binary relationship within NOS is the recent work by Raja et al. (2022) on the links between *colonial history* (a socio-institutional aspect of NOS) and *sampling strategies* (traditionally framed as an epistemic element of NOS) in the research field of palaeontology. In their study based on the Paleobiology Database, which looked into how fossil data are prospected, built and shared in this area across the world in the context of research into deep-time biodiversity, extinction and evolution, the authors found how particular colonial practices seem to be still informing sampling practices around fossil data, such as: material extractivism and epistemic injustice[8] (as seen in parachute science practices[9]), law and rights violation (in relation to access to and extraction from protected areas, etc.), and concentration of scientific knowledge and access to fossil samples from the global south in global north institutions. The authors then argue that such practices seem to be resulting in a skewed distribution of which kind of fossil data (vertebrate, invertebrate, etc.), and where they are recovered and reported from, is available to inform, for instance, further ecological meta-analysis and inquiries in the area of deep-time biodiversity, incurring specific sampling biases into the field.

Therefore, and as I recently argued elsewhere (Gandolfi, 2023b, p. 165; Gandolfi, 2024), in a scenario of NOS teaching that still treats epistemic and socio-institutional aspects of scientific development as disconnected (or that extensively privileges the former over the latter), we risk losing the potential of NOS within science education to contribute to people's informed and critical engagement with socio-scientific issues such as the COVID-19 pandemic or environmental issues by instead reinforcing *"essentially utilitarian, neoliberal and triumphalist views about science*

[8] "When the expertise of local researchers is devalued" (Raja et al., 2022, p. 145).

[9] "Where researchers, generally from higher-income countries, 'drop in' to other countries to conduct research and leave without any engagement with the local community, including local researchers" (Raja et al., 2022, p. 145).

[...] without any critical reflection or acknowledgement of its complexities, including political, economic and ethical commitments". Even NOS teaching approaches that attempt to foster a better understanding of the relationship between science and society still tend to prioritise students' learning how to navigate a diverse society based on scientific knowledge, such as making decisions about transportation options considering the current concerns about climate change. What I would like to emphasise here, inspired by Santos (2009), is that despite efforts to integrate the connections between science and society more effectively into education through NOS-based practices, most strategies continue to operate within an instrumental (or technical-rational) perspective which is primarily focused on *"preparing students to be in a society that is simply given to them"* (Gandolfi, 2023b, p. 165; Gandolfi, 2024).

Grounded on decolonial and Freirean ideas (Freire, 2004), I then propose a critical-decolonial (re)conceptualisation of NOS for addressing environmental (in)justices that involves engaging with the role of scientific development and work in these environmental injustices. This should be done not solely in relation to future generations or purely local viewpoints, as often seen in environmental discourses in the global North (e.g., DfE, 2022), but also from the socio-political and historical perspectives of communities that have been severely impacted by environmental issues for centuries, such as Indigenous and landless groups in global South. Here, I take initial inspiration from Paulo Freire's seminal work on Critical Pedagogies (Freire, 1972, 2004) because for him socio-political issues are central to education: *"the teacher must also be teaching in favor of something and against something. This 'something' is just the political project, the political profile of society, the political 'dream'"* (Shor & Freire, 1987, p. 46). In particular, the subsequent operationalisation of his work into the notion of *ecopedagogies*, as outlined by Misiaszek (2020, p. 359, my emphasis), can be of relevance to rethinking NOS for environmental (in)justices:

> [Freirean] ecopedagogies are grounded in critical thinking and transformability, with the ultimate goal being to construct learning with increased social and environmental justice (...) [they are] centered on *better understanding the connections between human acts of environmental violence and social violence that cause injustices/oppressions, domination over the rest of Nature, and planetary unsustainability*.

This Freirean socio-political framing can allow science educators, for instance, to challenge the overreliance on only *scientific facts* about the environment that, as I argued above, remains prominent in how science has been framed within EE and which, as argued by Sara Tolbert and Jesse Bazzul (Bazzul & Tolbert, 2019; Tolbert & Bazzul, 2017), only contributes to creating a picture of science and scientific practices with no ruptures, conflicts or injustices—that is, a naïve view of NOS as devoid of any social, political and cultural aspects. In addition, this Freirean approach to NOS could also support a better engagement with the concepts of *connections* and *scale* within science education, such as the links between local/global, national/international, here/there, individual/collective, now/future, human/nature surrounding scientific practices and knowledge related to environmental issues (Misiaszek & Torres, 2019), and to other socio-scientific issues, as exemplified by how such elements were central (implicitly or explicitly) to discourses about and experiences of the COVID-19 pandemic.

In particular, such work across the notions of *scale* and *connections* can support a more critical positioning of science—both as a human endeavour and as a body of knowledge and practices to be learned at school—within local-planetary thinking and practices around environmental injustices, including fostering widened notions of collectivity and collaborations within and among science and other areas of knowledge, practice and, importantly, life on this planet. This approach asks for embracing a collective reflection about the connections between, as Bruno Latour (2020) puts it, *"where we live in, and where we live from"*; that is, between different communities, and between humans and the environment within a local-global scenario of growing environmental emergencies and injustices. What can potentially emerge from this kind of work is then an acknowledgement of the links between environmental challenges and sociopolitical structures at diverse scales and levels of complexity, and the disruption of individualised views of (scientific) citizenship and literacy that are replaced by science educating for *widened collectivity* (Misiaszek, 2020).

Such Freirean approach can be illustrated by some of the NOS-related activities that were part of a project I developed with a science teacher and his year 8 students (aged 12–13) at a state secondary school in England, and which I have more thoroughly explored elsewhere (Gandolfi, 2021a, b). This project took place over a whole school year with the central aim of supporting the participant teacher (through a collaborative action-research approach) with embedding NOS into his school's existing curricular practices in science. Informing our approach to such embedding of NOS into his work was exactly the critical-decolonial perspective around NOS that I have been advocating across this section: one that does not separate NOS from scientific content, nor into a set of isolated epistemic and socio-institutional elements; one that is centred instead on recognising the socio-historical and political grounding of the scientific content to be taught and learned in contexts such as secondary schools in England. Across the project, several topics from the regular national science curriculum followed by the participant school were approached under this perspective, such as Medicines, Magnetism, Evolution and Biodiversity, and Earth's Resources. The sequence below (Fig. 4.5), taken from a larger set of resources employed during the work on the topic of Earth's resources, exemplifies the potential of the proposed Freirean approach to NOS in the case of environmental injustices:

Informing the work around the sequence displayed in Fig. 4.5 below there was also an engagement with decolonial thinking as a way for NOS teaching and learning to move beyond its usual overreliance on only epistemological aspects of scientific endeavours to also more critically engage with the socio-political and axiological aspects of science across our diverse and interconnected histories of lives in and from the environment. In particular, here I draw on several recent scholarship from fields of History, Philosophy and Sociology of Science (e.g., Ashby & Machin, 2021; Harding, 2008; Mitova, 2023) that have expanded our understandings of the deep socio-historical-political connections between the so-called modern (or

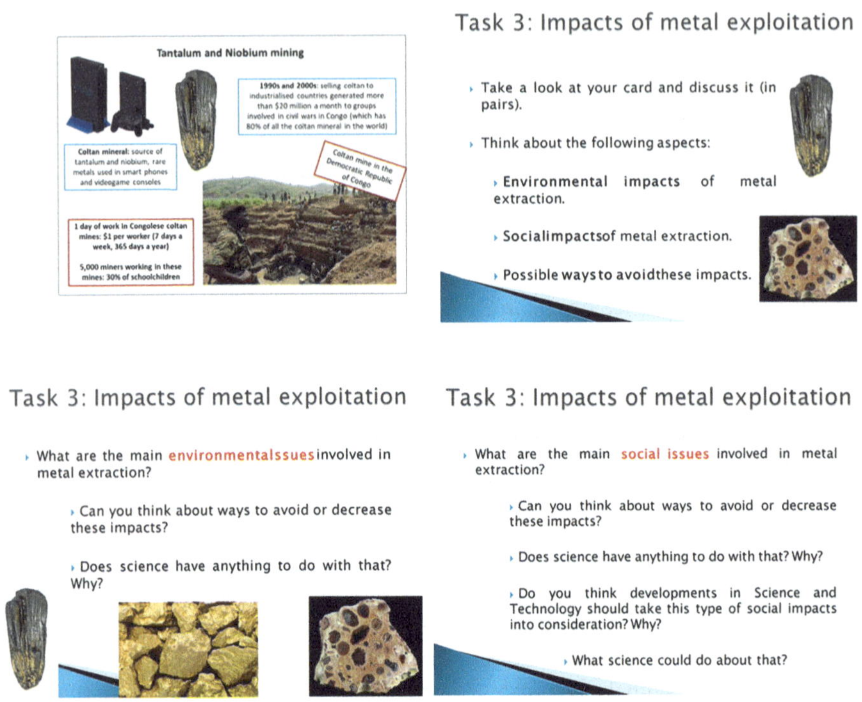

Fig. 4.5 Sample from teaching sequence on Earth's resources. (Source: own production)

Western) science and (neo)colonial endeavours, as I have already briefly touched upon earlier in chapter when drawing on Raja et al. (2022), Roy (2018), and Seth (2009), and on Quijano's (2007) concept of colonial matrix of power. The latter author, for instance, reminds us of the intertwined nature between environmental access and exploitation and scientific and technological development across the centuries, especially through practices grounded on biological and geological management, appropriation and exploitation of land and natural resources (such as minerals, plants, water, etc.), such as the example grounding the sequence displayed in Fig. 4.5.

Throughout the guided discussion activity outlined above, we then sought to reflect alongside the participant year 8 students on how the issues of environmental injustices connected to the example of niobium and tantalum exploitation cannot be solely attributed to some bad techniques or decisions in the development and/or application of scientific (and technological) knowledge, such as blaming the particular choice of niobium as a source for the electronics industry, and simply replacing it by a *greener* solution. Instead, we delved deeper—and critically—into unpacking the links between environmental injustices and systemic issues of over-exploitation and over-developmental approaches to our local and global relationships with land, natural resources and peoples working and living in certain areas of the world (e.g., Democratic Republic of Congo). As I argued elsewhere

(Gandolfi, 2021b, p. 512–515; Gandolfi, 2024), such decolonial approach to NOS allows science educators and their students to explore and reflect on how:

> [scientific development is] inherently tied to other elements of the decolonial endeavour: social, cultural and historical; that is, the development of scientific/technological knowledge cannot be disconnected from its socio-cultural and historical roots and, more importantly, from the often unequal 'dialogues' established as part of these processes (…) [This] recognition and deconstruction of the shared historical processes, interchanges and collaborative and exploitative systems behind the development of scientific/technological knowledges and practices are central to making explicit the holistic, dynamic and complex nature of the 'science' we teach in schools, including aspects of social justice, oppression and cross-cultural interactions.

A decolonial approach to NOS then involves recognising—in order to challenge and resist—the *oppressive* (and neocolonial) structural elements of science. That would involve acknowledging and reflecting on the implications to NOS and, via that, to environmental (in)justices of, for instance: the encroachment of scientific practices and communities by corporate power (e.g., Exxon, BP, Monsanto, Vale do Rio Doce)[10] (Carter, 2017; Eaton & Day, 2020; Tannock, 2020); the unequal and biased structures of funding within scientific and technological development (Carter, 2017); issues of violence, copyright, intellectual property, and regulation of access and violence to land, natural resources and local knowledge within scientific and technological development (e.g., Shebitz & Oviedo, 2018); among other aspects closely linked to NOS as I further expand in Gandolfi (2024), but which are often overlooked in NOS teaching. Such decolonial approach is also illustrated by Fig. 4.6 below, which displays another NOS-related sequence part of the same project depicted in Fig. 4.5, now about the topic on Extinction and Biodiversity:

As alluded to by the brief example displayed in Fig. 4.6 below, a decolonial approach to NOS also involves recognising that a diversity of knowledges, practices and ontologies about the natural world are central to scientific (and technological) developments surrounding issues of environmental (in)justice. As I argued elsewhere (Gandolfi, 2021b, p. 514), even the so-called 'modern Western' Science we often found in traditional science curricula is a result of diverse *"exchanges, appropriations and collaborations between different communities and [of] the circulation of diverse types of knowledges/resources, all promoted by historical and geopolitical contexts"*, such as colonialisms. As such, a wide variety of knowledges, practices and ontologies, such as those emerging from Indigenous and other land-based communities (Das & Lowe, 2018; Harding, 2008; Tuck et al., 2014), can play a central role in contributing to rethinking our engagement with the environmental (in)justices that such communities have already been facing for centuries. In this scenario, NOS teaching can position itself at the forefront of addressing environmental (in)justices and other socio-scientific issues in our post-pandemic world within science education by: firstly, helping students recognise the importance of diversity (of knowledges, practices, perspectives and lived experiences) to scientific work around such challenges; and, secondly, prompting students and teachers into

[10] See, for instance: https://www.theguardian.com/education/2023/mar/27/fossil-fuel-firms-us-universities-colonize-academia

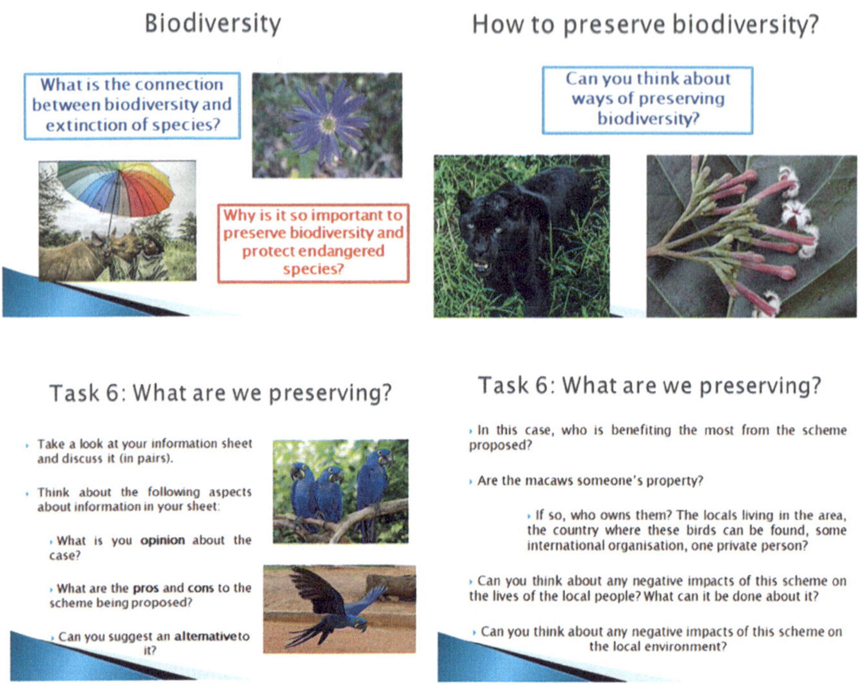

Task 6

Imagine yourself as an ornithologist working in the tropics. You meet a wealthy patron of preservation - someone who has purchased tens of thousands of acres to conduct research on innovative sustainable agriculture. This local magnate (whose fortune comes from owning a national fizzy pop company in England) is an avid birder. He wants to rescue the dwindling population of hyacinth macaws, whose habitat is shrinking due to deforestation of the rainforest in South America. These magnificent, impressive birds nest in the hollows of old trees, so that even if new trees are planted to replace the forests, the old trees and nesting sites are still lost. The situation is aggravated because the local people are poor enough that they are motivated to capture the birds and sell them to traders who smuggle them and market them to wealthy bird collectors. This patron wants a scheme to take macaws from the wild, raise them in captivity and release them on his own land, establishing a protected population on his "nature preserve".

Fig. 4.6 Sample from teaching sequence on Extinction and Biodiversity. (Source: own production)

more critical and social-justice informed explorations of science and scientific development that take into account this diversity when reflecting on ongoing socio-scientific issues and injustices.

4.3 Final Comments

In this chapter, I have argued for the importance of a socio-politically informed science education within a global-local landscape of environmental (in)justices, especially as we reconsider the kinds of relationships we seek to rebuild with the natural world and with each other at local and global scales in this apparent post-pandemic world. In particular, I have attempted to explore how a critical (in Freirean terms) and decolonial approach to NOS teaching and learning can support science educators' engagement with growing issues of environmental injustices across the world—and with other socio-scientific issues, like potential future pandemics—by challenging persisting views and practices within science education that are dissociated from socio-political elements, such as the role of corporate power and the over-exploitation of natural resources within the Global South. Adopting this critical-decolonial perspective around NOS would involve engagement with socio-political discussions around:

- The societal and axiological aspects of NOS;
- Recognising (to resist) the *oppressive nature of science*, that is, aspects related to: the encroachment of scientific practices and communities by corporate power; the unequal and biased structures of funding within scientific and technological development; issues of violence, copyright, intellectual property, and regulation of access and violence to land, natural resources, labour and local knowledge within scientific and technological development.
- The diversity of knowledges, practices and ontologies about the natural world (e.g., Indigenous communities and land-based knowledge), which can play a central role in rethinking our engagement with environmental (in)justices and with other kinds of socio-scientific issues.

Perpetuating purely celebratory and acritical views of *progress* within science education without considering how scientific endeavours have been historically linked to complex socio-political—and often oppressive—projects of land, natural resources and environmental knowledge exploitation (Gandolfi, 2021b, 2024) will certainly not be enough if our aim is to tackle environmental challenges and other current and future socio-scientific issues, like the next potential pandemic, in a socially just way. On the contrary, what is needed is the acknowledgement within science education practices, hopefully with the help of a critical-decolonial NOS perspective, of the links between environmental and other socio-scientific injustices and historical-socio-political structures and decision-making, as many other authors across this book have also argued.

References

Acosta, A. (2012). Extractivismo y neoextractivismo: dos caras de la misma maldición. *Más allá del desarrollo, 1*, 83–118.

Acselrad, H. (2010). Ambientalização das lutas sociais-o Caso do movimento por justiça Ambiental [The "Environmentalization" of social struggles—The environmental justice movement in Brazil]. *Estudos avançados, 24*, 103–119. https://doi.org/10.1590/S0103-40142010000100010

Ahmed, N., Marriott, A., Dabi, N., Lowthers, M., Lawson, M., & Mugehera, L. (2022). *Inequality kills: The unparalleled action needed to combat unprecedented inequality in the wake of COVID-19*. Oxfam GB.

Allchin, D. (2011). Evaluating knowledge of the nature of (whole) science. *Science Education, 95*(3), 518–542. https://doi.org/10.1002/sce.20432

Allchin, D. (2020). From nature of science to social justice: The political power of epistemic lessons. In H. A. Yacoubian & L. Hansson (Eds.), *Nature of science for social justice* (pp. 23–39). Springer International Publishing.

Alsop, S., & Bencze, L. (2020). Editorial: Reimagining science and technology education in the COVID-19 portal. *Journal of Activist Science & Technology Education, 11*(2), I–VII. https://doi.org/10.33137/jaste.v11i2.34530

Ashby, J., & Machin, R. (2021). Legacies of colonial violence in natural history collections. *Journal of Natural Science Collections, 8*, 44–54.

Bazzul, J., & Tolbert, S. (2019). Love, politics and science education on a damaged planet. *Cultural Studies of Science Education, 14*(2), 303–308. https://doi.org/10.1007/s11422-019-09913-2

Bencze, L. (2017). *Science and technology education promoting wellbeing for individuals, societies and environments: STEPWISE*. Springer International Publishing.

Bencze, L., & Carter, L. (2011). Globalizing students acting for the common good. *Journal of Research in Science Teaching, 48*(6), 648–669. https://doi.org/10.1002/tea.20419

Carter, L. (2017). A decolonial moment in science education: Using a socioscientific issue to explore the coloniality of power. *Revista Brasileira de Pesquisa em Educação em Ciências, 17*(3), 1061–1085. https://doi.org/10.28976/1984-2686rbpec20171731061

Das, S., & Lowe, M. (2018). Nature read in black and white: Decolonial approaches to interpreting natural history collections. *Journal of Natural Science Collections, 6*, 4–14.

Department for Education (DfE). (2022). *Sustainability & climate change: A strategy for the education & children's services systems*. Retrieved June 21, 2023, from https://www.gov.uk/government/publications/sustainability-and-climate-change-strategy/sustainability-and-climate-change-a-strategy-for-the-education-and-childrens-services-systems

Dunlop, L., & Rushton, E. A. (2022a). Putting climate change at the heart of education: Is England's strategy a placebo for policy? *British Educational Research Journal, 46*(6), 1083–1101. https://doi.org/10.1002/berj.3816

Dunlop, L., & Rushton, E. (2022b). A manifesto for education for environmental sustainability: Implications for science education. *School Science Review, 103*(383), 38–41.

Dunlop, L., Atkinson, L., & Turkenburg-van Diepen, M. (2021). "It's our future." Youth and fracking justice in England. *Local Environment, 26*(1), 110–130. https://doi.org/10.1080/13549839.2020.1867837

Dunlop, L., Rushton, E. A., Atkinson, L., Ayre, J., Bullivant, A., Essex, J., Price, L., Smith, A., Summer, M., Stubbs, J. E., Turkenburg-van Diepen, M., & Wood, L. (2022). Teacher and youth priorities for education for environmental sustainability: A co-created manifesto. *British Educational Research Journal, 48*(5), 952–973. https://doi.org/10.1002/berj.3803

Eaton, E. M., & Day, N. A. (2020). Petro-pedagogy: Fossil fuel interests and the obstruction of climate justice in public education. *Environmental Education Research, 26*(4), 457–473. https://doi.org/10.1080/13504622.2019.1650164

Eckstein, D., Künzel, V., & Schäfer, L. (2021). *Global climate risk index 2021 – Who suffers most from extreme weather events? Weather-related loss events in 2019 and 2000–2019*. German Watch. Retrieved June 21, 2023, from https://germanwatch.org/sites/germanwatch.org/files/Global%20Climate%20Risk%20Index%202021_1.pdf

Erduran, S., & Dagher, Z. (2014). *Reconceptualizing the nature of science for science education: Scientific knowledge, practices and other family categories*. Springer.

Ferguson, T., Roofe, C., & Cook, L. D. (2021). Teachers' perspectives on sustainable development: The implications for education for sustainable development. *Environmental Education Research, 27*(9), 1343–1359. https://doi.org/10.1080/13504622.2021.1921113

Freire, P. (1972). *Pedagogy of the oppressed*. Bloomsbury Academic.

Freire, P. (2004). *Pedagogy of indignation*. Paradigm Publishers.

Gandolfi, H. E. (2019). In defence of non-epistemic aspects of nature of science: Insights from an intercultural approach to history of science. *Cultural Studies of Science Education, 14*(3), 557–567. https://doi.org/10.1007/s11422-018-9879-8

Gandolfi, H. E. (2021a). "It's a lot of people in different places working on many ideas": Possibilities from global history of science to learning about nature of science. *Journal of Research in Science Teaching, 58*(4), 551–588. https://doi.org/10.1002/tea.21671

Gandolfi, H. E. (2021b). Decolonising the science curriculum in England: Bringing decolonial science and technology studies to secondary education. *The Curriculum Journal, 32*(3), 510–532. https://doi.org/10.1002/curj.97

Gandolfi, H. E. (2023a). Teaching in the age of environmental emergencies: A "utopian" exploration of the experiences of teachers committed to environmental education in England. *Educational Review*, 1–21. https://doi.org/10.1080/00131911.2022.2163378

Gandolfi, H. (2023b). Transnational conversations about science education: Paulo Freire, interculturality and socio-political transformation. *Cultural Studies of Science Education, 18*, 159–173. https://doi.org/10.1007/s11422-023-10163-6

Gandolfi, H. E. (2024). (Re) considering nature of science education in the face of socio-scientific challenges and injustices: Insights from a critical-Decolonial perspective. *Science & Education*, 1–27. https://doi.org/10.1007/s11191-024-00536-w

Gandolfi, H. E., Rushton, E. A., Silva, L. F., & da Silva Carvalho, M. B. S. (2024). Teacher educators and environmental justice: Conversations about education for environmental justice between science and geography teacher educators based in England and Brazil. *Cultural Studies of Science Education*, 1–30.

Glackin, M., & King, H. (2020). Taking stock of environmental education policy in England–the what, the where and the why. *Environmental Education Research, 26*(3), 305–323. https://doi.org/10.1080/13504622.2019.1707513

Grosfoguel, R. (2016). Del extractivismo económico al extractivismo epistémico y ontológico [From economic extractivism to epistemological and ontological extractivism]. *Revista Internacional de Comunicación y Desarrollo/RICD, 1*(4), 33–45. https://doi.org/10.15304/ricd.1.4.3295

Harding, S. G. (2008). *Sciences from below: Feminisms, postcolonialities, and modernities*. Duke University Press.

Hodson, D. (2014). Nature of science in the science curriculum: Origin, development, implications and shifting emphases. In M. R. Matthews (Ed.), *International handbook of research in history, philosophy and science teaching* (pp. 911–970). Springer.

Ideland, M. (2018). Science, coloniality, and 'the great rationality divide': How practices, places, and persons are culturally attached to one another in science education. *Science & Education, 27*(7–8), 783–803. https://doi.org/10.1007/s11191-018-0006-8

Kelly, G. J., Carlsen, W. S., & Cunningham, C. M. (1993). Science education in sociocultural context: Perspectives from the sociology of science. *Science Education, 77*(2), 207–220. https://doi.org/10.1002/sce.3730770208

Latour, B. (2020). *Critical Zones – An Exhibition at ZKM Karlsruhe 2020*. Retrieved June 21, 2023, from https://zkm.de/en/zkm.de/en/ausstellung/2020/05/critical-zones/bruno-latour-on-critical-zones

Layrargues, P. P. (2018). Subserviência ao capital: educação ambiental sob o signo do antiecologismo [Subservice to capital: environmental education under the sign of anti-environmentalism]. *Pesquisa em Educação Ambiental, 13*(1), 28–47. https://doi.org/10.18675/2177-580X.vol13.n1.p28-47

Le Grange, L. (2023). Decolonisation and anti-racism: Challenges and opportunities for (teacher) education. *The Curriculum Journal, 34*(1), 8–21. https://doi.org/10.1002/curj.193

Lederman, N. G., Abd-El-Khalick, F., Bell, R. L., & Schwartz, R. S. (2002). Views of nature of science questionnaire: Toward valid and meaningful assessment of learners' conceptions of nature of science. *Journal of Research in Science Teaching, 39*(6), 497–521. https://doi.org/10.1002/tea.10034

Levinson, R. (2006). Towards a theoretical framework for teaching controversial socio-scientific issues. *International Journal of Science Education, 28*(10), 1201–1224. https://doi.org/10.1080/09500690600560753

Levinson, R. (2018). Realising the school science curriculum. *The Curriculum Journal, 29*(4), 522–537. https://doi.org/10.1080/09585176.2018.1504314

Maldonado-Torres, N. (2007). On Coloniality of being: Contributions to the development of a concept. *Cultural Studies, 21*(2–3), 240–270. https://doi.org/10.1080/09502380601162548

Martin, S., Dillon, J., Higgins, P., Peters, C., & Scott, W. (2013). Divergent evolution in education for sustainable development policy in the United Kingdom: Current status, best practice, and opportunities for the future. *Sustainability, 5*(4), 1522–1544. https://doi.org/10.3390/su5041522

Mignolo, W. D., & Walsh, C. E. (2018). *On decoloniality*. Duke University Press.

Misiaszek, G. W. (2019). Ecopedagogy: Teaching critical literacies of 'development', 'sustainability', and 'sustainable development'. *Teaching in Higher Education, 25*(5), 615–632. https://doi.org/10.1080/13562517.2019.1586668

Misiaszek, G. W. (2020). *Ecopedagogy: Critical environmental teaching for planetary justice and global sustainable development*. Bloomsbury Publishing.

Misiaszek, G. W., & Torres, C. A. (2019). Ecopedagogy: The missing chapter of pedagogy of the oppressed. In C. A. Torres (Ed.), *The Wiley handbook of Paulo Freire* (pp. 463–488). Wiley.

Mitchell, G. (2019, July 17). *The messy challenge of environmental justice in the UK: Evolution, status and prospects*. Natural England Commissioned Reports, 273. Retrieved May 31, 2023, from http://publications.naturalengland.org.uk/publication/6558423485513728

Mitova, V. (2023). Why epistemic decolonisation in Africa? *Social Epistemology*, 1–14. https://doi.org/10.1080/02691728.2023.2184218

Morin, E., & Kern, A. B. (1999). *Homeland Earth: A manifesto for the new millennium*. Hampton Press.

Moura, C. B., Nascimento, M. M., & Lima, N. W. (2021). Epistemic and political confrontations around the public policies to fight COVID-19 pandemic: What can science education learn from this episode? *Science & Education, 30*, 501–525.

Osborne, J., Pimentel, D., Alberts, B., Allchin, D., Barzilai, S., Bergstrom, C., Coffey, J., Donovan, B., Kivinen, K., Kozyreva. A., & Wineburg, S. (2022). *Science Education in an Age of Misinformation*. Retrieved June 21, 2023, from https://sciedandmisinfo.sites.stanford.edu/sites/g/files/sbiybj25316/files/media/file/science_education_in_an_age_of_misinformation.pdf

Park, W., Cullinane, A., Gandolfi, H., Alameh, S., & Mesci, G. (2023). Innovations, challenges and future directions in nature of science research: Reflections from early career academics. *Research in Science Education*, 1–22. https://doi.org/10.1007/s11165-023-10102-z

Quijano, A. (2007). Coloniality and modernity/rationality. *Cultural Studies, 21*(2–3), 168–178. https://doi.org/10.1080/09502380601164353

Raja, N. B., Dunne, E. M., Matiwane, A., Khan, T. M., Nätscher, P. S., Ghilardi, A. M., & Chattopadhyay, D. (2022). Colonial history and global economics distort our understanding of deep-time biodiversity. *Nature Ecology & Evolution, 6*, 145–154. https://doi.org/10.1038/s41559-021-01608-8

Rede Brasileira de Justiça Ambiental (RBJA). (2001). *Manifesto de lançamento da Rede Brasileira de Justiça Ambiental [Launching manifest from the Brazilian Network of Environmental Justice]*. Retrieved 11 November 2023, from https://antigo.mma.gov.br/educacao-ambiental/pol%C3%ADtica-nacional-de-educa%C3%A7%C3%A3o-ambiental/documentos-referenciais/item/8077-manifesto-de-lan%C3%A7amento-da-rede-brasileira-de-justi%C3%A7a-ambiental.html

Rezende, F., & Ostermann, F. (2020). Hegemonic and counter-hegemonic discourses in science education from the perspective of a post-critical curriculum theory. *Cultural Studies of Science Education, 15*, 679–694. https://doi.org/10.1007/s11422-019-09969-0

Roberts, L. (2009). Situating science in global history: Local exchanges and networks of circulation. *Itinerario, 33*(1), 9–30. https://doi.org/10.1017/S0165115300002680

Roy, R. D. (2018, April 5). *Decolonise science—Time to end another imperial era*. The Conversation. https://theconversation.com/decolonise-science-time-to-end-another-imperial-era-89189

Running-Grass. (1995). Environmental education and environmental justice: A three circles perspective. *NYSOEA Pathways, 5*(1), 9–13.
Santos, W. L. D. (2009). Scientific literacy: A Freirean perspective as a radical view of humanistic science education. *Science Education, 93*(2), 361–382. https://doi.org/10.1002/sce.20301
Santos, A. H. D. (2018). A justiça ambiental e os novos direitos constitucionais: a função socioambiental dos territórios quilombolas do Parque Estadual da Pedra Branca [Environmental justice and the new constitutional rights: the socioenvironmental function of the State Park Pedra Branca's quilombo territories]. *Revista Brasileira de Estudos Urbanos e Regionais, 20*, 457–478. https://doi.org/10.22296/2317-1529.2018v20n3p457
Schilling Trein, E. (2018). An argument for vindicating a Marxist ontology in environmental education research. *Environmental Education Research, 24*(10), 1464–1475. https://doi.org/10.1080/13504622.2018.1545153
Schlosberg, D. (2013). Theorising environmental justice: The expanding sphere of a discourse. *Environmental Politics, 22*(1), 37–55. https://doi.org/10.1080/09644016.2013.755387
Seth, S. (2009). Putting knowledge in its place: Science, colonialism, and the postcolonial. *Postcolonial Studies, 12*(4), 373–388. https://doi.org/10.1080/13688790903350633
Shebitz, D., & Oviedo, A. (2018). Learning from the past: Reflecting on the Maya-ICBG controversy in the classroom. *Ethnobiology Letters, 9*(1), 59–66. https://doi.org/10.14237/ebl.9.1.2018.1095
Shor, I., & Freire, P. (1987). *A pedagogy for liberation: Dialogues on transforming education*. Greenwood Publishing Group.
Tannock, S. (2020). The oil industry in our schools: From Petro Pete to science capital in the age of climate crisis. *Environmental Education Research, 26*(4), 474–490. https://doi.org/10.1080/13504622.2020.1724891
Tolbert, S., & Bazzul, J. (2017). Toward the sociopolitical in science education. *Cultural Studies of Science Education, 12*(2), 321–330. https://doi.org/10.1007/s11422-016-9737-5
Tuck, E., & Yang, K. W. (2012). Decolonisation is not a metaphor. *Decolonisation: Indigeneity, Education & Society, 1*(1), 1–40.
Tuck, E., McKenzie, M., & McCoy, K. (2014). Land education: Indigenous, post-colonial, and decolonizing perspectives on place and environmental education research. *Environmental Education Research, 20*(1), 1–23. https://doi.org/10.1080/13504622.2013.877708
Voskoboynik, D. M. (2018, October 8). To fix the climate crisis, we must face up to our imperial past [Blog post]. *OpenDemocracy*. https://www.opendemocracy.net/en/opendemocracyuk/to-fix-climate-crisis-we-must-acknowledge-our-imperial-past/
Young, J. C. (2021). Environmental colonialism, digital indigeneity, and the politicization of resilience. *Environment and Planning E: Nature and Space, 4*(2), 230–251. https://doi.org/10.1177/2514848619898098
Zeidler, D. L., Herman, B. C., & Sadler, T. D. (2019). New directions in socioscientific issues research. *Disciplinary and Interdisciplinary Science Education Research, 1*(1), 1–9. https://doi.org/10.1186/s43031-019-0008-7

Haira Gandolfi is an assistant professor at the Faculty of Education, University of Cambridge/UK. She has a PhD in Education, an MSc in Science Education, and a BSc in Chemistry Teaching. She has experience in Chemistry teaching at secondary and postsecondary levels in Brazil, and her research interests are Decolonial Curricula and Pedagogies, Radical and Critical Education, Science Education, History, Philosophy, Sociology of Science, and Teacher Education.

Chapter 5
Critical Imagination and Conscientisation for Just Post-pandemic Science Education: Park Rangers as Public Science Educators

Betzabé Torres Olave and Gonzalo R. Guerrero

5.1 Introduction

Efforts to move towards just post-pandemic futures encounter a significant challenge: how to imagine and enact our transformative agency that works towards such a world? Education, and particularly educators, have a main duty in this '*conscientisation*' process and in advancing these efforts towards fairer worlds. A process of scientific conscientisation might imply understanding the role of science, in harmony with the position of human beings within nature: a transition towards understanding society in interdependence with nature and not as it is currently done traditionally in western modern and hegemonic science (Kato et al., 2023) through a false human-nature dichotomy (Jones, 2023). This approach to conscientisation could be articulated with dialogues between critical pedagogies (Torres-Olave et al., 2023), feminisms (Saeed, 2023), and framed in decolonisation of scientific education (Gandolfi, 2021). However, as Freire (1998/2018) has argued, even though our work towards conscientisation is essential, educators are not the only ones in charge of such a duty. Now, specifically in our field, science education, the role of scientists and other *public* science educators is also at the core since it shapes the bases on which science education is grounded: its values and practices. Therefore, in this chapter we also aim to learn with and from *public* science educators and their agentic practices towards imagining just post-pandemic worlds. In particular, we draw on a case of institutionalised park rangers as *public* (Giroux, 2004) and so-called

B. Torres Olave (✉)
University of Leeds, Leeds, UK
e-mail: b.torresolave@leeds.ac.uk

G. R. Guerrero
Pontificia Universidad Católica de Chile, Santiago de Chile, Chile

C. B. Moura (ed.), *A Sociopolitical Turn in Science Education*, Contemporary Trends and Issues in Science Education 62, https://doi.org/10.1007/978-3-031-78586-3_5

informal science educators to advance answers towards how to imagine fairer post-pandemic science and environmental education.

The role of our critical imagination as a "force that makes different possibilities visible" (Bazzul & Tolbert, 2019, p. 307) is scarce in science education literature (Bazzul, 2022). However, its essentiality as part of our agency as educators has been highlighted both in science education (Torres-Olave et al., 2023) and in futures studies (Facer, 2023). In the case of agency in science education, attention to the temporal dimension of it is crucial to how we imagine and reimagine possibilities to act (or not) across spaces (Torres-Olave, 2022). However, time for such a task is usually not considered in educators' work (Torres-Olave & Dillon, 2022), or agency is reduced to what educators do in practice without considering other elements of such practice, such as those political imaginative practices (Stetsenko, 2019). In that line, Facer (2022) has argued that "we now need to turn our attention towards understanding how temporal frames structure and delimit the nature of the problems we present to students as well as the as-yet unexplored potential for working with diverse temporal frames as a means of opening up new sites of dialogue and collective agency" (p. 47). On a similar line, Bonelli and Dorador (2021), Chilean scientists, have argued that we need to think *temporally* not just with other humans but also with non-human and more-than-human actors. Non-human entities refer to beings or entities that are not classified as human beings: this can include animals, plants, ecosystems, and other living organisms that exist independently of human presence. On the other hand, more-than-human entities are part of a larger web of life, and our actions and well-being are intertwined with the well-being of these diverse entities (Gudynas & Acosta, 2011). For many indigenous communities, more-than-human entities must be recognised to move towards a deeper understanding of the interconnectedness and interdependence of human beings with the natural world (Carvalho et al., 2020) which transcend larger temporal frames that our own existence (Bonelli & Dorador, 2021). More-than-human entities are part of our intergenerational stories as communities which are considered, for instance, by indigenous wisdom through nonlinear ontological understanding of time, perceiving past, present, and future simultaneously. In that sense, more-than-human actors might contribute to encompassing the relationships between humans, animals, plants, ecosystems, landscapes, geological formations, technologies, and other elements that shape our lived experiences in cyclical and multiple rather than linear perception of time. For instance, in the words of Bonelli and Dorador (2021),

> salares' (salt flats) deep-time history, which includes the periods of geological formation and biological evolution, the history of the Chilean neoliberal extractivist economy, its historical time, is extremely insignificant. It is only in recent decades that mining companies have successfully exploited the salares' subterranean waters and brines, an exploitation they typically justify by the current need to implement global energy transitions (p. 3)

As they notice, those ways of portraying extractivism as a technoscientific solution to justify their practices reduce our relationships with morethanhuman entities within short anthropocentric time scales. Those practices in the name of technoscientific progress undoubtedly affect how science education is understood (Kim &

Dionne, 2014; Kimmerer, 2012). There are, however, other stories where the scientific imagination has been linked to processes of conscientisation through, for example, learning with coastal communities caring relationships with the sea (George & Wiebe, 2020) or working with reparative environmental justice frameworks (Agathangelou & Killian, 2021) contesting the singularity and linearity of time and space moving away from narratives of 'science for domination' and exploitation. (see Kato et al., 2023).

Those past and present stories are important for highlighting and shaping how science, particularly the imaginaries, practices, and commitments of scientists and *public* science educators, are emphasised in science education. They serve as examples of public pedagogy, illustrating what the world has taught and is teaching us today (Giroux, 2004, 2006) regarding the roles of scientists, public science educators, the purpose of science, and the temporal scales and rhythms we engage with (Torres-Olave, 2024).

We begin this chapter by introducing the theoretical framework that informs our positionality and thinking –grounded in critical pedagogies, as well as futures studies. Subsequently, we contextualise this study within the backdrop of Chilean neoliberalism and its influence on science education, laying the groundwork for the case study we will delve into using this framework. To conclude the chapter, we engage in a discussion regarding how our example can serve as a form of public pedagogy, with potential inclusion in science curricula for both school and higher education contexts. Furthermore, we reflect on the implications of our example in shaping our conception of science and its role in envisioning just post-pandemic futures more broadly.

5.2 Theoretical Framing

5.2.1 Public Pedagogies and Public Science Educators

To envision fairer post-pandemic futures, we position ourselves within the framework of critical pedagogies, given their emphasis on equity, justice, and social change (Giroux, 2011). One of their central tenets is the pursuit of justice as a goal that education can contribute to through processes of emancipation and conscientisation (Darder et al., 2017), both on an individual and collective level. This process involves and demands reading and understanding the world in order to transform it (Freire, 1998).

When we examine the realm of science education through this critical lens, we uncover a history of oppression and inequalities that have been perpetuated in the name of science, as well as various socioscientific challenges (Tolbert et al., 2023). In that line, the journal *Nature*, through guest editors Nobles et al. (2022) recently published a special issue recognising the toxic legacy of science, and themselves as journal, in perpetuating systemic racism. Similarly, authors have engaged in unveiling stories where science discourses have helped to perpetuated racists views of certain bodies (Bernal-Munera, 2023), homophobia (Tilsen, 2023), and other intersected oppressions. However, we also encounter narratives of critical hope around science through the actions of individuals and communities (Torres-Olave, 2022), which demonstrate the potential for science education to advance towards epistemic, social, and environmental justice by engaging with diverse ways of knowing (Warren et al., 2020) while recognising the needed reparations of wrongs.

Within the context of the climate planetary emergency and from the perspective of critical pedagogies, questions have been raised regarding the content, attitudes, and skills that should be encompassed within the science curriculum (Gilbert, 2023). The actions and positions of different scientists and public science educators inform these attitudes and skills, thus serving as public pedagogy by teaching us something about the world: what holds value in science, how scientific inquiry is conducted, how science connects with real-world scenarios, and how science *can* and *should* contribute to the creation of fairer societies.

Public pedagogies as a framework, are rooted in critical pedagogies, encompassing the understanding that teaching and learning occur not only within formal educational settings but also through various cultural practices and spaces such as arts, media, museums, and parks, which impact knowledge about the world (Sandlin et al., 2011). Consequently, public pedagogies explore how different sites of practice function pedagogically, either directly or indirectly. In that way, public pedagogies hold the potential to address explicit and implicit narratives of science that have emerged in recent years, including the COVID-19 pandemic, women's rights discourses, science controversies, socio-environmental issues like the planetary climate crisis, and other topics that bridge scientific knowledge with society. When those narratives reproduce discourses of science that oppress or exploit others, they need to be unveiled and contested. For instance, using public pedagogies as a framework, Eaton and Day (2020) unveil the strategies employed by fossil fuel industries in their educational materials targeted at teachers. Their analysis reveals how these industries design learning resources that deflect responsibility for climate crises onto individuals. Eaton and Day (2020) demonstrate that through these pedagogical interventions, fossil fuel companies attempt to redirect attention away from corporate accountability and their industry's role in accelerating the effects of climate change. Hansson and Öhman (2022) have examined the role of museums as public pedagogies in promoting sustainability and sustainable development goals. Similarly, Park et al. (2022) have grounded their work on public pedagogies to develop 'food pedagogies' aimed at advancing social justice in the realms of health and food rights. These examples emphasise the significance of exploring diverse social issues as public pedagogies, explicitly informing learning and enhancing pedagogical values within public discussions. These discussions are shaping how various disciplines are learned and understood within the public sphere by either reproducing or contesting hegemonic narratives of science.

Following this line of thinking, Giroux (2004) argues that public pedagogies are about making the political *more pedagogical* Building upon this approach, Mayes and Center (2023) analyse students' climate activism, particularly their use of memes and humour on cardboard signs during protests, as a form of public pedagogy through political interventions, effectively acting as public climate educators. In another example, challenging the boundaries between techno-scientific knowledge and public knowledge of science, McEntee et al. (2023) contend that park rangers in Aotearoa/New Zealand possess a rich understanding of "science, place, multiple publics, and their limits" (p. 135), making them valuable public science educators and what the authors refer to as "environmental caretakers." Unfortunately, their voices are often marginalised due to epistemic hierarchies regarding who possesses, produces, and communicates scientific knowledge.

In the case of Chile, the country where we both grew up, different forms of public pedagogies and therefore public science educators have been in the spotlight in the lastyears due to ongoing discussion about the country's constitution. Those discussions have conquered the public imagination about different themes regarding science. We have seen ideas around women and LGBTQIA+ communities' rights (e.g. Hiner, 2022) rights of nature, indigenous lands (e.g. Green Rioja et al., 2022), and in general, how we relate to the other using scientific knowledge to frame such discussions. It is here where we set the context of this chapter, as we will further explore later.

5.2.2 *Agency, Temporality, Conscientisation, and What Is Imaginable*

Public pedagogies can also inform us about what the world is *being today*, and together with visions of futures we can explore what it *could be* (Facer, 2022). Futures studies deal with several issues but of particular interest for us are time, critical imagination, and conscientisation as part of what frames our sense of agency. Here, agency is understood as the power that lies within us to read the world for writing new ones (Torres-Olave, 2022) which have different dimensions: relational, practical, and temporal. It is that temporal dimension that we want to expand here through the concepts of critical imagination and conscientisation: how we think we shall live, how we imagine that potentiality, and what that means for our actions today.

In that line, Facer (2023) has argued that in neoliberal times certain *times* are of value with the tendency to put people "on track" and "behind". These neoliberal timescapes, rhythms of life (Adam, 1998), frame how we understand our relationships with human, non-human, and more-than-humans actors. In Facer's (2023) words "[w]hen timescapes are allied with power, as industrial and colonial histories show, arrhythmia becomes subject to intervention and control, bodies are forcibly and violently brought into dominant time structures" (p. 61); dominant time structures which then justify the exploitation of nature through accelerated environmental impacts (Kingsbury, 2022), unsustainable growth (Hickel, 2019) and scientific practices (Misiaszek & Rodrigues, 2023). They even define what is possible and what is not within and through science, delimiting our imagination and, therefore, our potential actions As Freire argued a while ago (Freire, 1997), the accelerated understanding of science and technology has been "the main bastion of capitalism" (p. 56) and its relationship and time dynamics. However, timescapes are not natural or given, nor are they singular; therefore, they can be rethought (Torres-Olave, 2024) in the same way we can reimagine our actions today and those futures we could advance towards.

But how do we imagine how shall we live and the role of science in such living? Imagination is essential to make sense of our actions and to go beyond limited situations (Torres-Olave et al., 2023). In this context, Bonelli and Dorador (2021) explored their imagination in terms of their work in microbiology. They developed what they call a "sympoetic imagination" which allows them to rethink and critique extractivist practices in the name of science, and to expand on other ways of scientific relations "through and by [their] affective engagements with salares" (p. 12). In these examples, their look at the more-than-human to allow themselves *to be taught* about the world: the *salares* can then be a *public pedagogue.* In that sense, public pedagogues of science are not just humans, but also the other actors we coexist with. As such, cho we recognise as a public science pedagogue needs to be widened. They are not just the formal and schooled science educator.

Putting together these ideas, we can crystallise a framework that allows us to see the world in different terms. Public pedagogies about science are out there and they need to be unveiled to reflect on how they contribute to notions of science and how some of these practices and values, such as thinking and imagining with nature and contesting neoliberal ways of relating, can be incorporated in science education to illuminate other ways of linking science and societies through conscientisation towards transformative agentic practices.

A critical scientific imagination and conscientisation are both essential in science education at school (Morales-Doyle & Frausto, 2021), in out-of-school contexts (Hecht & Nelson, 2022), and in higher education (Misiaszek & Rodrigues, 2023) levels because transformative actions are crucial to ending anti-environmentalism and unsustainability by disrupting what is thought as the mainstream practices of (neoliberal) science that falsely justifies acts of socio-environmental injustices in the name of progress and growth (Hickel, 2019; Liboiron, 2021).

5.3 A Case to Learn and Think with: Alto Maipo, the Monster in the Park

To explore these ideas and develop our argument, we will draw on a case study that forms part of an ongoing research project (Guerrero, 2024). This study focuses on socio-environmental conflicts and involves collaboration with science educators and park rangers who act as public science educators in El Morado natural monument, a national park in Chile.

Unfortunately, the national park is currently experiencing the adverse impacts of Alto Maipo hydroelectric project, referred to as Alto Maipo hereafter. This project has constructed underground tunnels that run directly beneath the El Morado natural monument, which is part of a glacier cirque, resulting in significant and irreparable damage and a lot of predicted dangerous impacts such as: threatening the safe supply of drinking water to 7 million people in Santiago (the capital of Chile); endangering irrigation waters for agriculture; accelerating the melting of glaciers and recharging of aquifers, also decertifying 100.000 ha of the basin and its area of direct influence (Folchi & Godoy, 2016).

The Alto Maipo project is being developed by AES Andes S.A. (previously AES Gener S.A.), a subsidiary of AES Corp, North America's second-largest energy company (Folchi & Godoy, 2016). STRABAG SpA Chile, an Austrian multinational with a workforce spanning over 20 countries, is responsible for the project's implementation. Situated near Santiago in San José de Maipo, the project remains under development. A 2011 agreement between AES Gener and Aguas Andinas, Chile's primary sanitary company, allows for the redirection of water originally destined for Santiago's water supply to the hydroelectric project. The plan aims to divert water from the Maipo River's main tributaries into a tunnel network, resulting in a severely limited ecological flow in the river that is insufficient to sustain life (Godoy, 2014). The infrastructure associated with the project includes two run-of-the-river centrals and 73 kilometres of abduction tunnels with diameters ranging from 6 to 8 meters. It is essential to note that this network of tunnels passes directly beneath the Andes Mountains, the El Morado National Park, and the San Francisco Glacier.

Like many high-cost projects, Alto Maipo depends on economic collaboration with foreign investors, relying on large investment groups and both national and foreign banks. Consequently, the project's financing is intricately intertwined with global capital. The network of investors involved comprises national and international institutions, banks, global corporations, subsidiaries, and associate companies. Information gathered from the Environmental Justice Global ATLAS platform[1] (Temper et al., 2015) reveals that Alto Maipo is co-financed by the International Finance Corporation and the US Overseas Private Investment Corporation (Folchi & Godoy, 2016). Through collaboration with a syndicate of international development banks, including the Overseas Private Investment Corporation (OPIC), the International Finance Corporation (IFC), as well as Chilean and international commercial banks, the project's financing has been increased to approximately US$3.05 billion, with 60 percent in debt and the remaining 40 percent in equity. Notably, despite significant flaws in the environmental impact assessment and inadequate consultation processes, Alto Maipo has received funding from nine banks (Temper et al., 2015). Paradoxically, Alto Maipo has been registered with the Clean Development Mechanism, enabling companies to earn carbon credits for initiatives aimed at reducing carbon emissions (UNFCCC, 2008). This demonstrates the

[1] https://ejatlas.org

involvement of international organisations and laws, such as the United Nations, within the network, and highlights how discourses surrounding carbon credits permeate the Chilean State within this framework. This phenomenon differs from traditional extractivism (Gudynas, 2018) and is now recognised as 'progressive' neo-extractivism (Svampa, 2019).

Neo-extractivism in the Global South constitutes a structural feature of capitalism (Svampa, 2019). Unfortunately, the dynamics of economic growth, based on historical extractivist, neoliberalism, and a particular mode of accumulation, have increased socio-environmental conflicts in Latin America. Therefore, analysing the successive economic cycles and historical or potential future dimensions of environmental conflicts are important to discuss the roles of science and environmental educators towards conscientisation (Guerrero, 2024). Consequently, new counter-hegemonic discourses and critical imagination are needed in science and environmental education to defend land, water, and territories that are also part of how we understand process of conscientisation in the name of science as a concrete and material socioscientific and political issue.

5.3.1 Park Rangers as Public Science Educators

The case we are presenting here is part of a larger project on envisioning opportunities for collective activism in science education –specifically within the context of extractivism in Latin America. The project is examining a critical approach to scientific and environmental literacies grounded in conscientization and drawing inspiration from outdoor science education. In the larger project, environmental conflicts are understood as potential scenarios/opportunities to analyse power dynamics and networks of actants (including humans, non-human, and more-than-human entities) involved in socio-environmental issues (Guerrero, 2022).The project develops a collaborative pedagogical approach to conscientisation with a critical focus, specifically tailored to address the Alto Maipo conflict shared above; and over the course of 2020–2022 interviews, conversations and focus groups were conducted with different actors, i.e., pre-and in-service teachers, scientists, and park rangers (Guerrero, 2024).

In this chapter, we focus on examining the discourses of awareness and critical imagination exhibited by specific group of park rangers who took part in the larger project positioning them as public science educators. Situated as public educators, park rangers could connect science, and particular science justice issues (Morales-Doyle, 2023) to local problems such as access to water and food security, environmental vitality and nature preservation. In that way, their discourses can make the political conflict of Alto Maipo more pedagogical. The Park Rangers site of work, El Morado natural monument is visited roughly by 15.000 visitors per year, including students and science teachers from different schools, mainly from Santiago, tourists, and general public. These park rangers' role in this national park is then relevant because of their proximity to significant events, theirs local knowledge and a special relationship with the natural monument. Moreover, they shape discourses of science and environmental education out of school and university for students, teachers, and general public. For anonymity purposes, park rangers' names and characteristics are not shared, however, it is important to highlight that they both have been working in the park over a decade. The excerpts of the interviews conducted by the second author of this chapter, are part of a research visit conducted by a participatory research group —which included 4 pre-service teachers, 2 in-service teachers, and 2 scientists.

5.3.2 A Monster in the Park, a Precious Resource in Danger: Two Storylines to Learn with

Through interviews and conversations, the park rangers shared different stories of the tensions navigated while doing their work. Here, the focus is on excerpts from two stories that we would like to think with to expand on the role of public pedagogies for imagining just post-pandemic science education and working towards conscientisation processes.

> (...) well, we usually teach about the impressiveness of the glacier at a low altitude, how the glacier was formed, and the importance of the water cycle. We also reflect on why the water cycle is essential, along with teachers. The glacier is *a fundamental part of everything*, serving as the base for the commune of San José de Maipo and meeting the water needs of all of Santiago. We intertwined all this information so that the children could understand where the river originated, the significance of its tributaries, the importance of water to the community, and why it was crucial to take care of this precious resource.
>
> (...) nowadays, protected areas are an opportunity for social, cultural, economic, or scientific development, and obviously, everything is interconnected. That's why I was saying that El Morado is a small laboratory, *a scientific area, a very small scientific laboratory*, where you can benefit greatly, not only in terms of chemistry, physics, and biology. Many visitors, such as schools or universities, also come to talk about history, geology or landscape geomorphology, painting, photography, physical education, mathematics, and they link it all with what the protected area is about. (Park Ranger 1)

During the park visit, the park rangers narrate the natural history of the location, including the stories of pre-Columbian peoples who inhabited and protected the area. They also highlight the sensitivity of insects to the impact of Alto Maipo, for instance, the changes in sound and temperature of the ground which in turn has cascading effects on the entire trophic chain of the park, including the glacier. These examples demonstrate the limitations of relying solely on reports or environmental regulations that fail to capture such nuanced effects and also several concerning observations:

> (...) however, in this entire sector, there are noticeable changes. For instance, the mountains and the glacier in the background, situated in the San José volcano, have undergone significant alterations. Comparing the present to a decade ago, the glacier was much wider and larger. The glaciers in the central zone are estimated to melt up to 6 meters per year. Unfortunately, it has become apparent that the dust emitted by the Alto Maipo trucks, along

> with the exhaust pipe emissions and the heat produced by the truck engines, have a negative impact. The dust, coupled with increased temperatures, is being carried by the wind and affecting the glaciers. These changes in temperature and dust accumulation are influencing the glaciers' melting process. For instance, if you shine a torch at night and realise that there is dust in the environment, I have done the test, it is like, I do not know. It is like the light from Star Wars, it's very similar.
>
> (...) the consequences are also affecting the local wildlife. Species are migrating away from their usual habitats due to the lack of food caused by the changes in the environment. Additionally, the direct pollution of rivers, noise pollution, and the rise in water temperatures due to the altered water flow are concerning issues. The project seems to be releasing heavy minerals and materials into the rivers, further aggravating the situation. Moreover, atmospheric impacts are being observed. The large number of vehicles and trucks moving in the area contributes to the dispersion of dirt and dust particles. These particles adhere to the glaciers, causing them to warm up and accelerate the melting process. (Park Ranger 2)

However, as shared in the next excerpt, there are some tensions between public science and environmental educators and unfortunately, when it comes to addressing more political or critical issues, the voices of park rangers are often silenced by institutional constraints:

> (...) well, I could talk about Alto Maipo a lot, but, probably, we are conditioned... We cannot talk about a project of which we are prohibited from issuing judgments because, of course, technically, we have the expertise, but here the one who has the voice are those who participated in evaluating the project. (Park Ranger 1)

Consequently, their agency is diminished, leaving the task of engaging with these political issues primarily to teachers or those fortunate enough to possess deeper knowledge about the happenings beneath the glacier and specifically about Alto Maipo.

5.4 Discussion: A Glacier as a Transdisciplinary Actor to Develop Critical Scientific Imagination and Conscientisation

A fundamental part of everything is how one of the park rangers portrays the Glacier. For both of them, the Glacier represents not just nature, but also brings with its humanity and fairer conditions for living for human and non-human actors. As noted by them, "the consequences are also affecting the local wildlife" and the presence of Alto Maipo brings questions about "the importance of water to the community". In that way, in their discourses, we can see how the threat to the Glacier has become a transdisciplinary actor to think with. It brings and facilitates discussions not just about science, but also about basic human rights such as water, and how knowledge and voices are distributed and valued. Moreover, it fosters discussions about what is essential for living, either a National Monument, or an industry, and the need to bring new relationships through science that are not exploitative and extractivist for the sake of development for a few at the expenses of others. In Kimmerer's (2012) words, the environmental crises we are suffering today are also part of "the degradation of our relationship with the living world and the extinction of an ethical responsibility for the land which sustains us" (p. 317). Expanding what counts as science education to think with and within nature as the park rangers do, is a way of developing a critical scientific imagination that can help us, researchers, educators, and students, develop conscientisation of what is of value, what is essential, and whose knowledges we bring to learn with when learning science in different settings.

However, there are actors, such as the park rangers and the Glacier itself, whose voices have been erased and silenced. When a voice is silenced, that silence also carries out discourses of science (Torres-Olave & Dillon, 2022). What is untold can be presented as of non-value or not essential for the experience within the park and what the park in itself represents for the local community, not only today but at larger scales. As noted by George and Wiebe (2020), learning with more-than-human entities and thinking time with other rhythms and scales can present challenges to Western modern science, particularly shifting "possessive, property-driven forms of citizenship and instead center reciprocal relations of abundance and care with and among human and more-than-human lifeworlds" (p. 519). Learning through the relationship between the park rangers and the Glacier has meant learning stories that travel across generations. Stories that can be of value to science and other disciplines. Stories that can advance caring, ethical, and conscious relationships through science for intergenerational justice.

Even though this process of developing conscientisation is neither simple nor solely the responsibility of schoolteachers, there are ways of distributing such ethical tasks. There are different discourses of science to whom we also need to demand ecological responsibility and, consequently, actions. For example, using the park, as the same park rangers stated, as a *scientific laboratory* to learn and cultivate imagination. An imagination that expands the relationships we build with others, that is ecological, ethically responsible and that centres learning experiences not just in students but also within the world of which we are also part and the relationship we build in the process (Hecht & Nelson, 2022). This is necessary not just for school experiences but also for higher education, such as through land-based pedagogies (Hage et al., 2022) for science teacher education or other science-related professions, as the park rangers also shared. If this goes beyond a visit to a sustainable practice, park rangers' knowledges can be set in an ecology of knowledges needed for science education, overcoming epistemic hierarchies of who owns scientific knowledge and the different ways of knowing science. As argued by McEntee et al. (2023), these stories also show that park rangers "have a nuanced knowledge of place" (p. 127). The movements, changes, losses and impacts of the park highlighted by them, require a rich understanding of the relationships, the geographies, and the diversity of habitats of the park. Through their knowledge of affective, structural, and content-related dimensions, they advance notions of science and ecological ethics essential for conscientisation processes.

Park rangers' stories also highlight the cruciality of the space to learn with and the value of oral stories of those who have been in the space for, as them, over two decades. When they say "Comparing the present to a decade ago, the glacier was much wider and larger", they bring with it other stories and dimensions of science to which we usually do not pay the necessary attention: non-linear, more broad temporal dynamics and rhythms. As such, park rangers play a crucial role in introducing students to different temporal and spatial scales. Similarly to Bonelli and Dorador (2021), park rangers are thinking not only in human but also in non- and more-than-human time scales. One of our problems today is that we are having temporal gaps between imagination and socio-cultural-environmental conditions. Those temporal thinkings have political effects because in the case of the Glacier, shapes what is of value *today* for *a few.* Thinking within extractivist and immediatism temporal scales frames our beliefs, our need, and to what we pay attention to or not (Torres-Olave, 2024). As stated by Facer (2023) we need to "cultivate such a temporal imagination as a resource for possibility thinking...start[ing] from the assumption that the temporal imagination, as with any other imaginative activity, is resourced by social and cultural resources that offer tools, images and concepts to scaffold and mediate our ideas of the world" (p. 62). Park rangers' stories of and with the Glacier can help to scaffold those ideas of the world realising the impacts that monsters such as Alto Maipo have on it.

In their stories, park rangers also share about other *starting* points. It is not just the Glacier and what it is *today*, but also *what it was*, and *what it could be* in the future with its influences on life more broadly. Those stories of science are vital because they problematise when time "starts" to be counted, which as a consequence, impacts our ways of thinking and the limits of our imagination. If we continue thinking in the scales we inhabit, or in the short term, conscientisation process will be hindered because moving towards fairer worlds is not just about *us:* it also needs to include the past and futures to which we do not belong. Those temporal scales strongly influence how we live today. For example, in the case of Chile, neoliberal logic has shaped neo-extractivist practices, and notions of techno-scientific futures as linear with infinite accumulation and growth.

In the case of Alto Maipo, the project poses a serious threat to the water supply of over 7 million Chileans. Regrettably, the potential consequences of this undertaking have not been adequately considered. The importance of addressing this issue cannot be underestimated. The impact on local flora and fauna, as well as the fragile habitats in the vicinity, demands serious consideration and measures to mitigate its effects. Therefore, it is vital to recognise that a climate crisis and the impact of global warming is already exerting considerable stress on the Andean ecosystems, and any additional disturbances from large-scale projects like Alto Maipo may only worsen the situation. Collective, critical, and relational responsibility from critical science educators is vital to contesting this monster in the park. As stated elsewhere, "[s]eeking just futures, pedagogically, is an attempt to enable students and educators to see, understand and transform unjust forces in our world that are temporally rooted both in the past and the present" (Torres-Olave, 2024, p. 13). Bringing different visions, such as those of the park rangers, can offer different rhythms and disrupt what is considered a problem to reflect and act upon through both the seen and unseen in our science classrooms.

5.5 Conclusions

We started this chapter by proposing a challenge to move towards just post-pandemic futures through and with science education. In such a challenge, we believe in the need to cultivate critical imagination and conscientisation to enact our transformative agency towards such a worlds. However, for just post-pandemic futures, we believe in the essentiality of at least two issues: (1) widen who counts as science educators and where we set the responsibility of educating the scientific imagination, and (2) delinking temporal thinking from our own tempos and rhythms to think with and through other scales.

Hegemonic public scientific voices have shown extractivist logic in the name of scientific progress, which will not help us move towards those fairer worlds we want to build. However, there are silenced voices, of human, non-human, and more-than-human actors that have always been there as public science educators. Attention is needed to those voices, an attention that can move us away from the self (Torres-Olave, 2024) to start seeing those actions and stories of science that can cultivate a scientific imagination which is critical, relational, and environmentally responsible with larger time scales. When thinking of who counts as a science educator, particularly in public spaces, we can then move towards asking who produces and communicates science and the role epistemic hierarchies play in diminishing certain actors' agency. Learning with park rangers has taught us how other actors are thinking and imagining with non-human and more-than-human actors. Such is the case of the river, of the salares, of the Glacier. Those ways of imagining need to be incorporated to the practices and values of science within science curriculum.

With such learning, we can move to the second task to cultivate our imagination and conscientisation: delinking temporal thinking from our own tempos and rhythms to think with other scales. An accelerated time does not allow us to look beyond our own temporal frame and therefore, leaving human, non-human, and more-than-human actors behind under the accelerated machine of capitalism. It also delimits the border of our imagination, of what is possible, and therefore our transformative agency. Thinking with other time scales and frames within science curriculum is also crucial. As such, we hope this chapter contributes to how we think and experience time, with others, through institutions, ways of thinking, and how time, its directions and rythms, are taught (or not) in science.

This whole book has prompted us to imagine several aspects of post-pandemic science education. Along these lines, we firmly believe that science and environmental education should be acknowledged as a collective endeavour that demands collaboration, dialogue, and attention to the position of human beings in the world. In a post-pandemic era, a transition towards understanding and reflecting on the interdependence and interconnectedness of human, non-human, and more-than-human entities (such as rivers or glaciers) seems to be imperative. And, in our specific context in Latin America, by taking these factors into account, we can promote a science and environmental education for conscientisation, aiming at protecting both the invaluable water supply and the delicate biodiversity of the Andes, paving the way for a harmonious coexistence between humans and nature.

To continue our exploration, cultivating our critical imagination and conscientisation processes, we ask ourselves to persist in challenging the borders of science education, but also its tempos and rhythms and the actors we bring along to our science classrooms. This includes not only to problematise whose voices count as science education, but also to problematise the temporal and geographical spaces in which we think. It includes asking what stories we value and want to tell of *origins* and *futures* for and within science. What temporal scales are we going to engage with in our pedagogical and research practices? What is the role of temporal pauses for that attention we highlighted as essential? We do not know yet what all these practices entail, but we believe temporality needs our attention not just in scientific terms, but also in its social dimension to contest those hegemonies of accelerated capitalist times.

References

Adam, B. (1998). *Timescapes of modernity*. Routledge.

Agathangelou, A. M., & Killian, K. D. (2021). About time: Climate change and inventions of the decolonial, planetarity and radical existence. *Globalizations, 18*(6), 821–838. https://doi.org/10.1080/14747731.2021.1945400

Bazzul, J. (2022). Politics, imagination, and the problem of antiquation: Embracing old and new materialisms. *Critical Education, 13*(1), 97. http://ojs.library.ubc.ca/index.php/criticaled/article/view/186366

Bazzul, J., & Tolbert, S. (2019). Love, politics and science education on a damaged planet. *Cultural Studies of Science Education, 14*(2), 303–308. https://doi.org/10.1007/s11422-019-09913-2

Bernal-Munera, M. (2023). A Freirean liberatory perspective of community colleges education: Critical consciousness and social justice science issues in the biology curriculum. *Cultural Studies of Science Education, 18*, 41–55. https://doi.org/10.1007/s11422-023-10152-9

Bonelli, C. & Dorador, C. (2021). Endangered Salares: Micro-disasters in northern Chile, *:* Latin American Science, Technology and Society, 4(1). https://doi.org/10.1080/25729861.2021.1968634

Carvalho, I. C. d. M., Steil, C. A., & Gonzaga, F. A. (2020). Learning from a more-than-human perspective. Plants as teachers. *The Journal of Environmental Education, 51*(2), 144–155. https://doi.org/10.1080/00958964.2020.1726266

Darder, A., Torres, D., & Baltodano, M. (2017). *The critical pedagogy reader* (3rd ed.). Routledge.

Eaton, E., & Day, N. A. (2020). Petro-pedagogy: Fossil fuel interests and the obstruction of climate justice in public education. *Environmental Education Research, 26*(4), 457–473. https://doi.org/10.1080/13504622.2019.1650164

Facer, K. (2022). *Anthropocenes, crises and sustainability: Educating the temporal imagination in the era of climate change*. 47–49. Paper presented at the 15th Invitational Seminar on 'Challenges for environmental and sustainability education research in times of climate crisis', Gent, Belgium. https://biblio.ugent.be/publication/8768733

Facer, K. (2023). Possibility and the temporal imagination. *Possibility Studies & Society, 1*(1–2), 60–66. https://doi.org/10.1177/27538699231171797

Folchi, M., & Godoy, F. (2016). La disputa de significados en torno al Proyecto Hidroeléctrico alto Maipo (Chile, 2007-2015) [The dispute of meanings surrounding the Alto Maipo hydroelectric project (Chile, 2007-2015)]. *Historia Ambiental Latinoamericana Y Caribeña (HALAC) Revista De La Solcha, 6*(1), 86–104. https://doi.org/10.5935/2237-2717.20160005

Freire, P. (1997). *Pedagogy of the heart*. Continuum.

Freire, P. (1998). *Pedagogy of freedom. Ethics, democracy, and civic courage*. Rowman & Littlefield.

Gandolfi, H. E. (2021). Decolonising the science curriculum in England: Bringing decolonial science and technology studies to secondary education. *The Curriculum Journal, 32*(3), 510–532. https://doi.org/10.1002/curj.97

George, R. Y., & Wiebe, S. M. (2020). Fluid Decolonial futures: Water as a life, ocean citizenship and seascape relationality. *New Political Science, 42*(4), 498–520. https://doi.org/10.1080/07393148.2020.1842706

Gilbert, J. (2023). Re-thinking science education for the Anthropocene. In S. Tolbert, M. Wallace, M. Higgins, & J. Bazzul (Eds.), *Reimagining science education in the Anthropocene* (Vol. 2, pp. 9–22). Springer.

Giroux, H. (2004). Public pedagogy and the politics of neo-liberalism: Making the political more pedagogical. *Policy Futures in Education, 2*(3–4), 494–503. https://doi.org/10.2304/pfie.2004.2.3.5

Giroux, H. (2006). Cultural studies, public pedagogy, and the responsibility of intellectuals. *Communication and Critical/Cultural Studies, 1*(1), 59–79. https://doi.org/10.1080/1479142042000180926

Giroux, H. (2011). *On critical pedagogy*. Bloomsbury.

Godoy, F. (2014). *Conocimiento y medio ambiente. Aproximación social del conocimiento científico en conflictos socioambientales el caso del Alto Maipo* [Knowledge and the environment. Social approach to scientific knowledge in socio-environmental conflicts: The case of alto Maipo]. [Master dissertation, University of Chile]. University of Chile theses online https://repositorio.uchile.cl/handle/2250/133916

Green Rioja, R., Loncon, E., & Carrillo Vidal, A. (2022). Transforming Chile from the ground up. *NACLA Report on the Americas, 54*(4), 382–388. https://doi.org/10.1080/10714839.2022.2145114

Gudynas, E. (2018). *Extractivismos y corrupción. Anatomía de una relación íntima [Extractivism and corruption. Anatomy of an intimate relationship]*. Editorial Quimantú.

Gudynas, E., & Acosta, A. (2011). La Renovación de la Crítica al Desarrollo y el Buen Vivir Como Alternativa [The renewal of criticism of development and good living as an alternative]. *Utopía y Praxis Latinoamericana, 16*(53), 71–83.

Guerrero, G. (2022). Community-based outdoor science education in Chile: A contribution to expanding networks within pro-eco-justice Dispositifs. *Journal for Activist Science and Technology Education, 12*(1) Retrieved from https://jps.library.utoronto.ca/index.php/jaste/article/view/38141

Guerrero, G. (2024). *Critical scientific and environmental literacies: Exploring opportunities for conscientisation from outdoor science education* [Doctoral thesis, University College London].

Hage, S. M., Castagna Molina, M., & McCowan, T. (2022). Climate change and the role of universities: The potential of land-based teacher education and agroecology. *Revista Brasileira De Política E Administração Da Educação, 38*(00) https://doi.org/10.21573/vol38n002022.122906

Hansson, P., & Öhman, J. (2022). Museum education and sustainable development: A public pedagogy. *European Educational Research Journal, 21*(3), 469–483. https://doi.org/10.1177/14749041211056443

Hecht, M., & Nelson, T. (2022). Getting 'really close': Relational processes between youth, educators, and more-than-human beings as a unit of analysis. *Environmental Education Research*. https://doi.org/10.1080/13504622.2022.2074378

Hickel, J. (2019). The contradiction of the sustainable development goals: Growth versus ecology on a finite planet. *Sustainable Development, 27*(5), 873–884. https://doi.org/10.1002/sd.1947

Hiner, H. (2022). We didn't see it coming. *NACLA Report on the Americas, 54*(4), 415–423. https://doi.org/10.1080/10714839.2022.2145121

Jones, N. (2023). Reimagining Freire: Beyond human relations. *Cultural Studies of Science Education, 18*(1), 205–216. https://doi.org/10.1007/s11422-023-10154-7

Kato, D. S., Galamba, A., & Monteiro, B. A. P. (2023). Decolonial scientific education to combat 'science for domination'. *Cultural Studies of Science Education, 18*, 217–235. https://doi.org/10.1007/s11422-023-10165-4

Kim, E. A., & Dionne, L. (2014). Traditional ecological knowledge in science education and its integration in grades 7 and 8 Canadian science curriculum documents. *Canadian Journal of Science, Mathematics and Technology Education, 14*(4), 311–329. https://doi.org/10.1080/14926156.2014.970906

Kimmerer, R. W. (2012). Searching for synergy: Integrating traditional scientific ecological knowledge in environmental science education. *Journal of Environmental Studies of Science, 2*, 317–323. https://doi.org/10.1007/s13412-012-0091-y

Kingsbury, D. V. (2022). Energy transitions in the shadow of a dictator: Decarbonizing neoliberalism and lithium extraction in Chile. *The Anthropocene Review, 10*(2), 555–572.6. https://doi.org/10.1177/20530196221087790

Liboiron, M. (2021). *Pollution is colonialism*. Duke University Press.

Mayes, E., & Center, E. (2023). Learning with student climate strikers' humour: Towards critical affective climate justice literacies. *Environmental Education Research, 29*(4), 520–538. https://doi.org/10.1080/13504622.2022.2067322

McEntee, M., Medvecky, F., MacBride-Stewart, S., et al. (2023). Park rangers and science-public expertise: Science as care in biosecurity for kauri trees in Aotearoa/New Zealand. *Minerva, 61*, 117–140. https://doi.org/10.1007/s11024-022-09482-9

Misiaszek, G. M., & Rodrigues, C. (2023). Teaching justice-based environmental sustainability in higher education: Generative dialogues. *Teaching in Higher Education, 28*(5), 903–917. https://doi.org/10.1080/13562517.2023.2214879

Morales-Doyle, D. (2023). Putting science education in its place: The science question in social justice education. *Cultural Studies of Science Education, 18*(1), 81–94. https://doi.org/10.1007/s11422-023-10151-w

Morales-Doyle, D., & Frausto, A. (2021). Youth participatory science: A grassroots science curriculum framework. *Educational Action Research, 29*(1), 60–78. https://doi.org/10.1080/09650792.2019.1706598

Nobles, M., Womack, C., Wonkam, A., & Wathuti, E. (2022). Ending racism is key to better science: A message from Nature's guest editors. *Nature, 610*(7932), 419–420. https://doi.org/10.1038/d41586-022-03247-w

Park, S. J., Yeatman, H., Russell, J., & MacPhail, C. (2022). Creating food learning opportunities for adults within urban settings: A framework for food pedagogies. *Community Health Equity Research & Policy, 44*(2). https://doi.org/10.1177/0272684X221118791

Saeed, T. (2023). Reimagining education: Student movements and the possibility of a critical pedagogy and feminist praxis. In *Decolonizing education for sustainable futures*. Bristol University Press. https://doi.org/10.51952/9781529226119.ch004

Sandlin, J. A., O'Malley, M. P., & Burdick, J. (2011). Mapping the complexity of public pedagogy scholarship 1894–2010. *Review of Educational Research, 81*(3), 338–375. https://doi.org/10.3102/0034654311413395

Stetsenko, A. (2019). Radical-transformative agency: Continuities and contrasts with relational agency and implications for education. *Frontiers in Education, 4*, 1–13. https://doi.org/10.3389/feduc.2019.00148

Svampa, M. (2019). *Neo-Extractivism in Latin America*. Cambridge University Press.

Temper, L., del Bene, D., & Martinez-Alier, J. (2015). Mapping the frontiers and front lines of global environmental justice: The EJAtlas. *Journal of Political Ecology, 22*(1), Article 1. https://doi.org/10.2458/v22i1.21108

Tilsen, J. (2023). "The freshness of irreverence": Learning from ACT UP toward sociopolitical action in science education. *Cultural Studies of Science Education, 18*(1), 143–158. https://doi.org/10.1007/s11422-023-10162-7

Tolbert, S., Frausto, A., & Torres-Olave, B. (2023). From false generosity to true generosity: Theorizing a critical imaginary for science education. In S. Tolbert, M. Wallace, M. Higgins, & J. Bazzul (Eds.), *Reimagining science education in the Anthropocene: Volume II* (pp. 163–184). Palgrave Macmillan.

Torres-Olave, B. (2022). *Strengthening science teacher agency: A Freirean perspective of two teacher-led communities in Chile* [PhD thesis, University of Bristol]. British Library EThOS – Search and order theses online https://ethos.bl.uk/OrderDetails.do?uin=uk.bl.ethos.852589

Torres-Olave, B. (2024). Trains: Attention, and an ethics of the other. *Review of Education, Pedagogy, and Cultural Studies*, 1–15. https://doi.org/10.1080/10714413.2024.2314867

Torres-Olave, B., & Dillon, J. (2022). Chilean physics teacher educators' hybrid identities and border crossings as opportunities for agency within school and university. *Journal of Research in Science Teaching, 59*(10), 1737–1900. https://doi.org/10.1002/tea.21774

Torres-Olave, B., Tolbert, S., & Frausto Aceves, A. (2023). Reflecting on Freire: A praxis of radical love and critical hope for science education. *Cultural Studies of Science Education, 18*(1), 1–20. https://doi.org/10.1007/s11422-023-10168-1

United Nations Framework Convention on Climate Change UNFCCC. (2008). *CDM: Alto Maipo Hydroelectric Project*. Clean Development Mechanism. https://cdm.unfccc.int/Projects/Validation/DB/2CB7GPPZM1GN0OG90BPS2R3LL4O5LR/view.html

Warren, B., Vossoughi, S., Rosebery, A. S., Bang, M., & Taylor, E. V. (2020). Multiple ways of knowing*: Re-imagining disciplinary learning. In *Handbook of the cultural foundations of learning* (pp. 277–293). Taylor and Francis. https://doi.org/10.4324/9780203774977

Betzabe Torres Olave (she/her/ella) is a Physics teacher and science education researcher. Her research interests are centered around educators' agency, collective praxis, and justice-oriented pedagogies in the context of [science] education. Her research critically debates how neoliberal agendas have shaped science education, looking for counternarratives to such a project in educators' daily practices through the concept of critical hope. Philosophically, she is interested in Paulo Freire's philosophy of education and feminist philosophies of science. She is also interested in qualitative methodologies, physics education research, and inter- and transdisciplinary praxis.

Gonzalo R. Guerrero is a Physics and Maths Teacher who graduated from Universidad de Santiago, Chile. He holds a master's degree in Research and Innovation in Curriculum and Teacher Education, specializing in Didactics of Experimental Sciences, from Universidad de Granada, Spain. Currently, he is a PhD researcher in Science Education at the Curriculum, Pedagogy, and Assessment Department, Institute of Education, University College London, United Kingdom. Additionally, he works as a researcher at Centro de Acción Climática, Pontificia Universidad Católica de Valparaíso, focusing on critical scientific and environmental literacies, activism, and teacher training.

Part III
Experiences, Tales and Local Perspectives

Chapter 6
The Unfinished Sociopolitical Turn of Science Education in Mexico in the New Curricular Framework for Public Education: A Critical Consciousness Without Action?

Liliana Valladares

6.1 The Current Context of Science Education

We are living in a context of great social and technological transformation and a growing environmental crisis that locally and globally questions us: How and why are we teaching science? What goals should science education pursue?

According to Molla (2021), educational systems currently face various tensions that can be summarized as follows:

- Intensification of market rationality, which erodes public investment in education.
- Increasing dataism, "digital panopticon" (Han, 2017, p. 37) and a surveillance capitalism (Zuboff, 2019), endangering the privacy of educational communities.
- Post-truth and infodemic, which generates greater public mistrust in science.
- Development of industry 4.0, forcing educational systems to respond to the hyperautomation of jobs.
- Alarming levels of inequalities, strongly visible with the pandemic.
- A global context of environmental and climate emergency, with a permanent risk of experiencing future pandemics and other catastrophes that threaten the survival of humanity and our continuity on Earth.

What implications do these realities have on educational systems? How is science education responding creatively to current needs arising from these global tensions?

If education as a social process always responds to the social, historical, cultural, political and economic context in which it takes place, there is no doubt that the

L. Valladares (✉)
Universidad Nacional Autónoma de México, Mexico City, Mexico
e-mail: lilianavalladares@filos.unam.mx

C. B. Moura (ed.), *A Sociopolitical Turn in Science Education*, Contemporary Trends and Issues in Science Education 62,
https://doi.org/10.1007/978-3-031-78586-3_6

current challenges demand the configuration of new pedagogical practices that allow societies to respond to the current planetary crisis and navigating the risks and tensions that the postpandemic worlds impose on us in an extraordinary way (Hauberg-Lund, 2019). Responding to these challenges, which have a particularly strong scientific component, requires a fundamental rethinking of the values, purposes and methods of science education (Laherto & Rasa, 2022).

A new way of thinking about education—its policies, plans, programs and practices—is urgent, and it should translate into the necessary sociopolitical actions so that we all participate creatively in the social, economic, cultural, political, and technological transformations that society requires to become more inclusive and more committed to social and educational justice.

As Joffe (2020) points out, similarly to when the Sputnik satellite was launched in 1957, the COVID-19 pandemic represented an opportunity to seriously transform science and technology education. The current planetary crisis demands imagining new pedagogies so that human beings transcend anthropocentric visions, developing socio-ecological processes of qualification, socialization and subjectivation (Biesta, 2013), in which we could assume our limitations (ecological, sociological, economic, ontological, among others), as well as our vulnerability to and our co-constitution with the environment (Hauberg-Lund, 2019).

We need to develop and learn new educational practices, based on innovative languages, which rewrite new types of intersubjective and interobjective relationships of respect and solidarity. We need to imagine for the next generations new cultures of socio-environmental rationality and ecocentrism; new, productive and sustainable economies; and new forms of fair and democratic public, community and political life (Petrie et al., 2020). This requires redefining the knowledge and practices that are necessary for education, to reach and maintain this future horizon, starting from our realistic circumstances.

In what follows, it is argued that science education is a programmatic concept that has responded to the context of social crisis by outlining a "sociopolitical turn" in its ultimate sense. This sociopolitical turn is expressed as an increasing socioscientific activism carried out by science teachers and students, supported by different pedagogical frameworks. With the purpose of observing some of the common traits underlying sociopolitical action in science education, five of the most relevant pedagogical frameworks reported in literature are characterized. Among them, special attention is paid to the Culturally Relevant Pedagogy (CRP) approach and in its three tenets: (i) academic success, (ii) cultural competence, and (iii) critical consciousness. The three different components of the last tenet, critical consciousness, are detailed, as parts of a sociopolitical development, to show how a critical analysis of reality is not enough to achieve sociopolitical action, because political efficacy is also required as another precursor of the action for social transformation. The recently proposed curricular framework in Mexico is described and analyzed in light of the CRP approach and the development of critical consciousness. It is argued that the so-called New Mexican School (NMS), although it describes an important shift towards critical social analysis, as a previously absent curricular objective in Mexican education, still has the chance to provide teachers of pedagogical principles to ensure science education for sociopolitical action. Finally, some reflections and recommendations are proposed to consolidate this sociopolitical turn in Mexican science education.

6.2 The Programmatic Nature of Scientific Literacy and Its Sociopolitical Turn

The aims and pedagogies in the field of science education are taking new directions (Laherto & Rasa, 2022), and understanding them is essential because they both configure the horizons towards which societies develop as well as their results. Norris et al. (2014) have characterized scientific literacy as a programmatic concept that points to desirable objectives for educators, scientists and educational policy makers, defining the horizon and the type of citizens and society that are wanted to build through education.

Due to its programmatic nature, each sociohistorical context should renegotiate the definition of scientific literacy, since it always responds to the legitimized needs expressed by a group of individuals who wish to become scientifically literate and the social communities of which they are a part. The political and axiological commitments of each definition must be made explicit to clearly decide the course that a particular society or community decides to undertake through its scientific literacy processes (Norris et al., 2014).

Two different programmatic senses printed in science education process have been unveiled by Roberts (2007), who has distinguished two main visions of scientific literacy to which almost all definitions respond (Sjöström & Eilks, 2018; Bybee, 2016).

On the one hand, Vision-I focused on the learning of scientific content and processes for their subsequent application (Bybee, 2016), which emphasizes science as a discipline that demands propositional and procedural knowledge, metacognition, and disposition (Liu, 2013). On the other hand, Vision-II groups the definitions of a scientific literacy focused on understanding the usefulness of scientific knowledge in life and society and in promoting its learning from significant contexts (Bybee, 2016), contextualizing it and relating it to technology, environment and society (Liu, 2013).

Vision-II is developed as a product of what Mansour and Wegerif (2013) call the "sociocultural turn in science education", which questioned the decontextualization of knowledge in Vision-I. This sociocultural turn places science in a context of cultural practices and interests that demand attention from the school, which could no longer be blind to the students' cultures, and which seeks to understand their personal/collective ideas about science.

The study of Odden et al. (2021), who reviewed the published articles in the *Science Education* journal between 1922 and 2019, corroborates the presence of this sociocultural turn. These authors found a historical displacement in the science education field from topics that focus primarily on particular types of science content (1950–1960), to topics focused on science students (1970–2019). The approach to issues related to race, gender, identity and critical discourse analysis, distinctive of the sociocultural turn, has had a high prevalence in the mid-1980s and after 2010.

As a result of the contributions of authors such as Santos (2009), Yore (2012) and Liu (2013), a third view of scientific literacy has been added to these two orientations: Vision-III, which assumes that school science should go beyond its contextualization and involve, at a greater extent, a social component and higher external impact (Sjöström & Eilks, 2018), since the crisis scenario we are facing requires training students to act and transform the social communities and contexts in which they live.

Vision-III integrates three innovations: a fusion of the fundamental and derivative senses of scientific literacy (Yore, 2012); the introduction of the notion of science engagement and participation (Liu, 2013); and the inclusion of a political and emancipatory agenda, aligned with equity and social justice (Santos, 2009).

This last trait has become increasingly relevant nowadays (Laherto & Rasa, 2022), since, as Galamba and Matthews (2021) describe, we are witnessing how various countries are facing a rise in fascism (and its religious intolerance, racist, sexist and homophobic practices), which threatens progressive groups that advocate for equity, freedom of expression, sexual orientation, and religion. These authors take Gottfried's (2017) definition of fascism as: "…something cruel, unscrupulous, arrogant, obscurantist, anti-liberal and anti-working-class…Fascism is also a political and economic system…" (p. 316). Galamba and Matthews (2021) remind us of how throughout history, in different fascist, authoritarian and anti-democratic regimes, science education was used as an "apolitical" tool to establish, justify, perpetuate and spread violent and intolerant ideologies.

Given the growing trend of contexts similar to fascism, Galamba and Matthews (2021) warn that some science education practices, especially those focused on neoliberal possessive individualism, may be adequate to support and deepen fascist-like ideologies, overlooking the development of sympathy, altruism and interpersonal relationships.

Though authors like Hadzigeorgiou (2015) have criticized the conception of science education as sociopolitical action, arguing that it downplays the importance of knowledge for its own sake and totally neglects the personal/aesthetic dimension of science, reinforcing a technocratic attitude, others as Galamba and Matthews (2021) argue that science education should develop a sociopolitical agenda that challenges the oppression of minorities and other fascism-related traits, engaging the youth in some sort of activism related to ethics and social responsibility.

Similarly, Tolbert and Bazzul (2017) are among the critical scholars who advocate for more politicized approaches in science education, as the major crises worsen (climate change, environmental destruction, growing social inequalities, and new apartheids). This qualitative change in the goals of science education delineates a "sociopolitical turn" in which science education should provide the educational communities with more opportunities for political engagement, becoming a "site of contestation" (Tolbert & Bazzul, 2017, p. 323) in relation to these crises. As a part of this slowly emerging sociopolitical turn in science education, Tolbert and Bazzul (2017) identify the implication of three elements: the rise of the status of politics, the development of political consciousness, and an engagement with activism. Likewise, they recognize that more and more science teachers are prepared to engage their students politically, designing educational experiences that challenge the hegemonic and inequitable education policies and practices.

How is this sociopolitical turn expressed in science teaching? In what follows, some of the main pedagogical frameworks and categories that have been developed around Vision-III of scientific literacy are summarized, showing how science education has been imbricated with sociopolitical action.

6.3 Pedagogical Frameworks of the Sociopolitical Turn of Science Education

Among the main frameworks and categories that have been developed around Vision-III of scientific literacy and that commit it to sociopolitical action are: (1) Socially Responsible Science Education, (2) Civic Science Education, (3) Future-oriented Science Teaching, (4) A Curriculum for Sociopolitical Activism and (5) Justice-Centered Science Pedagogy. This last one is particularly relevant for the Latin American context and the so-called New Mexican School.

6.3.1 Socially Responsible Science Education

Fuchs and Tan (2022) propose grouping under the category of Socially Responsible Science Education (SRSE) the set of visions for school science that seek to foster social responsibility in learners, encouraging them to take a critical stance towards products of science and technology, while also empowering them to reduce the potential harm these products have on individuals, societies, and environments.

SRSE, in a consistent way with Vision III, aspires to develop fully engaged citizens, that apply their understanding of scientific concepts and processes in issues that affect their lives, their culture and the environment, to act morally and ethically to reduce potential harm, all as a means of caring for themselves, others and their community.

According to Fuchs and Tan (2022) there are four educational frameworks which can support SRSE, and each one provides opportunities and challenges to foster social responsibility in science education. These frameworks share among their premises that sociopolitical action is explicitly promoted and modeled from school, linked to scientific content, without separating it from its social and humanistic matrix, and paying special attention to the place and context in which learning occurs: (1) Science, Technology, Society and Environment (STSE): based on critical and place-based pedagogies on socio-scientific issues and reflective actions; (2) Socioscientific Issues (SSI): aimed to fostering students' ability to reason morally and ethically about SSI through dialogue and debate; (3) Science & Technology Education Promoting Wellbeing for Individuals, Societies & Environments (STEPWISE): focused on producing Research-informed and Negotiated Actions (RiNA), where students conduct their own primary and/or secondary research to inform and take action around SSI (Bencze, 2017); (4) Critical pedagogy of place: oriented to encouraging social actions from teachers and students to re-inhabit their places, for improving their social and ecological life (Gruenewald, 2003).

6.3.2 Civic Science Education

Levy et al. (2021) proposed a Civic Science Education (CSE), to recover the insistence that science education should result in shaping a better society. CSE includes those "…experiences that have been intentionally designed to foster or enhance individuals' interactions with and/or engagement in science-related public matters…" (Levy et al., 2021, p. 1056). That is, those educational schemes that support people's ability to understand science issues are under this umbrella, explore them and take informed actions and decisions.

From the point of view of Levy et al. (2021), CSE has the enormous potential to link students' scientific and civic learning by highlighting the relevance of science-related public matters for the lives of students and by providing opportunities for them to explore and contribute to their communities, while enhancing the motivation for scientific learning. According to the level of participation in science and technology matters, these authors recognize three CSE sub-categories or levels: foundational, exploratory, and purposefully active. The last one prioritizes participation in the public sphere through providing powerful, memorable experiences; for example, promoting awareness about a specific public problem related to science, or designing and enacting plans to affect/influence it.

According to Levy et al. (2021), among the current educational practices related to CSE are: Socioscientific Issues (SSI); Science, Technology & Society (STS); Science, Technology, Societies & Environment (STSE); Citizen Science; Environmental Education; Education for Sustainable Development.

6.3.3 Future-Oriented Science Teaching

According to Laherto and Rasa (2022), for science education to attain these new aims toward transformative action, projective and transformative agentic orientations are required. Following the argument of Lloyd and Wallace (2004), Laherto and Rasa (2022) discuss how the typical approaches in science curricula either pass over the temporal dimension, are commonly oriented towards the history of science, or have been criticized as superficial, politically naive and, therefore, ineffective in bringing about the necessary individual and societal change, as is the case of some approaches based on SSI and STSE connections. These authors argue that a future-oriented teaching can help students assess the positive and negative impacts of science and technology on society and deepen their understanding of the complexity of these interactions. From their point of view, the future is inherent to the nature of science; modelling and prediction, such as creativity and imagination, play a key role in both science and its learning, so there are several reasons for explicitly discussing futures in science education, just as environmental education has done. Given that how one perceives the future affects how one acts in the present, these future-oriented pedagogies in science education could contribute to the transformativity and transgressiveness of learning that are necessary for a new critical approach to science education.

6.3.4 A Curriculum for Sociopolitical Activism

According to Hodson (2010, 2011, 2020), an important promoter of the sociopolitical turn in science education, the transition from Vision-II to Vision-III consists mainly of providing opportunities for a science education that is much more radical and politicized, in which students not only address complex and controversial socio-scientific issues (SSI), but also formulate their own position concerning and mainly prepare for, and engage in, sociopolitical actions "… that they believe will "make a difference"..." (Hodson, 2020, p. 596), both individually and collectively. Following Hodson (2020), it is not enough for students to learn that social, political, and economic forces influence science and technology, because they need to learn how to participate, to experience participation and, moreover, they need to encourage others to participate.

For Hodson (2020), among the main four goals for science education—i.e., learning science; learning about science and scientific practice; doing science and addressing SSI—it is the last one that allows finding effective, appropriate and socially responsible ways to respond to the current crisis through developing the critical skills to confront the scientific, personal, social, economic, environmental and moral-ethical aspects of SSI.

6.3.5 Justice-Centered Science Pedagogy

For Morales-Doyle (2017), science education can be a catalyst for transformation and social justice through the implementation and design of educational experiences under a framework that this author calls "Justice-Centered Science Pedagogy" (JCSP). JCSP encompasses curriculum, teaching practices, and classroom structures, and it is the result of combining the contributions of critical pedagogy (developed in Latin America) and culturally relevant pedagogy (rooted in the United States' Black freedom struggle) as a way of approaching the inequalities present in larger oppressive systems.

The Culturally Relevant Pedagogy approach was developed by Ladson-Billings (1995) to synthesize the actions of effective teachers of African American students. According to Ladson-Billings, teachers should attend three tenets to achieve social transformation: (i) students must experience academic success, (ii) must develop and/or maintain cultural competence, and (iii) they must develop a critical consciousness to challenge the status quo of the current social order.

In contexts of oppression, as Morales-Doyle (2017) points out, it is essential to attend to this third criterion, since it is the conscientization that Freire spoke of in Latin America, and which broadens the goals of education towards social change, and not only concentrating them on the development of technical and labor skills of interest to the global competitive economy. Conscientization allows people to understand the historical and political conditions that created the present inequitable social circumstances and realize that they can transform those realities, thus eliminating oppressions.

Morales-Doyle (2017) sustains that science education, which usually omitted the praxis of social transformation and the development of critical consciousness, under JCSP framework is now committed to sociopolitical action, by combining critical pedagogy with culturally relevant pedagogy. To study any scientific content of the curriculum, the JCSP stops working with SSI decontextualized to the oppressive realities that are lived in many spaces, to instead work with "Social Justice Science Issues" (SJSI), a term that refers to SSI that are also generative themes, as Freire proposed, but situated in each context. That is, culturally-relevant-SSI, or problems of social injustice that are not only important to students and their communities, but also are the starting point for scientific literacy, because they contextualize science learning within real world problems without denying the structures of social domination and inequality that are present in local contexts, and that are often suppressed or treated uncritically by traditional school science. SJSI allow to organize a curriculum that is more critical and locally relevant. With this original proposal to further politicize the traditional SSI and to engage SJSI, Justice-Centered Science Pedagogy contributes to reimagining science classrooms, seeking "...to subvert schools' historical roles as reproducers of social inequality..." (Morales-Doyle, 2017, p. 1057).

6.4 Culturally Relevant Pedagogy (CRP) and the Development of Sociopolitical Action in Science Education

The approach of Culturally Relevant Pedagogy (CRP), as shown by Morales-Doyle (2017), makes it easier to realize that in achieving social transformation in schools, it is necessary to understand that the experiences of inequality, violence, and oppression that students experience within the school are very similar to the experiences they deal with outside of school, especially the ways in which economically/racially/ethnically marginalized students in society are perceived and treated by those with greater authority and power (Elmesky, 2018). Hence, we cannot ignore the need to interrupt science teaching practices that reproduce the same unbalanced perspectives and unfair treatment that young people experience in extracurricular contexts (Elmesky, 2018).

The view of schools as safe sites that protect children from oppressive structures must be challenged; instead, schools must begin to stop reacting and take responsible action (Biesta, 2013), such as addressing issues of social justice (Elmesky, 2018). As Biesta (2013) warns, the school should stop thinking of itself as just a function of and thus functional for society; the school also needs to be dysfunctional in relation to society because, following Meirieu (2020), education always also comes with a "duty to resist" (p. 157).

How is sociopolitical action achieved in schools? Why is CRP a strategic approach to achieve this? As suggested by Hipolito-Delgado et al. (2021), education literature could benefit from more engagement with the literature from community psychology about sociopolitical development; I would make this recommendation extensive to delve into the sociopolitical turn of science education, but this requires first understanding the CRP approach and its three main tenets.

Ladson-Billings (1995) defines CRP as a form of teaching where teachers use a student's culture to guide student learning: "...Culturally relevant pedagogues are those who challenge the construction of knowledge that rarely gives students access to knowledge and information about how we make decisions about what counts as valuable..." (Ladson-Billings, 2021, p. 70). As Jones and Taylor (2022) point out, CRP teachers must guide students on this journey to become more aware and critical of systems—at every level—that create and perpetuate oppression.

Ladson-Billings (2009, 2021) has argued that teachers engaged with a CRP approach produced Black student outcomes that were greater than their colleagues working in the same school and at the same grade level. Harrell-Levy (2018) warns that, although there are some gaps in the literature, most research suggests that CRP may elevate levels of sociopolitical participation; that means that the culturally relevant class does contribute to the likelihood of students' sociopolitical involvement.

According to Kolovou (2022), CRP introduces two important cornerstones: the empowerment of the collective in contrast to simply the individual, and the establishment of a school culture of caring, not only for students' academic success, but for their overall well-being. These cornerstones are encapsulated in three educational outcomes or propositions upon which rest the pedagogical practices of teachers that are aligned with a CRP: (i) Academic achievement/student learning, (ii) Cultural competence, and (iii) Sociopolitical/critical consciousness (Ladson-Billings, 2021). Culturally Relevant Science Teaching (CRST) is a mirror of CRP but in the science fields, so it also includes (i) Academic success in science, (ii) Cultural competence in science, and (iii) Sociopolitical consciousness in science. CRST assumes that a better understanding of the sociopolitical nature of science would allow students to be active stakeholders in their communities. In that sense, the development of sociopolitical consciousness in science implies the teaching about how issues of power and oppression intersect science practices, for example, through the analysis of the underrepresentation of minority groups in science, technology, engineering, and mathematical fields, or identifying dominant, exclusionary narratives that are overly masculine and Eurocentric within science texts and diverse learning resources (Jones & Taylor, 2022).

Regarding (i) Academic success, Ladson-Billings (1995) argues that teachers and schools must make it their priority to provide students with many opportunities to experience academic success through achieving authentic learning, relevant to students' lives. That includes developing their academic skills, such as reading, writing, mathematical, technological, and intrapersonal and interpersonal social and political skills; acquiring and creating new knowledge individually and collectively; regulating personal emotions, which are constantly impacted by the social, economic, and political inequalities and inequities around them; and recognizing that learning is a lifelong journey (Cataldo, 2022).

In regard to (ii) Cultural competence, Ladson-Billings (1995) argues that culturally competent teachers provide their students with the opportunities to affirm and value their culture, while acquiring knowledge and fluency in, at least, another culture. Such is the case of minoritized students, for whom another culture is generally the mainstream one, but in this approach, it is important to challenge existing power structures that break through these cultures (Ladson-Billings, 2021).

Finally, with respect to (iii) Critical Consciousness, this tenet is, according to Ladson-Billings (2021), the most ignored, and according to Jones and Taylor (2022) the most understudied and underutilized, despite being fundamental to the entire approach. Critical Consciousness refers to the fact that schools must teach students how to develop their critical and sociopolitical consciousness, that is, their ability to critically analyze the world, explicitly the societal inequalities and inequities that affect their lives. All of the pedagogical frameworks in Sect. 6.3 share the assumption that academic success could be favored by promoting cultural competence and sociopolitical action, and vice versa, however the interrelation among these components is often minimized or disarticulated. Critical Consciousness is central to our analysis because it focuses on the achievement of sociopolitical action as a goal of science education. Let's see.

6.4.1 Critical Consciousness and Sociopolitical Action: The Third Tenet of CRP

According to Anyiwo et al. (2020), critical sociopolitical action refers to the efforts to influence or transform communities and societal systems to be more equitable and just. This action could be individual and interpersonal, collective, or digital and can occur through conventional politics and community engagement, such as writing letters to political officials about sociopolitical concerns, voting, volunteering, protesting, boycotting, blocking traffic, occupying buildings, social media campaigns, among others. However, this action can also manifest in subtle and seemingly mundane strategies, such as wearing clothing with cultural and political messages (Anyiwo et al., 2020).

Anyiwo et al. (2020) identify as core factors that shape and influence sociopolitical action and youth's participation in activism to the sociocultural factors, the school context and the sociopolitical development. Sociocultural factors include variables that influence or shape activism and the desire for social change, for example, racially discriminatory experiences. School context mainly refers to the perception of school climate as a space for diversity and sociopolitical growth and an increase in civic engagement (Anyiwo et al., 2020).

Finally, critical sociopolitical development (SPD) is a "…process by which individuals acquire the knowledge, analytical skills, emotional faculties, and the capacity for action in political and social systems necessary to interpret and resist oppression…" (Watts et al., 2003, p. 185). In other words, it is the process through which the oppressed or marginalized people learn to critically analyze their social conditions and act to change them (Watts et al., 2011) or improve them (Watts & Hipolito-Delgado, 2015). Another definition of SPD characterizes it as: "… the process by which people: develop knowledge of self, social inequality, and social action; develop efficiency in engaging in action; and use this knowledge to claim control of their sociopolitical reality…" (Hipolito-Delgado et al., 2021, p. 5).

Watts et al. (2011) outline a theoretical formulation of the development of critical consciousness with three distinct components that are combined in a fluid and recursive way (Hipolito-Delgado et al., 2021): (A) Critical social analysis or reflection, (B) Political efficacy, and (C) Participation in civic or political action.

(A) Critical social analysis or reflection refers to a social analysis, recognition, and questioning of economic, racial/ethnic, social and gender systemic inequities that constrain wellbeing and human agency and its moral rejection. (B) Political efficacy is the perceived ability to effect social and political change through individual and/or collective activism, that means, the sense of agency or the feeling that people have about their capacity to create a perceived change to address injustices. (C) Critical action refers to individual or collective action taken to change some aspects or conditions of society (practices, decisions, or institutional policies) perceived as unjust (Watts et al., 2011; Diemer et al., 2016). While critical reflection or awareness represents the cognitive component of the SDP, political efficacy embodies the emotional one, and critical action represents the behavioral component (Nicholas et al., 2019).

There is a complex and non-linear relationship between these three components, since the social or political action may influence critical reflection, and vice versa. Critical social analysis is seen by some authors as a necessary component (but not sufficient), and as a precursor for effective political action since people do not participate blindly without some understanding or awareness of why they are engaging in political action. Neither do people participate if they do not have a sense of agency for effective political action: "…That is, people may understand structural inequalities, but do not feel compelled to act on their insights unless they believe their efforts will yield a desired outcome…" (Watts et al., 2011, p. 45).

Even though it is unclear how the shift from critical social analysis to sociopolitical action could be fostered, Watts and Hipolito-Delgado (2015) have identified the prominence of fostering an awareness of sociopolitical circumstances, encouraging critical questioning, and cultivating collective identity. For these authors, the sense of collectivity is thought to inspire action or change, helping participants see beyond their personal experience, so community organizing is a promising strategy for bridging the gap between critical social analysis and sociopolitical action. Murray and Milner (2015) also highlight the centrality of developing collective identity in ways that can increase the ability to fight for social change. Harrell-Levy (2018) found a positive relationship between service learning and sociopolitical involvement, and mainly between pedagogical experience and sociopolitical efficacy and involvement.

Although all the components of SDP are important, sociopolitical action is crucial because it involves engaging others towards social justice: "…it is the only component of critical consciousness with the potential to change social and structural aspects of society…" (Watts & Hipolito-Delgado, 2015, p. 864). However, these authors have found that theorists expend most of their intellectual energy describing the component of critical social analysis, its features, and its importance for liberation, as if sociopolitical action arises without other preconditions such as leadership or mobilization skills. CRP practitioners do the same by developing a wide range of techniques and exercises for advancing critical analysis, but: "…ironically, the sociopolitical action phase received the least attention from theorists…" (Watts & Hipolito-Delgado, 2015, p. 853).

An example can be found in Murray and Milner (2015). These authors outline four pedagogical principles that can be useful to increase youths' sociopolitical development. The four principles contribute clearly to the development of critical social analysis, but even the authors ask themselves, how much do they add up for the deployment of social change actions? Many curricular proposals reproduce a similar pattern: they provide valuable elements for critical analysis but leave aside political efficacy and action for social change. As Jones and Taylor (2022) warn, in the CRST there should be clarification on how to use the knowledge that results from critical social analysis associated with science to really transform oppressive realities in ways that are liberating and empowering.

Figure 6.1 synthesize the importance, for a science education committed to social action, of exposing the different elements of CRP, since this dissection facilitates a more precise identification of one of the main challenges of a transformative science school: we need not just achieving a critical analysis of science, but also making a pedagogical effort to attain a better political efficacy that be more easily translated into concrete action for social change.

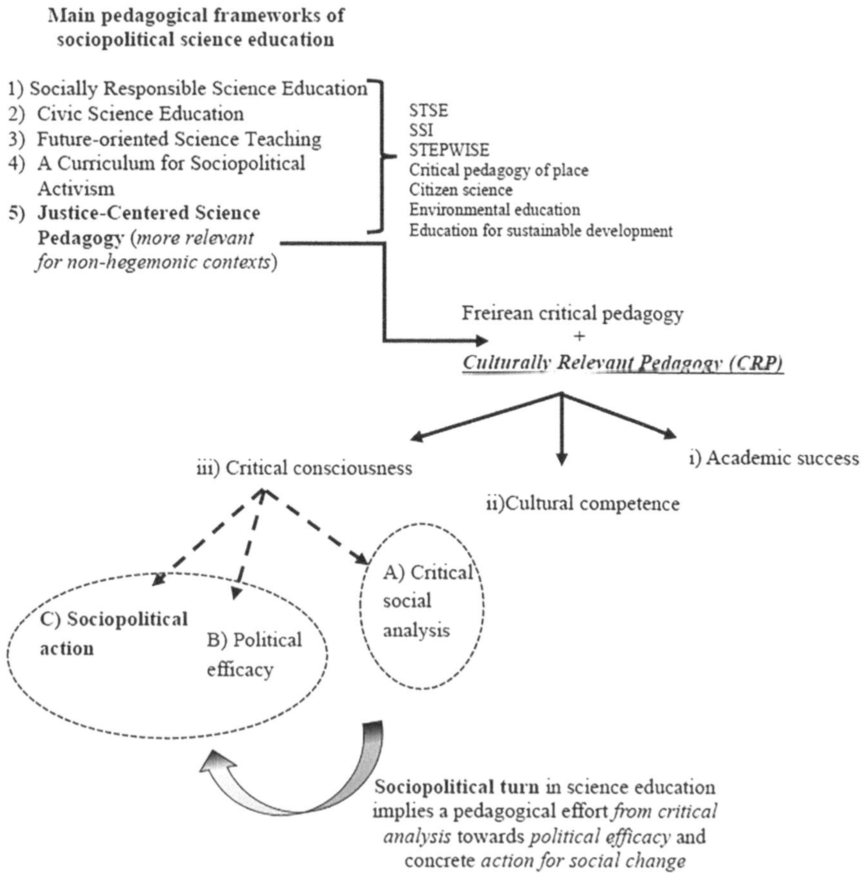

Fig. 6.1 Sociopolitical turn in science education: Culturally Relevant Pedagogy (CRP) components underlying sociopolitical action

6.5 The Mexican Curriculum as an Unfinished Example of a Sociopolitical Science Education

The New Mexican School (NMS) is the name given to the process of transformation of the hegemonic educational trends and of implementing a new national educational policy, which began in 2019, by the current Mexican government (2018–2024). The NMS began with direct modifications to the Mexican Constitution, substantially changing articles 3rd, 31st and 73rd (Tiburcio & Jiménez, 2020), by proposing a democratic, equitable, comprehensive, inclusive and intercultural education, that questions and transforms the structural inequalities that have configured this Latin American country.

This educational proposal arose as a response to a diagnosis of Mexican education in its last 50 years, in which it was identified (SEP, 2022):

- That in the different reforms that have been applied in the history of basic education in Mexico (including elementary and secondary levels), a neoliberal orientation has prevailed.
- This neoliberal orientation reproduces inequality, racism and classism as structural mechanisms on which the great educational problems of this country are based, such as school desertion, repetition, low academic performance in standardized tests.
- Many of the problems in education are individualized and blamed on the subjects of education.
- Historically, diversity has been ignored as an instituting element of the national curriculum, so contents that are culturally relevant, pertinent, and significant for the students are not prioritized.
- Previous curricular proposals that promoted the demerit of teaching and the teaching figure as a subject that only prescribes the contents of its subject, the objectives, skills or key learnings, and the planning and evaluation provided.

To respond to these deficiencies and gaps, the NMS, in its new common curricular framework, proposes (SEP, 2022):

- The human right to education as a fundamental principle of the national educational policy in Mexico. Such right could only be ensured if the school stops tolerating inequalities, racism, sexism, classism and violence.
- A recentering of the community as the one that constitutes and articulates the educational processes, so that the students have the possibility of being effectively educated as subjects with rights, within a framework of interaction between the school and the community.
- A teaching revaluation in which the curricular autonomy of the teachers is recognized, to decide on their didactic practice and the syllabus and to establish a pedagogical dialogue in the teaching and learning processes.
- A structure of formative fields, instead of subjects, that favors the integration of knowledge and the links among contents and the diverse territories, understood as "…socially and culturally symbolized spaces, which are constantly resignified in a framework of tensions…[…] where reading, arithmetic and science are equally important, like the land, the rivers, the home, the body, ancestral knowledge, architecture, community festivals, spaces for coexistence…" (p. 69).
- Learning based on projects as a strategy to identify the curricular content involved in different fields and to carry out relevant articulations. These projects should be aimed at social justice and solidarity with the environment, respecting diversities.
- Transition from school grades to the establishment of content by learning phases, which would allow more time to be devoted to the cognitive, emotional, social, and cultural processes that students require for the optimal development of their abilities.

In the formative field dedicated to the sciences, named "Knowledge and scientific thought", NMS aims to develop a scientific attitude to question, investigate, think and interpret natural and social phenomena and processes, generating new meanings and strategies to solve problematic situations at nature, school, personal, family and community level.

Analyzed under the light of the CRP approach, in this new curricular framework the school is characterized as a sociopolitical space, where the common is built from the diverse, always in tension and conflict. At school: "… converge and manifest, in their tension and complexity, the effective exercise of the rights of children and adolescents, and the inequalities that articulate social exclusions based on class, gender, sexuality, nationality, ethnicity, ability and age…" (SEP, 2022, p. 9). The distributive function of the school is questioned, interrogating the human capital paradigm, in which school success and failure are individualized; then, justice and social transformation are incorporated as ends of education.

Regarding Academic success, central to the CRP approach, NMS is committed to offering students "…an education that allows them to transform the world as profoundly as they consider necessary. This task implies providing them with the intellectual, affective, emotional, social and political tools so that they can proceed with this transformation, [...] protecting them during this formative time, creating bridges with other institutions, and supporting them to be affectively, intellectually and morally free and self-employed…" (SEP, 2022, p. 12). NMS is based on social humanism (Tiburcio & Jiménez, 2020); as students are conceptualized as subjects with rights, it is expected that through education students can develop all their potential, individually and collectively.

In relation to Cultural competence, NMS adopts this tenet as part of its guidelines, recognizing that: "... the participation of children and adolescents in their schools and communities around different aspects of life allows them to interact, learn and dialogue with people of different ages, creeds, ethnic groups, languages, abilities, sexual preferences, with the view of transforming their reality and participating in the generation of common benefits for all..." (SEP, 2022, p. 12).

Finally, regarding Critical consciousness, NMS conceives education as a political act and the curriculum as a "…battlefield that reflects other struggles: business, political, economic, religious, identity, cultural..." (SEP, 2022, p. 32). Dialogue and a necessary transformation are assumed as basic premises, because we live in an unequal society; so the state, the community and the school must be engaged with "… the oppressed, women, indigenous people, Afro-descendants, people with disability, sexual diversity, and not engaged with a type of education that removes children and adolescents from their historical and social conditions, reducing them to repositories of information whose social and economic value translates only into the form of human capital …" (SEP, 2022, p. 13).

NMS recognizes education as a condition of possibility of emancipation, and is therefore considered as an enabling right, "... perhaps the most important right of the social..." (Martínez, 2022, p. 104). Likewise, learning is understood as a fact situated in a specific social, historical and territorial context that links students with their reality in order to transform it, identifying relevant problems and collaboratively taking action to build solutions.

Consequently, it is possible to assume that NMS is in accordance with the three CRP tenets and, therefore, it is also engaged in sociopolitical action. It is important to note that NMS was not conceptualized from the CRP perspective, but I argue this is a relevant approach to understand the sociopolitical turn of Mexican educational change and it could be also useful for similar analysis in another contexts. The NMS sociopolitical turn is manifested in the explicit promotion of actions among students, such as: building projects of service and solidarity exchange between the school and the community-territory from the reappropriation of knowledge, experiences, wisdom and practices learned during the educational processes, which occurs inside and outside the school space; participating in caring for the environment in an active, supportive and resilient way in the face of climate change; applying critical thinking as a basis for free, conscious and responsible decision-making, based on the exercise of the students' rights, for healthy eating, more physical activity, sexual and reproductive health.

However, even though NMS provides teachers with "learning progressions" oriented towards concrete actions, and despite NMS supplying some didactic guidelines to trigger the teachers' creativity for teaching critical social analysis, the new Mexican curricular framework still has the chance to provide guidelines for the teacher to teach sociopolitical action. In other words, CRP offers new light and possibilities to NMS fully respond to what the sociopolitical turn of science education demands.

If we remember the three components of critical consciousness: (i) critical social analysis or reflection, (ii) politics efficacy (the perceived ability to effect sociopolitical change), and (iii) participation in civic or political action (Hipolito-Delgado et al., 2021), the NMS proposes school contents for science education that clearly address the first component, and perhaps the second, since its emphasis is on: (1) knowing and interpreting "…the natural and sociocultural phenomena and processes from the exploration, observations and identification of environmental regularities…" (SEP, 2022, p. 145); (2) recognizing the interactions among elements of the natural environment, biodiversity, matter and energy; (3) analyzing "...costs and benefits for the family and the environment..." (SEP, 2022, p. 147) of certain consumption habits, as well as the causes and consequences of environmental degradation, pollution and climate change in order to "... propose actions for the common benefits..." (SEP, 2022, p. 148).

The predominant verbs in the learning progressions in science fields are: recognize, analyze, interpret, explain and value, so, although necessary for sociopolitical action, they will not immediately translate into acts and actions of sociopolitical change. The NMS formative field dedicated to knowledge and scientific thought contains didactic guidelines that suggest to the teacher some class activities to identify and characterize SSI, associated with the quality of air, water or soil and the form, as well as their associations with the prevalence of respiratory and gastrointestinal diseases in their locality. However, to achieve greater impact on practices and decisions committed to deeper sociopolitical change, NMS could include guidelines aimed at explicitly promoting political efficacy or critical action, as CRP suggest.

Of the four total contents and of the twelve learning dialogues (interactions between contents of physics, chemistry, biology, mathematics) that organized secondary education, all include activities related to the identification, investigation, analysis, explanation and interpretation of related data to local SSI, but only one proposes the translation of these analyzes into concrete action (which in this case consists of participating in the design of a community vaccination action plan, such as launching a local campaign). The rest of the science contents are not grounded in actions of social change, but rather promote social analysis and argumentation in students. The same happens at the elementary level.

The great affinity that the NMS' theoretical framework shows with the critical content of the CRP approach, reveals some of the possibilities that these new curricular guidelines have to consolidate a science education for a sociopolitical action, and in this way to contribute to social transformation.

Hipolito-Delgado et al. (2021) argue that critical reflection and sociopolitical efficiency are independent predictors of sociopolitical action. These authors also found evidence of how participation in transformative student voice (TSV) and classroom leadership opportunities positively influenced sociopolitical efficacy. Political efficacy is important because people will act when they believe they can make an impact. According to Hipolito-Delgado et al. (2021), there is evidence that CRP is positively related to sociopolitical efficacy, but additional research is needed to better understand educational practices that promote sociopolitical efficacy and action.

For Diemer et al. (2016), educational interventions that focus only on critical social analysis or reflection would be incomplete because, although it is a necessary step, it is insufficient for sociopolitical action, so it is very important to design educational experiences that foster among students an engagement in action while developing critical reflection, critical motivation, and collective identity simultaneously. Hipolito-Delgado et al. (2021) warn that critical pedagogy, with its emphasis on critical consciousness, while necessary, is rarely accompanied by opportunities to develop and practice skills required for activism. Critical awareness of oppression is a crucial first step; for people with a history of oppression, the ability to think independently and critically about dominant discourse, which typically marginalizes communities, is a prerequisite for liberation, but it is not enough. Developing a sociopolitical science education requires also to pay attention on how promoting sociopolitical efficacy and action, in such a way that teaching scientific contents to attain academic success could be accompanied by the deployment of cultural competence and critical consciousness, as CRP sustains.

6.6 Conclusion: Promoting Sociopolitical Science Education

We are facing a rise in fascism. Consequently, social tensions and demands for justice are increasing among vulnerable groups, historically excluded from and oppressed by the benefits of human development. This postpandemic context makes unavoidable the necessity to reconsider schools as sociopolitical spaces. This implies that professional development programs need to address teachers' sociopolitical consciousness and equip them with strategies and practices to foster the critical consciousness of their students (Kolovou, 2022). This has been partially attained by NMS, but there is a chance to fortify this advance. CRP approach offers valuable theoretical elements that potentially could inspire and strengthen teachers' pedagogical guidelines aimed to contribute to larger social changes.

Usually, mathematics and science teachers do not feel confident in using CRP approaches (Kolovou, 2022), even though disparities stratified by race and social class are intensified in these subjects, in which the curriculum is mainly Eurocentric and the perspectives of Indigenous, Latinx, and Black people's contributions are erased. Kolovou (2022) argues how, despite the promising results associated with CRP, such implementations remain relatively low and sporadic in science classes, partly because teachers feel unfamiliar or not confident with using CRP practices, so it is urgent to support teachers and provide them with pedagogical guidelines and inclusive tools for equitable education.

Jones and Taylor (2022) agree that teachers need dedicated time and space to succeed in the implementation of each CRP tenet. Most importantly, making the shift to teach science in this way should have a lasting impact on our world, giving students the power to be agents of critical change. These authors share practical suggestions, ideas and examples of how teachers can practically implement the third tenet of Culturally Relevant Science Teaching to engage and empower students in science classrooms, fostering the students' sense of sociopolitical consciousness.

It is our hope that these general recommendations may encourage science teachers to materialize a sociopolitical science education that is able to respond to the challenges facing our world. As expressed by the voices of some teachers, the NMS has many challenges, since it is not an easy task to create a democratic and participatory society in Mexico, where diversity and equity are addressed, as well as a school that transcends beyond the cognitive to civic science education (Tiburcio & Jiménez, 2020). Some of the principles and axes of the NMS are still far from the science teachers' cultures and from the students and their families. This poses the question of how it would be possible to implement what discursively remains written in the official documents of the NMS. Throughout this chapter, I have argued the importance of knowing some of the different frameworks for a sociopolitical science education, paying special attention to the CRP contributions and the three components of sociopolitical development (SDP) in contexts such as the Latin American ones. May this work serve as an invitation to continue the analysis of the challenges that scientific literacy has for social transformation in non-hegemonic contexts, such as Mexico.

References

Anyiwo, N., Palmer, G. J., Garrett, J. M., Starck, J. G., & Hope, E. C. (2020). Racial and political resistance: An examination of the sociopolitical action of racially marginalized youth. *Current Opinion in Psychology, 35*, 86–91.

Bencze, J. L. (Ed.). (2017). *Science & technology education promoting wellbeing for individuals, societies & environments*. Springer.

Biesta, G. (2013). Responsive or responsible? Democratic education for the global networked society. *Policy Futures in Education, 11*(6), 733–744.

Bybee, R. (2016). Scientific literacy. In R. Gunstone (Ed.), *Encyclopedia of science education* (pp. 944–946). Springer.

Cataldo, K. (2022). Culturally relevant pedagogy in the post COVID-19 era. *Current Issues in Comparative Education, 24*(2), 115–118.

Diemer, M., Rapa, L., Voight, A., & McWhirter, E. (2016). Critical consciousness: A developmental approach to addressing marginalization and oppression. *Child Development Perspectives, 10*(4), 216–221. https://doi.org/10.1111/cdep.12193

Elmesky, R. (2018). Science education reform: Can students learn science while navigating oppressive schools within an oppressive society? In L. Bryan & K. Tobin (Eds.), *Thirteen questions: Reframing education's conversation: Science* (pp. 391–400). Peter Lang.

Fuchs, T., & Tan, Y. S. M. (2022). Frameworks supporting socially responsible science education: Opportunities, challenges, and implementation. *Canadian Journal of Science Mathematics and Technology Education, 22*, 9–27. https://doi.org/10.1007/s42330-022-00200-x

Galamba, A., & Matthews, B. (2021). Science education against the rise of fascist and authoritarian movements: Towards the development of a *pedagogy for democracy. Cultural Studies of Science Education, 16*, 581–607. https://doi.org/10.1007/s11422-020-10002-y

Gottfried, P. (2017). The uses and misuses of "Fascism". *Society, 54*(4), 315–319. https://doi.org/10.1007/s12115-017-0144-7

Gruenewald, D. A. (2003). The best of both worlds: A critical pedagogy of place. *Educational Researcher, 32*(4), 3–12.

Hadzigeorgiou, Y. (2015). A critique of science education as sociopolitical action from the perspective of liberal education. *Science & Education, 24*, 259–280.

Han, B.-C. (2017). *Psychopolitics: Neoliberalism and new technologies of power*. Verso.

Harrell-Levy, M. (2018). Sociopolitical efficacy as a mediator of pedagogical experience and sociopolitical involvement. *Citizenship, Social and Economics Education, 17*(2), 100–117.

Hauberg-Lund, M. (2019). Dark pedagogy in the Anthropocene. In J. Lysgaard, S. Bengtsson, & M. Hauberg-Lund (Eds.), *Dark pedagogy: Education, horror and the Anthropocene* (pp. 103–141). Palgrave Macmillan.

Hipolito-Delgado, C. P., Stickney, D., Kirshner, B., & Donovan, C. (2021). Fostering youth sociopolitical action: The roles of critical reflection, sociopolitical efficacy, and transformative student voice. *Urban Education, 59*, 1132–1158. https://doi.org/10.1177/00420859211068468

Hodson, D. (2010). Science education as a call to action. *Canadian Journal of Science, Mathematics and Technology Education, 10*(3), 197–206.

Hodson, D. (2011). *Looking to the future: Building a curriculum for social activism*. Sense Publishers.

Hodson, D. (2020). Going beyond STS education: Building a curriculum for sociopolitical activism. *Canadian Journal of Science, Mathematics and Technology Education, 20*, 592–622.

Joffe, A. (2020). *Coronavirus: A sputnik moment for science education. The COVID-19 crisis: Impact and implications*. https://www.jstor.org/stable/resrep26356.23

Jones, B., & Taylor, J. (2022). Within the walls of the classroom: How science teachers' instruction can develop students' sociopolitical consciousness. *Science Education, 106*(5), 1021–1031.

Kolovou, M. (2022). Embracing culturally relevant education in mathematics and science: A literature review. *The Urban Review, 55*, 133–172. https://doi.org/10.1007/s11256-022-00643-4

Ladson-Billings, G. (1995). Toward a theory of culturally relevant pedagogy. *American Educational Research Journal, 35*, 465–491. https://doi.org/10.3102/00028312032003465
Ladson-Billings, G. (2009). *The Dreamkeepers: Successful teachers of African American children*. Jossey Bass.
Ladson-Billings, G. (2021). I'm here for the hard re-set: Post pandemic pedagogy to preserve our culture. *Equity & Excellence in Education, 54*(1), 68–78.
Laherto, A., & Rasa, T. (2022). Facilitating transformative science education through futures thinking. *On the Horizon, 30*(2), 96–103. https://doi.org/10.1108/OTH-09-2021-0114
Levy, B., Oliveira, A., & Harris, C. (2021). The potential of "civic science education": Theory, research, practice, and uncertainties. *Science Education, 105*, 1053–1075.
Liu, X. (2013). Expanding notions of scientific literacy: A reconceptualization of aims of science education in the knowledge society. In N. Mansour & R. Wegerif (Eds.), *Science education for diversity* (pp. 23–39). Springer.
Lloyd, D., & Wallace, R. (2004). Imaging the future of science education: The case for making futures studies explicit in student learning. *Studies in Science Education, 40*(1), 139–177.
Mansour, N., & Wegerif, R. (2013). Why science education for diversity? In N. Mansour & R. Wegerif (Eds.), *Science education for diversity* (pp. ix–xx). Springer.
Martínez, M. A. (2022). Descolonizar la escuela: educación por la justicia en comunidades originarias del sureste de México. *Revista Internacional de Educación para la Justicia Social, 11*(1), 101–117.
Meirieu, P. (2020). Pedagogía: El deber de resistir. UNAE. http://repositorio.unae.edu.ec/bitstream/56000/1373/1/PhilippeMeirieu%20completo.pdf
Molla, T. (2021). Critical policy scholarship in education: An overview. *Education Policy Analysis Archives, 29*(2). https://doi.org/10.14507/epaa.29.5655
Morales-Doyle, D. (2017). Justice-centered science pedagogy: A catalyst for academic achievement and social transformation. *Science Education, 101*, 1034–1060.
Murray, I., & Milner, R. (2015). Toward a pedagogy of sociopolitical consciousness in outside of school programs. *The Urban Review, 47*, 893–913. https://doi.org/10.1007/s11256-015-0339-4
Nicholas, C., Eastman-Mueller, H., & Barbich, N. (2019). Empowering change agents: Youth organizing groups as sites for sociopolitical development. *American Journal of Community Psychology, 63*(1–2), 46–60. https://doi.org/10.1002/ajcp.12315
Norris, S. P., Phillips, L., & Burns, D. (2014). Conceptions of scientific literacy: Identifying and evaluating their programmatic elements. In M. Matthews (Ed.), *International handbook of research in history* (pp. 1317–1344). Springer.
Odden, T., Marin, A., & Rudolph, J. (2021). How has science education changed over the last 100 years? An analysis using natural language processing. *Science Education, 105*, 653–680.
Petrie, K., Kemmis, S., & Edwards-Groves, C. (2020). Critical praxis for critical times (163–178). In K. C. Mahon et al. (Eds.), *Pedagogy, education, and praxis in critical times*. Springer.
Roberts, D. (2007). Scientific literacy/science literacy. In S. Abell & N. Lederman (Eds.), *Handbook of research on science education* (pp. 729–780). Lawrence Erlbaum.
Santos, W. (2009). Scientific literacy: A Freirean perspective as a radical view of humanistic science education. *Science Education, 93*(2), 361–382.
SEP. (2022). *Marco Curricular y Plan de Estudios 2022 de la Educación Básica Mexicana*. Secretaria de Educación Pública.
Sjöström, J., & Eilks, I. (2018). Reconsidering different visions of scientific literacy and science education based on the concept of bildung. In Y. Dori (Ed.), *Cognition, metacognition, and culture in STEM education* (pp. 65–88). Springer.
Tiburcio, C., & Jiménez, V. (2020). Concepciones docentes sobre la interculturalidad en la Nueva Escuela Mexicana. *Revista Entrerios, 3*(1), 84–99.
Tolbert, S., & Bazzul, J. (2017). Toward the sociopolitical in science education. *Cultural Studies of Science Education, 12*, 321–330. https://doi.org/10.1007/s11422-016-9737-5

Watts, R. J., & Hipolito-Delgado, C. (2015). Thinking ourselves to liberation? Advancing sociopolitical action in critical consciousness. *The Urban Review, 47*, 847–867. https://doi.org/10.1007/s11256-015-0341-x

Watts, R. J., Williams, N. C., & Jagers, R. J. (2003). Sociopolitical development. *American Journal of Community Psychology, 31*, 185–194. https://doi.org/10.1023/A:1023091024140

Watts, R. J., Diemer, M. A., & Voight, A. M. (2011). Critical consciousness: Current status and future directions. *New Directions for Child and Adolescent Development, 134*, 43–57.

Yore, L. (2012). Science literacy for all: More than a slogan, logo, or rally flag! In K. Tan & M. Kim (Eds.), *Issues and challenges in science education research* (pp. 5–23). Springer.

Zuboff, S. (2019). *The age of surveillance capitalism*. Profile Books.

Liliana Valladares is Doctor in Philosophy of Science, Associate Professor (Faculty of Philosophy and Letters of the National Autonomous University of Mexico). Main research interests include: Social and cultural contexts in science education; Public pedagogy and education in non-school contexts; Epistemology and science education; Contemporary debates in the humanities, social sciences, and educationg.

Chapter 7
"You're a Boy; You're a Girl": Witnessing Anti-queer Violence in and Imagining Queer Futures for Post-COVID Science Education

Matthew Weinstein and Alysa M. Schafer

7.1 Introduction

Alysa's Story

While setting my high school science class up one day: fixing the slides, and getting my class on track, I was thrown off by a terrifying yell. I swung my head towards the left door to my classroom after hearing the commotion when I heard one of my students yell, "I am going to beat that tranny's ass!". I was stunned. That sort of violent rhetoric spoken into the room felt like it set the mood for the rest of my day, even the rest of my week. I knew I had transgender students in my class that period, and I knew that they were watching my reaction to that statement. He then threw himself down, fuming. I told him to "please not use that kind of language in my room, it is unacceptable". The student retaliated saying "I don't care; they are a tranny and they deserve it".

This is a paper about the dangers that have emerged for queer youth and teachers after the reopening of face-to-face schooling in the U.S. in the wake of their closures–to inhibit the spread of COVID-19. We do not define queer here, and we use it interchangeably with LBGTQ+ to mean those gender/sex* identities and practices that challenge heteropatriarchal/cis-het norms and hegemony. Furthermore, while we use the term queer inclusively, there is a central concern for us for trans youth who have been the most explicit target of attacks in recent alt-right legislation,

M. Weinstein (✉)
University of Washington, Tacoma, WA, USA
e-mail: matthew@uw.edu

A. M. Schafer
Independent Scholar, Tacoma, WA, USA

C. B. Moura (ed.), *A Sociopolitical Turn in Science Education*, Contemporary Trends and Issues in Science Education 62,
https://doi.org/10.1007/978-3-031-78586-3_7

media, and organizing. The return to normal urged on schools and communities produced environments that had new and explicit dangers for queer youth and teachers. Reactionary social movements singled out queer youth as targets of erasure, making that part of their identities outlaw in the literal sense—trans and other queer identities were not even medicalized, but made to disappear (Casper et al., 2022).

This paper is also about the role of science classrooms both as problematic spaces for queer youth, but also about their potential to support queer identities. As we will explore, the science classroom, in addition to all the other "usual" ways classrooms reinforce cis-hetero-patriarchy, are involved in the additional ideological work of naturalizing binary and material (biologically determined) narratives of sex/gender. In re-reading the science classroom as a place to complicate the binary story, we are helping envision the science classroom as a place that is especially well situated to produce queer well-being. Clearly this is not a neutral paper; it is a direct call for action on the part of straight and queer teachers and youth to transform classrooms and larger cultures to support queer identity in all its complexities. In concerning itself with queer identities and power as they intersect and materialize in schools and science classrooms we identify with the sociopolitical project described by Gutiérrez (2013) in analyzing her subject matter field (mathematics) to "imagine new possibilities for relationships between people, mathematics [or science], and the globe."

In many ways we are simply producing an accounting of the circumstances in which queer youth and teachers find themselves now, shortly after the reopening of schools in 2022. But we will also use narrative and reflection to examine the experiences of queer teachers and youth in this moment and the potential of science classrooms as a space of resistance and well-being.

Note that this paper is primarily grounded in our own narratives. It is not autoethnography; it is not narrative analysis. It is a witnessing to the suffering wrought by the "torquing" that youth and teachers must do to survive in binary sex/gender categories. Torquing refers to the turns one takes when biography meets social categories (Bowker & Star, 1999): here the categories of nature and culture, male and female, as well as sex and gender.

This is also **not** an intersectional analysis. We know that there is further complexity to this story when we consider the ways that class and race intersect with sexgender. The authors are both positioned as white and middle class (now, if not in their childhoods), we are exploring our experiences, and sharing what we have witnessed. This is a kind of testimony fraught with all the complexity such an act entails (Felman & Laub, 1991). As such we are employing the use of counter narrative to claims of progress, but also to the pessimism of powerlessness (Miller et al., 2020).

Finally, we will be intentionally sloppy with sex versus gender. Traditionally this distinction was made where one (sex) was seen as material and the other (gender) was seen as sociocultural (see for instance, Keller, 1992); in other words, one was nature and the other was culture. As will be clear, we do not make such a division. We do not see sex as "pre-cultural"; rather, central to our framework is the idea that sex is the gendering of nature, as much as gender is the culturing of sex. Our commitments are to queering both sides of that binary. This also reveals the intellectual tradition we both think from: critical queer feminist science studies (Cipolla et al., 2017). This is a stance that allows us to critically engage with science, society (e.g., schooling), and their respective hierarchies unbeholden to either while never being able to be fully separable from them. We are, after all, trying to push the bus we are riding on (Martin quoting Berger and Luckmann, 1998, p. 25).

7.2 Queer Tellings

This paper builds on previous scholarship on queering science education (e.g., Snyder & Broadway, 2004; Fifield & Letts, 2014; Letts & Fifield, 2019). Queer theory provides a critical lens for understanding and challenging binary assumptions, emphasizing the diversity of gender and sexual identities and the ways in which these identities are shaped by social, cultural, and historical factors. It is also a political project of ensuring the well-being of people who fall outside the cisheteronormative dominant culture and structures (Lundin, 2014; Mattheis et al., 2022). It is an explicitly political scholarship, working towards imaginings, theories, narratives, and histories that might engender societies and institutions welcoming of diversity, consent, and emergent ontologies and identities. In this we take a lesson from Julian Gill-Peterson's (2018) *Histories of the Transgender Child* which narrates a history of trans children both subject to medicalization yet in no way defined by medicine's cruel engagement with queer bodies. We are also both deeply informed by Judith Butler's implosion/explosion of sex and gender, in which the performance and speech act invert the conventional (even within academic circles) of biological sex and social-constructed gender (1990).

Science has played a major role in reifying the conventional order. Science and science education have often invested themselves in binary categories, such as male/female, heterosexual/homosexual, living/nonliving, and natural/unnatural (synthetic or artificial), that reinforce normative assumptions about gender and sexuality. These assumptions can limit the ability of science to fully apprehend the diversity and complexity of human (and non-human!) experience and can also contribute to the marginalization and exclusion of LGBTQ+ individuals in scientific fields.

The work of Donna Haraway is also centrally important to us as a way to open up science as a critical field. We draw on Donna Haraway's notion of imploded culture and nature (1994, 1997, p. 149) or *natureculture* to recognize the ways that appeals to the natural are tangled up with presumptions of hierarchies, social orders, and presumptions of the natural. We argue that it is critical to apprehend biological science as a discourse on nature rather than nature itself (1991a, b, c). Following Haraway, we seek to point to ways that biological (and more generally scientific) discourse can support queer students and both own up to and resist its own reactionary histories.

7.3 COVID and the Normal

In the wake of the COVID-19 partial closure of our state from spring 2020 through the fall of 2021, there was a political demand that schools as anchor institutions return to normal. Normal was in no way theorized or even given substantive description. As the protests in the wake of George Floyd's murder pointed to, normal was not a safe or acceptable state of affairs—George Floyd was a black resident of the state of Minnesota killed by police after he was suspected of using a counterfeit $20 bill. Certainly, for students and teachers marked as queer, i.e., differing in their sexual practices, identity, presentation, or desires from a cis, heterosexual norm, "the normal" was not a location that supported wellbeing. Research provides a complex picture of what the COVID-19 interregnum, i.e., the time between the pre-COVID and return to "normal" regime (though we note COVID is still rampant, one of the authors is recovering as we finish this article) has meant for the queer community. On the one hand in the U.S. many queer youth were more or less trapped in households that posed emotional and physical harms, which is not to deny the many supportive families (Gill & McQuillan, 2022). Other research notes that many who identified as straight prior to the interregnum, embraced queer identities during it. Given time to reflect on their wants and desires, there is some indication that more people shifted their self-identification during the solitude that COVID established for many (Klein, 2022).

The terms of normality to which public space "returned" after the COVID interregnum were largely set by the Trump administration's, i.e., Donald Trump the 45th president of the U.S., broad attack on queer, especially trans, people even as he lost the election. The Trump administration had set a tone for right wing erasure and attacks on trans youth. His administration removed references to sexuality in policy, they protected doctors who refused care for LGBTQ patients, they rolled back Obama era anti-discrimination protections, and in schools gave guidance that schools should not acknowledge trans students identities, worked hard to ban LGBTQ youth participating in 4-H (the premier youth development program of the U.S. Department of Agriculture, started in the early twentieth century; the 4 "h"s are head, heart, hand and health), and even used Title IX funds. (Title IX is the section of the federal law governing education that bars descrimination on the basis of sex) to discriminate against trans students. (HRC Staff, 2020). They set the tone and policy path that would eventually be taken up by a variety of states to basically outlaw trans youth–disallowing teachers, doctors, mental health workers, and even parents to support them–though often these policies were often carried out in the name of supporting parents.

At a more popular level, a variety of reactionary social media personalities (often referred to in press coverage as "influencers") such as Andrew Tate (arrested in Romania in December 2022 on rape and sexual trafficking charges) articulated explicit misogynist and anti-queer messages and garnered large audiences especially with teen and preteen boys left socially isolated by the epidemic. One British survey found that 80% of 16–17 aged youth had consumed his videos, and 45% of boys ages 16–24 had a positive view of him (Oppenheim, 2023). Tate, however, has not been alone. He is part of a long-standing network of social media personas pushing anti-queer and misogynist messages, largely targeting teens and tweens boys and attacking women (verbally harassing them) who the sphere identifies as threats in the spaces they value, e.g., online gaming (Bezio, 2018). Often this network is called the *manosphere* (Ribeiro, et al., 2021).

Alysa's Story
In my own classroom, I teach boys who talk about Tate nonstop, calling him a player and "rizzler" ("rizz", coming out of the word charisma means to have play or game, another way to commodify and objectify women). The same boys that idolize Tate in my classroom are the same boys I hear homophobic and transphobic rhetoric from, the same boys who speak lowly of women, the same boys who participate in toxic masculinity.

Tate's discourse provides a particularly clear example of the rigid *natureculture* that permeates reactionary thinking, one in which male and female are genetically or—in some other way—essentially defined and hierarchically ordered. That rigidity takes all challenges to this order as perversions that have to be destroyed or repressed. Thus womens' control of their bodies, the existence of trans people, queer relationships, and the expansion of rights for LGBTQ+ people are seen as one and the same: violations of a natural order in which men (by some inherent qualities) are legitimated in their power. In short, all challenges to either the hegemonic position of males, or the gender binary on which it rests are met with symbolic, legal, or physical violence.

Alysa's Story
I feel the impact of the manosphere misogyny on my high school students. My opening story provides one clear instance. In another, a group of boys made cruel jokes while I taught about HeLa cells and the life of Henrietta Lacks (Skloot, 2010), a poor black cancer patient whose cells were taken and used to test treatments for small pox, AIDS, etc. When I stated that I was skipping over parts of the movie version of Henrietta Lacks's story because they recounted particularly the abusive treatment of Henrietta's children after she died, one of these boys stated, "that is the best part". I had already set a tone that this movie was to be taken seriously as many people's lives were negatively affected by the choices made to take and use HeLa cells. The group of boys kept validating each other every time they made racist, sexist or homophobic comments about the movie or spoke in ways that affirmed violence on AFABs (assigned female at birth). The language and attitude seemed taken directly from the right-wing media.

The increase in alt-right policy decisions at the state and national levels and the impact of the social media discourses of the "manosphere", have resulted in limiting the civil liberties of queer people and others. It has curtailed the ability of individuals to express themselves both at home and in the public sphere (Campbell-Fox et al., 2024). The success of these reactionary moves has led to many LGBTQ+ youth losing gender-affirming services, access to peer-to-peer events, social supports, and other resources necessary to support their mental health (Hawke et al., 2021; Gattamorta et al., 2022). Studies have shown that due to being isolated at home, LGBTQ+ youth have experienced increased threats to their mental health due to family abuse, homophobic and transphobic ideologies of family members, and isolation. While the abuse of being kept home is not directly caused by policy decisions regarding legislation such as the Don't Say Gay Or Trans Bill, it did legitimate the harm of LGBTQ+ youth.

Matthew's Story

In January 2022 I was teaching a course on culture of secondary schools to a cohort of math and science masters students at UWT. Alysa was a part of that science cohort, and I had previously been their instructor in two methods classes, a multiculturalism class, and a couple of social foundations classes—we are a small institution.

The class was timed so that it occurred simultaneous with the start of our Master's students' student teaching. That moment, early 2022, was a time when schools were transitioning to "normalcy," a term freighted with meaning. Ostensibly the normal longed for in this return was of pre-COVID connection; but the COVID interregnum, as I like to view it, was also a time of social upheaval, of the Black Lives Matter protests, etc. So the normal was an evocation of the past, a kind of nostalgia for a society that felt ordered for the dominant classes, even if that past was violent and oppressive, for others. Afterall, the reference point for normal was January 2020, a time when Trump was still in office.

It was in the context of a return to normal, of a generation of traumatized and socially isolated students, that Alysa shared accounts of explicit homophobic and transphobic language as well as accounts of a student population completely emotionally disconnected from schools and community.

My own reaction to Alysa's reports was to try to intervene. I was already inviting one of the counselors at Alysa's school to come in and speak about their role in the school system and how teachers might partner with them, and it seemed like a perfect opportunity to push for some sort of "cultural intervention" (Layne, 2000). The counselor was very clued into the lives of students, including ones not conforming to hegemonic sexual identities. Later, I had planned to invite representatives from the state GLSEN chapter (GLSEN, 2023). GLSEN is an organization of teachers supporting queer youth (GLSEN no longer is an acronym, but when founded stood for the Gay, Lesbian, and Straight Education Network; GLSEN will be discussed later in this document).

(continued)

The fact is, both of these guests whom I had reached out to provided little support. The counselor seemed overwhelmed and couldn't fathom the breadth of the issue. GLSEN, in a tactical mistake, sent an elementary school expert who was able to provide statistics and general information about support and its import, but the class was already aware of the depth, breadth, and urgency of the problem.

My reach out to Alysa for this paper was inspired by our sense of frustration at the time and my need to understand what we were facing, to understand what was motivating student homophobia, and how our field of STEM both was a site of reinforcing anti-trans identities as well as a possible ground for reimagining them.

7.4 Queer Lives During and After COVID-19

Part of this story has to be understood in the context of the long struggle over the rights of LGBTQ+ people including expansion of privacy, marriage, and protected status right decisions both at state and federal levels. The protections regarding sexual orientation and gender were found to be included in Title VII of the Civil Rights Act (Thoreson, 2020). That is, the queer community had felt a sense of rapid progress and acceptance backed up by legal and cultural transformations. This sense of progress was evident legally and culturally.

Matthew's Story

I want to point to real, queer-positive progress in a rather surprising space. I am an avid contra dancer. Contra dance is a social dance form in which, together with a partner, one dances with an entire line of other couples, almost always to live music: a fusion including Celtic, Eastern European, and swing. The dance is called while the music plays with a similar vocabulary of moves as square dance.

The dance, in the past, was highly gendered. Everything was gents do this, ladies do that. But by the 1990s that started to change. There had already been forming a parallel dance culture of "gender neutral dances"—going back as far as the 1970s (Ricciotti, n.d.). These queer friendly dance events used a variety of terms to designate the binary roles of the dance (e.g. blue and red) or even translated the calls of dances to use no role references at all (e.g., the caller would just say, "person on the left," etc.) However, in "cis-het" dance events, the genders of gents and ladies remained.

A wide variety of acts challenged the cis-heteronormativity of the dance community. Men had long enjoyed wearing skirts at dance; also, a kind of cross-dressing occurred to make the dances gender balanced, when the numbers were off. Sometimes people would dance gender backwards, if they were

(continued)

really feeling subversive ("gent" in lady role and vice versa). But the serious change happened when young dancers in the twenty-teens let the older leadership know that they needed the dances to move towards gender neutrality.

In response to those pressures, contra dance has moved to being queer friendly. Gone are ladies and gents, in their place two new gender roles which anyone could occupy: larks and robins. Almost every dance in our region has become a queer dance. Gender play is encouraged. This is a community that has fully embraced (at least here in the Pacific Northwest) queering as a positive value and a norm against the reactionary tide. My story is highly abbreviated, but I wanted to point to a space that, despite fetishizing tradition and gender, has transformed itself into a queer positive space and that schools could reflect on as a positive model of changing sexgender culture.

More often, however, the COVID shutdown has been felt as a reversal. The sequestering of individuals lead to both a loss of personal and social well-being. Queer youth were taken away from friend groups that created community for them and were confined to rooms or houses that often were not safe for them (Hawke et al., 2021; Gattamorta et al., 2022). Within schools themselves, supports that teachers needed to make a positive environment, were often put on hold. GLSEN trainings for teachers stopped occurring at my (Alysa's) school site for two years. This coupled with the power of the manosphere to articulate radical discourses of hate through the internet (the only window of sociality left in the pandemic) created a wake of real loss and decline in the queer community. The hope of achieving a full sense of humanity and equality seemed lost or hidden from LGBTQ+ youth after COVID-19. This is not to say that the advocacy and the struggle for justice by queer folks had disappeared—but the dominant voices of those in power created a legal and cultural landscape where LGBTQ+ youth were not safe to be themselves. Our sense has been that safe spaces were smaller, more segregated and dispersed; LGBTQ+ youth found themselves isolated, even oppressed.

Alysa's Story

In the school year of 2021–2022 I completed my student teaching experience to become certificated. During this time, I witnessed myriad incidents of teachers and students being transphobic, homophobic, and violent or hateful toward LGBTQ+ community members. For example, teachers policing bathrooms, teachers calling people who use pronouns snowflakes, and purposeful misgendering of transgender students. This violence was coming from school staff toward LGBTQ+ students, from students to other LGBTQ+ students, from students to LGBTQ+ school staff, and from students simply brining violent rhetoric into the space of the classroom or the school. (e.g., see the

(continued)

opening story). I was also told by many of my colleagues that identify as allies at my school site that they were also witnessing these acts simultaneously. I was not alone. My colleagues recalled LGBTQ+ hate existing at the school, but not to this extent for many years prior to COVID-19. These acts, while fewer than last year, have continued into the current school year of 2022–2023. My colleagues have agreed that the culture feels better, but not back to how it was before the COVID-19 pandemic.

Contrary to last year, I am out now to my colleagues and students as nonbinary: something I did not do last year due to the social tensions and fear of being harassed by school staff or students. While I do not feel specifically targeted by my colleagues or administration for being nonbinary, I do feel unsupported. I gathered from the recurring transphobic and homophobic language, comments, and opinions that it was not a safe place for someone like me to come out when I was student teaching. There’re a few reasons for this. One of them being that I was not protected by a union if I were to voice complaints on the basis of my identity. I felt I was on the bottom of the totem pole. Next, I felt that I was an alien entity at my school site for I was not one of the staff members, therefore I felt as if I was an Other. Thus, I grappled with the possibility of exposing a vulnerable part of myself to a community that I did not feel close to or even part of. Furthermore, if I did expose myself, I felt it was possible that I could lose my student teaching position. Due to this possibility, I stuck it out. Sticking it out was what I had to do to feel the most protected. Unfortunately, I struggled with this endurance and my mental health took a strong hit from this experience.

Prior to the COVID lockdown there had been a long struggle resulting in the creation of queer-progressive spaces, representations, policies, and discourses. The COVID-19 crisis has resulted in the foreclosing of much of that space. The community, the advocacy, the footing that the queer community began to create with LGBTQ+ youth was broken by the lock down with it a loss of social well-being. This is not to say that the advocacy and the struggle for justice by queer folks has disappeared, but that the dominant voices of those in positions of power created a landscape where LGBTQ+ youth were not safe to be themselves. Safe spaces felt smaller, more segregated and dispersed.

7.5 Schooling as a Site of Queer Struggle

This paper is about the science classroom, but the science classroom is co-extensive with schools and school systems. The struggle of queer students and teachers and the policing of both take place in this larger context that we can only gesture at here; the contestation over science education is part of a larger contestation happening at all levels of the institution.

It is useful to think about the ways that cis-heteronormativity is built materially into the school buildings themselves. Gendered bathrooms, often too expensive to remodel, force binary gender identities on the student and sometimes staff bodies. Students are also gendered on attendance sheets and other bureaucratic forms in ways that often misgender queer students. The use of pronouns tied to gender has become a point of enormous conflict, and a means by which staff and administrators can either acknowledge students' identities or erase them, leading the classroom to become a site which reproduces erasure and misrecognition experienced elsewhere (e.g., the home).

On the other hand, resisting these attempts to fix or impose gender are student and community organizations committed to queer students' wellbeing. Organizations like GLSEN and, PFLAG (which formerly stood for Parents, Families and Friends of Lesbian and Gays) provide adult support for queer children/students. Gay Straight Alliances (GSAs) are student organizations that support queer students and allow them to tell their own stories about what they are experiencing in schools.

Finally, the movement to diversify the workforce itself, including expanding the queer teacher workforce, supports queer students by providing both empathetic adults and models for students working to understand how to live in a cis-heteronormative world.

Science teachers are involved in these extra-science classroom dimensions of both supporting and marginalizing queer students. Schools are sites where rigid *naturecultures* that cannot see beyond compulsory heterosexuality and binary sex are confronted by the complexities of both biology and identity.

Too often, the ways schools have responded to the terror that queer students and teachers experience is through ineffective gestures that leave the basic power structure intact. These take the form of diversity statements, anti-bullying programs, and social-emotional learning. In some way, each of these place the responsibility on the targeted person: either to change, e.g., in what are packaged as Social and Emotional Learning Curricula (SEL), or to report victimization, e.g., anti-bullying programs (Meyer, 2008).

Instead of a liberal response that a priori sees all as equal in the political space of the school, we are calling for something more explicit, a discourse that names the links between patriarchy and queer bashing, and that sees compulsory cis-partriarchal-heterosexism as structural in nature and that strategizes for a more radical vision of queer wellbeing.

In considering the impacts of architectures and policies on queer teachers and students, there is a slippage between the materiality of the building with its binary gyms and bathrooms and the materiality of the body with its complex protean expressions. The biology pointed to in these policies and architectures are sites of struggle between radically differing *naturecultures*. These are conflicting discourses on the meanings and organization of bodies. As Haraway in her interview with Penley and Ross (1991) explains,

> [...Remember] that biology is not the body itself, but a discourse. When you say that my biology is such-and-such–or, I am a biological female and so therefore I have the following physiological structure–it sounds like you're talking about the thing itself. But, if we are committed to remembering that biology is a *logos*, is literally a gathering into knowledge, we are not fooled into giving up the contestation for the discourse. (p. 11)

Haraway is asking for a sociological imagination of the biological here. Biology is a site of struggle not a fixed and final truth. Of course, this struggle takes its most explicit form in the biology classroom, where these tacit presumptions in policy and geography are put into explicit discourse, but also where the history and contestation over these ideas can be made most clear. It is to the biology curriculum and classroom which we now turn.

7.6 The *NatureCultures* of Extant Science Education

The science classroom is often complicit in legitimating the anti-queer *natureculture* examined above. The historical narrative surrounding biological sex has been shaped by a combination of cultural, social, and scientific factors. For a long time, it has followed a binary perspective, categorizing individuals strictly as male or female based on reproductive anatomy and chromosomes. This binary reading of the body then gets mapped onto the cells and material of the body through some combination of genes (XX vs. XY) or hormones (androgens vs. the others, see Montañez, 2017; Richardson, 2013). This reading by design or effect reinforces existing heteropatriarchal hierarchies and social relations by making invisible or pathological queer people and bodies. Importantly, science texts (or curriculum more broadly) naturalize these hierarchies and categories by reading them into nature. As Haraway notes (1997) in her examination of the social function of science texts that (according to Eric Engels) "Content with the great divide between nature and culture, biology textbooks tend to explain the" social "in terms of the 'natural'" (p. 103). In short, the curriculum of schools can be particularly harmful by reinforcing rigid *naturecultures* that pathologize queer youth and thus make them "natural" objects of attack.

Alysa's Story

I witnessed a truly tragic event last year. I watched a biology teacher teaching a lesson on genetics. In one part of the lesson, when explaining how chromosomes work, she pointed at a girl and stated "she has XX, so they are a girl." Then, she pointed at a boy and said "he has XY, so he is a boy." I was truly appalled. There were a few transgender students in that class, and I knew they were uncomfortable. Many transgender students in the class confided in me later and wrote comments on their activity lesson worksheet stating that they didn't like how the lessons were "transphobic" and "made" [them] uncomfortable. What they were objecting to was a compulsory (in the sense of offering no other options) cis-heternormative pedagogy. I did indeed concur that the lessons being taught were problematic and perpetuated the compulsory cis-heteronormative ideals of the patriarchy. This is just one of so many transphobic moments those students face, and sadly one of those students, was Scott, whose story I have already told, who dropped out because they had put up with so much erasure and denial from their teachers.

The most thorough analysis of heteronormativity in biology textbooks to date in the U.S. was conducted by Snyder and Broadway (2004) who analyzed biology textbooks and the absence of discussions and representations of sexuality outside the heterosexual norm. While this research is slightly dated, similar findings have been documented in Canada (Bazzul & Sykes, 2011), Finland (Aivelo et al., 2022), and Sweden (Junkala et al., 2022). The study by Snyder and Broadway found that the textbooks erased queerness in a variety of ways. Texts, they found, primarily omitted information rather than actively promoted biased content. For instance, the absence of visual representations of alternative family structures, such as those headed by gay or lesbian parents, was highlighted. The only mention of homosexuality was in relation to AIDS counseling. The textbooks primarily focused on reproduction and anatomy, disregarding other aspects of human sexuality. The presentation of sexuality within a heteronormative framework implied that only heterosexual behaviors were acceptable. In textbooks that mention homosexuality at all, it was often linked with drug use, reinforcing negative stereotypes. One textbook explained the historical and social contexts of AIDS, opening the possibility for teachers to engage students in discussions about sexual orientation. The general disappearance of terms like "homosexuality" and "gay" from the textbooks raised questions for the authors about the pressure from reactionary activist groups or a heteropatriarchal normalizing decision to emphasize only heterosexuality in biology classrooms. The emphasis on heterosexual intercourse and the silence on diverse sexuality aligned with conservative agendas that have prioritized heterosexual marriage as the only acceptable life.

The science of sexual development and genetics, however, undermines every part of the cis-heterosexual just-so story that Snyder and Broadway identified. The entire theory of sex hormones has been deconstructed. Nelly Oudshoorn's (1990) article "On the Make of Sex Hormones," analyzed how gender bias impacted scientific research and knowledge of the role played by hormones in sexual development. Oudshoorn argued that the dominant narrative in scientific literature had often overlooked the complexities and variations of hormonal functioning in different sexes, perpetuating a biased and oversimplified understanding. This gender bias had led to a hyperfocus on so-called "male" hormones while underemphasizing the role of "female" hormones in research and medical practices. Oudshoorn's work highlighted the importance of critically examining the historical narrative surrounding sex hormones and challenging gender bias in scientific studies to foster a more inclusive and accurate understanding of the diverse manifestations of biological sex (See Helm, 2019, for a short overview).

Nehm and Young (2008) critique contemporary biology for adhering to outdated views on sex hormones, emphasizing the flawed link between hormones and specific sexual characteristics. They highlight how sex hormones are not solely tied to distinct sexes, challenging the oversimplified mapping of hormones to gendered bodies. Current scientific understanding acknowledges a spectrum of biological sex, encompassing chromosomal, gonadal, and hormonal variations, defying strict male-female categorizations. This nuanced perspective recognizes the intricate interplay of genes, hormones, and identity, offering a complex understanding that surpasses conventional binary notions seen in both popular culture and past scientific narratives, expanding the realms of possibility beyond previously envisioned limits. These new models of sex are part of the natureculture that values empiricism, complexity, surprise, and change. It is a natureculture opposed to orthodoxy or fixity, and thus invites queer readings of nature. It is, afterall, a natureculture interested in peeling back seemingly infinite layers of the onion that is (their) nature, meaning the nature that the culture of science imagines through its ("power charged") dialogues with its experimental subjects (Haraway, 1991b, p. 198). But, we are too aware of the long history of pathologization of queer folx (Terry, 1999) and the effects of the medicalization of trans and gender-nonconforming bodies (Gill-Peterson, 2018), to not question the science as well, and to recognize we must be alert to the ways that naturecultures generally render some folx unnatural.

Feminist queer science studies allows us to understand the nature of science as a social process and the ways that nature and the social interweave. Queer theorists have argued that science is not a neutral, objective enterprise but is instead shaped by cultural and social factors, including heteronormative assumptions about gender and sexuality. By highlighting the social and cultural dimensions of science, queer theory can help to challenge the ways in which scientific research is often used to reinforce social hierarchies and promote the interests of dominant groups. In this way, queer theory can contribute to a more critical and reflexive approach to science that takes into account the ways in which scientific knowledge is produced and the broader social and political contexts in which it operates.

But the topic and content of science is not just presented in classrooms that are formally titled science (or biology, chemistry, geology, or physics). Science knowledge, i.e., discourses on nature also emerge in the course of allied subjects such as health and sex education. These too are sites where the dogma (Kuhn, 1970) of reactionary *naturecultures* is proclaimed and contested over.

Alysa's Story

In my school, sex education is incorporated in my chemistry class, simply because that is what sophomores are required to take. All chemistry teachers were mandated to take the Sex Education teacher's workshop for the "FLASH" (Family Life and Sexual Health) curriculum that the district uses–adopted and distributed through a near-by district.

During the COVID-19 pandemic, steps were taken to reform sex education in Washington state and other similarly liberal states like New York, New Hampshire, New Jersey and Wisconsin. In my home state of Washington, the department of education proclaimed that "By the 2022–23 school year, all schools must provide comprehensive sexual health education (CSHE) to all students. 'All students' includes any student enrolled in a district or public charter school, such as those in special education programs, online learning or alternative learning experiences, etc." The state department of education website explicitly instructed school districts to be inclusive: "acknowledge that

(continued)

some students identify as transgender (their gender identity is not consistent with the gender assigned to them at birth) or non-binary (they don't identify as either male or female or they identify as both male and female) and may use the pronouns they/them rather than she/her or he/his."

I recently took the sex education training to teach sex ed. at my school. Planned Parenthood's website says that "Family Life and Sexual Health (FLASH) is a comprehensive science- based sexuality education curriculum developed by Public Health Seattle-King County and designed to prevent unintended pregnancies, STDs and sexual violence" (Planned Parenthood, n.d.). The training focussed on responding neutrally and factually to students' questions and the research basis for the training. The training completely bypassed any instruction on how to educate students on queer youth's sexual experiences. Also, the training encouraged deferring difficult questions; instructors were to tell the student to reach out to a trusted family or care providers. But for queer youth there may not be an adult that is not going to harm them. For example, according to the TREVOR Project (2022), a nonprofit working on suicide prevention among LGBTQ youth, reported that "Fewer than 1 in 3 transgender and nonbinary youth found their home to be gender-affirming." and "60% of LGBTQ youth who wanted mental health care in the past year were not able to get it." (p. 4) The FLASH lessons included resources that educators could access to support queer youth, however, this assumed that instructors would access them; there should have been explicit training on how to teach sex education that is inclusive for queer youth.

Furthermore, the push for abstinence has not yet been purged from the sex ed. curricula pushed on districts. The FLASH websites page declares that "It is designed to support young people in making healthy choices: abstain from sex…" (King County, n.d., Theory of planned behavior tab). Abstinence-only sex ed. is problematic in general. However, this has specific implications for queer youth. Fisher (2009) has pointed out that "Heteronormativity is a central tenet of abstinence-only-until-marriage sexuality education. Curricula operate from the presuppositions that all students are or should be heterosexual, that all students will and shall marry someone of the other sex, and that all students should engage in heterosexual relations only within the context of marriage." (p. 68). In short, this perpetuates compulsory cis-heterosexuality and further harms queer youth

We have tried to capture the ways that existing science education collaborates with and naturalizes the cisheteronormative structure of schooling. For instance in sex education's stress on abstinence, it becomes merely the extension of a general ethos that assumes heteronormativity. In the case of binary sex myths, however, science does the ideological work of locating the cultural assumption of dual sex identities in nature and thus plays an essential role in marking queer students as deviants or broken. In response to this, we want to outline elements of a framework that could position the science classroom otherwise.

7.7 Imagining a Better *NatureCulture* for Science Education and Its Fellow Travelers

In this section we propose components of a science education space that is in its construction queer friendly. This is not a curriculum, i.e., a specific discourse designed through techniques such as scope and sequence to educate students, but the criteria needed for a transformed curriculum. It takes the wellbeing of queer students and teachers as the central demand and organizing kernel, rather than, say engineering a la the Next Generation Science Standards.

7.7.1 Critical Rejection of Out of Date Science

Today our understanding of the biology of sex has significantly evolved since the early 1900s, revealing a stark contrast between the knowledge of that era and present-day research. In the early 1900s, the prevailing view of sex biology was rooted in a binary framework, emphasizing clear and distinct categories of male and female based on external genitalia and reproductive functions. This limited perspective failed to acknowledge the complexities and variations that exist in human biology. Today, research has unveiled a more comprehensive understanding of sex biology, challenging the binary model and recognizing a continuum of variations and spectrums. We now understand that sex is not solely determined by external genitalia but is influenced by a combination of genetic, hormonal, anatomical, and neurological factors. It has become increasingly clear that biological sex exists on a spectrum or even polymorphously rather than a rigid dichotomy (Fausto-Sterling, 1993). Contemporary research has shed light on intersex variations, chromosomal anomalies, and the complexities of gender identity. The advancements in our understanding have highlighted the importance of embracing diversity and inclusivity in our scientific frameworks. We must continue to explore, question, and challenge existing paradigms to foster a more accurate and comprehensive understanding of the intricacies of human biology and reject long outdated science surrounding sex, gender, and biology.

7.7.2 Complexity and Uncertainty

The known historical research on sexgender is giving way to a universe of uncharted territories, challenging our preconceived notions, and expanding the boundaries of scientific understanding. This expansion reveals the complexity and diversity in the biology of gender and sex, and need to embrace a multitude of possibilities and perspectives that were previously overlooked. Embracing uncertainty and unknowns is pivotal when it comes to queering science and paving the way for a more nuanced understanding of gender and sex. By acknowledging the limits of past and present science, we create space for exploration and discovery. This opens possibilities for new sexgender beings and becomings.

7.7.3 Deobjectification of People

Science education is about an objectified nature. It reduces people to organs, material to molecules, and phenomena to variables. The suffering of queer people in medicalization of their bodies, including the psychologicization of their non-heteronormativity, is part of that history of objectification, in which knowledge is dyadic: there is a knower and there is the known. To queer science education requires confounding that dichotomous relationship. It was, after all, the demands of queer AIDS activists that forced the The U.S. Food and Drug Administration and others to develop new ways to gain knowledge of medicine and to work with the objects who were also the consumers of that knowledge (Epstein, 1996). Queer, feminist, and anti-colonial science studies have imagined/described other ways of being in science (Haraway, 1991b; Harding, 1991, 2011; Keller, 1983). Central to these other ways is an analysis of power between those collecting knowledge and the observed. A queer friendly science education has to make that relational vision of knowledge, i.e., its embedded location within matrices of power, desire, and resistance (its situatedness), manifest and open to reflection and discussion.

7.7.4 Representation of Queer Folx Everywhere

To improve queer representation in science education and science, it is essential to go beyond token hiring and instead focus on systemic changes. One key aspect is updating science curricula to incorporate queer perspectives. This can be achieved by including content that highlights the contributions of queer scientists throughout history and exploring the intersection of queerness and scientific fields. By integrating these discussions into the curriculum, we can provide a more comprehensive understanding of scientific enterprise. In addition to inclusive curriculum development, it is important to incorporate queer stories and narratives into science education (Lundin, 2014). Textbooks, case studies, and classroom discussions should feature diverse experiences and voices within the scientific community. This inclusion allows for a broader representation of queer individuals and their contributions, inspiring and validating queer students' aspirations in scientific fields. Representation in educational materials is also crucial (Snapp et al., 2015). By ensuring that illustrations, diagrams, and photographs reflect diverse gender identities and sexual orientations, we can challenge stereotypes and provide positive role models for all students (Lundin, 2014). Using inclusive language and imagery promotes a sense of belonging and acceptance within the scientific learning environment (Fredman et al., 2015). To further emphasize the significance of queer representation, it is valuable to create displays or exhibits that highlight the achievements of queer scientists and science advocates. These displays can showcase their research, activism, and impact on the scientific community, providing visible role models and empowering queer students to envision themselves in scientific careers.

Integrating queer history into science education is another important step. By exploring the impact of societal attitudes, discrimination, and prejudice on the progress of queer scientists throughout history, teachers can shed light on the challenges they faced (Lundin, 2014). This could foster empathy and a deeper appreciation of the contributions made by queer individuals to science.

Finally, collaboration with LGBTQ+ organizations is key to fostering inclusivity in science education. These organizations can be a site of resources, workshops, and programs that can connect the science classroom to the rest of the school and the broader work to provide a continuous space of inclusivity (Schey, 2021) by helping students, staff, and administration understand the evolving meaning of queerness, confront bias, develop the skills to engage in dialogue, and build the skills of queer students to engage successfully with science.

7.7.5 Acknowledgement of Queer Relations Across the Natural World

A queer friendly science education must locate non-heteronormative relationships across species and materialities (Russell, 2021). It is too easy to anthropomorphize and hetero-morphize the animal world and think that it mirrors and thus legitimates extant power orders. The world is filled with sexualities alien to our own, from hermaphroditic worms, self-impregnating reptiles, matriarchal bonobos and hyenas as well as all the asexually reproducing beings. Biology education, especially given the role of evolution with its engine of diversity, invites sexual curiosity, i.e., amazement at the sheer variety and complexity of reproductive choreographies (Roughgarden, 2004). This includes more than biology; all sciences challenge our taken for granted rules of nature, scales of time and space, and relations of being (Barad, 2007).

7.7.6 Critical History and Sociology

While science is a critical partner in the production of *naturecultures* that can make queer convivial worlds, it has also been implicated in the oppression of queer people through multiple forms of medicalization (Terry, 1999; Townsend, 2011) and psychological labeling (Wolfe & Fogwell, 2022) as well as explicit discrimination (Else, 2021). Students need to have discussed the ways that homophobia has been institutionalized rather than having the topic repressed, especially since transphobia is currently being remedicalized in Florida and elsewhere (Yurcaba, 2022). Not acknowledging the continued ways that science is "conjured" (Toumey, 1996) to justify anti-queer policy (see our conclusion), means denying the lived realities of our students, thus alienating them.

This grappling requires engaging with both critical histories and sociologies of science. The history of science provides cases of science's institutions of power, its role in supporting ideologies of race and gender, as well as its many breaks with those ideologies (Haraway, 1997). A science education that is queer friendly has to rehearse the ways that trans bodies have been integral to the production of gender (Gill-Peterson). The critical sociology of science (e.g., Cipolla et al., 2017) allows students to understand the ways science makes, excludes, and circulates people, materials, and ideas. By comprehending the "game" of science, i.e., "science in action" rather than "ready made" (Latour, 1987, Introduction), science education offers the ability of students who are queer or otherwise marginal to join effectively, knowing when to participate and when to refuse, when to resist, and when to call for radical transformation.

7.7.7 Sex Positive

To make the science classroom–and the school–a positive place for queer students a radically different framework for sexuality is needed. The sex-positive movement is a cultural shift towards embracing pleasure and moving away from clinically diagnosing and medicalizing bodies. It promotes the idea that all consensual sexual activities and expressions are valid and should be respected, challenging societal stigmas and fostering healthier attitudes towards sexuality (Cruz et al., 2017). This movement aims to reclaim from abstinence the narrative around sexuality, advocating for comprehensive sex education that prioritizes pleasure, consent, and well-being rather than focusing solely on risks and diseases (Manning, 2021). It also seeks to deconstruct rigid gender roles and expectations, emphasizing inclusivity and empowerment for all individuals. By promoting a holistic understanding of human sexuality, the sex-positive movement envisions a world where people can freely explore, express, and celebrate their sexuality without shame or judgment.

Sex positive science, health and sexual education are essential for queer students in that it foregrounds not the biochemistry—not unimportant—but the experience of sexuality in all of its diversity. By not fearing sexuality, by emphasizing the cooperative nature of sexuality, and by embracing the breadth of sexual practice, it decenters the implicit cis-heteronormativity of extant science educations which focus on heterosexual intercourse (reproduction) and legal relations (abstinence until marriage). These shifts allow the content to speak to queer students, but also have the added benefit or removing shame (labelling valid and invalid sexual ways of interacting) and irrelevance (discussing things so abstractly and biochemically that the link to human experience is missing) for all students.

This is not an exhaustive list of directions that science education needs to take, but it does mark a distance to travel to accomplish a science education in which queer students are included and allowed to have their curiosity flourish (Gilbert & Gray, 2019), which should afterall be the ethos of science education.

7.8 Conclusion

Queer students have faced a sudden and seemingly surprising swell of violence as well as trans and homophobic discourse occuring at the level of interpersonal relations and state, violence–the latter reified in exclusionary policy/legislation. The contestation between those seeking expanded freedoms and those seeking to erase those identities seems encapsulated in a photo in a recent article on the myths underlying the reactionary attack on trans youth (and adults) by anthropologist and human rights activist Alex Hinton (2023). In the photo there is a sign posted in the hallways of Congress by conspiracy theorist, reactionary leader, and state representative Marjorie Taylor Greene. It is displayed across the hall from a LGBTQ+ flag raised by DC's representative Marie Newman. Greene's sign says, "There are TWO genders: MALE & FEMALE 'Trust the science!'" (emphasis original).

The science that Greene referenced in her sign, the one that states that biology provides two clear and mutually exclusive genders, seems far from the science of genetics, zoology, botany, and evolution. It is, to use Chris Toumey's (1996) evocative phrase, "conjured science," science evoked rather than reflecting the working models and practices of those who endure biology, chemistry, geology, and physics coursework. It is also a science that too conveniently fits the rigid *natureculture* that Greene wants to impose on the world. Science for her is a political tool to naturalize hierarchies and differences that she would rather merely presume. This reactionary *natureculture* is belied however by the work of scientists of many stripes who find nature remarkably queer and infinitely complex.

We have tried to witness the right's attack and reflect on the *natureculture* that these attacks are premised upon. We have also tried to witness the ways that teachers, curriculum, school building, administrators, school boards, universities, and education policy conspire with this reactionary *natureculture*. But we also see in science the potential for a more honest, radical, queer, just, and healing science education and schooling writ large. However, as we have argued, this takes imagining science education and schooling through the lenses of queer theory, science studies and allied histories—as well as the work of contemporary science itself. This is the curriculum work we are calling for as a sociopolitical project.

References

Aivelo, T., Neffling, E., & Karala, M. (2022). Representation for whom? Transformation of sex/gender discussion from stereotypes to silence in Finnish biology textbooks from 20th to 21th century. *Journal of Biological Education*, 1–15. https://doi.org/10.1080/00219266.2022.2047099

Barad, K. M. (2007). *Meeting the universe halfway: Quantum physics and the entanglement of matter and meaning*. Duke University Press.

Bazzul, J., & Sykes, H. (2011). The secret identity of a biology textbook: Straight and naturally sexed. *Cultural Studies of Science Education, 6*, 265–286. https://doi.org/10.1007/s11422-010-9297-z

Bezio, K. M. (2018). Ctrl-Alt-Del: GamerGate as a precursor to the rise of the alt-right. *Leadership, 14*(5), 556–566. https://doi.org/10.1177/1742715018793744

Bowker, G. C., & Star, S. L. (1999). *Sorting things out: Classification and its consequences*. MIT.

Campbell-Fox, K., Kimball, D., Puckett, J. A., Dunn, T., & Rabine, S. (2024). "Stop playing politics with my kid:" Thematic analysis of a trans-exclusionary sports legislation hearing in Texas. *International Journal of Transgender Health*, 1–17. https://doi.org/10.1080/26895269.2024.2373233

Casper, A. M. A., Rebolledo, N., Lane, A. K., Jude, L., & Eddy, S. L. (2022). "It's completely erasure": A qualitative exploration of experiences of transgender, nonbinary, gender nonconforming, and questioning students in biology courses. *CBE Life Science Education, 21*(4), ar69. https://doi.org/10.1187/cbe.21-12-0343

Cipolla, C., Gupta, K., Rubin, D. A., & Willey, A. (2017). *Queer feminist science studies: A reader*. University of Washington Press.

Cruz, C., Greenwald, E., & Sandil, R. (2017). Let's talk about sex: Integrating sex positivity in counseling psychology practice. *The Counseling Psychologist, 45*(4), 547–569. https://doi.org/10.1177/0011000017714763

Else, H. (2021). Largest-ever survey exposes career obstacles for LGBTQ scientists. *Nature*. https://www.nature.com/articles/d41586-021-00221-w

Epstein, S. (1996). *Impure science: AIDS activism, and the politics of knowledge*. University of California Press.

Fausto-Sterling, A. (1993). The five sexes: Why male and female are not enough. *The Sciences, 33*(2), 20–26.

Felman, S., & Laub, D. (1991). *Testimony: Crises of witnessing in literature, psychoanalysis, and history*. Routledge.

Fifield, S., & Letts, W. (2014). [Re]considering queer theories and science education. *Cultural Studies of Science Education, 9*, 393–405. https://doi.org/10.1007/s11422-013-9509-4

Fisher, C. M. (2009). Queer youth experiences with abstinence-only-until-marriage sexuality education: "I can't get married so where does that leave me?". *Journal of LGBT Youth, 6*(1), 61–79. https://doi.org/10.1080/19361650802396775

Fredman, A. J., Schultz, N. J., & Hoffman, M. F. (2015). "You're moving a Frickin' big ship": The challenges of addressing LGBTQ topics in public schools. *Education and Urban Society, 47*(1), 56–85. https://doi.org/10.1177/0013124513496457

Gattamorta, K. A., Salerno, J. P., & Roman Laporte, R. (2022). Family rejection during COVID-19: Effects on sexual and gender minority stress and mental health among LGBTQ university students. *LGBTQ+ Family: An Interdisciplinary Journal, 18*(4), 305–318. https://doi.org/10.1080/27703371.2022.2083041

Gilbert, A., & Gray, E. (2019). Wonder in the science classroom. In S. Fifield & W. Letts (Eds.), *STEM of desire: Queer theories in science education* (pp. 109–123). Sense.

Gill, E. K., & McQuillan, M. T. (2022). LGBTQ+ students' peer victimization and mental health before and during the COVID-19 pandemic. *International Journal of Environmental Research and Public Health, 19*(18), 11537. https://doi.org/10.3390/ijerph191811537

Gill-Peterson, J. (2018). *Histories of the transgender child*. University of Minnesota Press. https://doi.org/99979044586.

GLSEN. (2023). *About us*. Retrieved June 22, 2023 from https://www.glsen.org/about-us#snt%2D%2D1

Gutiérrez, R. (2013). The sociological turn in mathematics education. *Journal of Research in Mathematics Education, 44*(1), 37–68.

Haraway, D. J. (1991a). *Simians, cyborgs, and women: The reinvention of nature* (1st ed.). Routledge.

Haraway, D. J. (1991b). Situated knowledges: The science question in feminism and the privilege of partial perspective. In D. J. Haraway (Ed.), *Simians, cyborgs, and women: The reinvention of nature* (pp. 183–201). Routledge.

Haraway, D. J. (1991c). The biopolitics of postmodern bodies: Constitutions of self in immune system discourse. In D. J. Haraway (Ed.), *Simians, cyborgs, and women: The reinvention of nature* (pp. 203–230). Routledge.

Haraway, D. J. (1994). A game of cat's cradle: Science studies, feminist theory, and cultural studies. *Configurations, 2*(1), 59–71.

Haraway, D. J. (1997). *Modest_Witness@Second_Millennium.FemaleMan©_Meets_OncoMouse™*. Routledge.

Harding, S. G. (1991). *Whose science? Whose knowledge?: Thinking from women's lives*. Cornell University Press.

Harding, S. G. (2011). *The postcolonial science and technology studies reader*. Duke University Press. https://doi.org/10.1215/9780822393849

Hawke, L. D., Hayes, E., Darnay, K., & Henderson, J. (2021). Mental health among transgender and gender diverse youth: An exploration of effects during the COVID-19 pandemic. *Psychology of Sexual Orientation and Gender Diversity, 8*(2), 180–187. https://doi.org/10.1037/sgd0000467

Helm, R. R. (2019, December 19). *Friendly neighborhood biologist here*. Retrieved June 26, 2023 from https://twitter.com/RebeccaRHelm/status/1207834357639139328

Hinton, A. (2023, April 23). Two myths fueling the conservative right's dangerous transphobia. *Sapiens*. https://www.sapiens.org/culture/transphobia-myths/

HRC Staff. (2020, June 11). *The real list of Trump's "unprecedented steps" for the LGBTQ community*. Human Rights Campaign. https://www.hrc.org/news/the-list-of-trumps-unprecedented-steps-for-the-lgbtq-community

Junkala, H., Berge, M., & Silfver, E. (2022). Diversity in sex and relationship education—Limitations and possibilities in Swedish biology textbooks. *Sex Education, 22*(5), 521–537. https://doi.org/10.1080/14681811.2021.1966407

Keller, E. F. (1983). *A feeling for the organism: The life and work of Barbara McClintock*. W. H. Freeman.

Keller, E. F. (1992). *Secrets of life, secrets of death: Essays on language, gender and science*. Routledge.

Klein, J. (2022, February 25). *Why sexual identities shifted during the pandemic*. BBC. https://www.bbc.com/worklife/article/20220224-why-sexual-identities-shifted-during-the-pandemic

Kuhn, T. S. (1970). The function of dogma in scientific research. In B. A. Brody (Ed.), *Readings in the philosophy of science* (pp. 356–373). Prentice-Hall, Inc.

Latour, B. (1987). *Science in action*. Harvard University Press.

Layne, L. L. (2000). The cultural fix: An anthropological contribution to science and technology studies. *Science, Technology and Human Values, 25*(3), 352–379.

Letts, W. J., & Fifield, S. (2019). *STEM of desire: Queer theories and science education*. Brill Sense.

Lundin, M. (2014). Inviting queer ideas into the science classroom: Studying sexuality education from a queer perspective. *Cultural Studies of Science Education, 9*(2), 377–391. https://doi.org/10.1007/s11422-013-9564-x

Manning, J. (2021). Communication studies about sex: Implications for relationships, health, culture, and identity. A review. *El Profesional de la Informacion, 30*(1), e300114+. Gale OneFile: Informe Académico, link.gale.com/apps/doc/A654208730/IFME?u=wash_main&sid=bookmark-IFME&xid=84db89b8. Accessed 6 July 2023.

Martin, E. (1998). Anthropology and the cultural study of science. *Science, Technology, & Human Values, 23*(1), 24–44.

Mattheis, A., Lovos, J., Humphrey, C., Eichenberger, L., & Nazar, C. R. (2022). Queering the common core (and the NGSS): Challenging normativity and embracing possibility. *Journal of Homosexuality, 69*(12), 2007–2026. https://doi.org/10.1080/00918369.2021.1987748

Meyer, E. J. (2008). Gendered harassment in secondary schools: Understanding teachers' (non) interventions. *Gender and Education, 20*(6), 555–570. https://doi.org/10.1080/09540250802213115

Miller, R., Liu, K., & Ball, A. F. (2020). Critical counter-narrative as transformative methodology for educational equity. *Review of Research in Education, 44*(1), 269–300. https://doi.org/10.3102/0091732x20908501

Montañez, A. (2017, September). Beyond XX and XY: The extraordinary complexity of sex determination. *Scientific American*.

Nehm, R. H., & Young, R. (2008). "Sex hormones" in secondary school biology textbooks. *Science & Education, 17*(10), 1175–1190. https://doi.org/10.1007/s11191-008-9137-7

Oppenheim, M. (2023, February 16). Figures that lay bare the shocking scale of toxic influencer Andrew Tate's reach among young men. *Independent*. https://www.independent.co.uk/news/uk/home-news/andrew-tate-influence-young-men-misogyny-b2283595.html

Oudshoorn, N. (1990). On the making of sex hormones: Research materials and the production of knowledge. *Social Studies of Science, 20*(1), 5–33. http://www.jstor.org/stable/285100

Penley, C., & Ross, A. (1991). Cyborgs at large: Interview with Donna Haraway. In C. Penley & A. Ross (Eds.), *Technoculture* (Vol. 3, pp. 1–20). University of Minnesota Press.

Planned Parenthood. (n.d.). *Family life and sexual health (FLASH)*. https://www.plannedparenthood.org/planned-parenthood-south-east-north-florida/education-programs/flash#:~:text=Family%20Life%20and%20Sexual%20Health%20(FLASH)%20is%20a%20comprehensive%20science,pregnancies%2C%20STDs%20and%20sexual%20violence

Ribeiro, M. H., Blackburn, J., Bradlyn, B., De Cristofaro, E., Stringhini, G., Long, S., Greenberg, S., & Zannettou, S. (2021). *The evolution of the manosphere across the web AAAI conference on web and social media, Virtual*.

Richardson, S. S. (2013). *Sex itself: The search for male and female in the human genome*. University of Chicago Press. http://ebookcentral.proquest.com/lib/washington/detail.action?docID=1431257

Ricciotti, C. (n.d.). *Gender free dancing: A brief historical perspective*. Retrieved June 26, 2023 from https://contra.maiamccormick.com/assets/pdfs/gender-free-history_brief.pdf

Roughgarden, J. (2004). *Evolution's rainbow: Diversity, gender, and sexuality in nature and people*. University of California Press.

Russell, J. (2021). *Queer ecopedagogies: Explorations in nature, sexuality, and education*. Springer. https://doi.org/10.1007/978-3-030-65368-2

Schey, R. (2021). Fostering youth's queer activism in secondary classrooms: Youth choice and queer-inclusive curriculum. *Journal of Adolescent & Adult Literacy, 64*(6), 623–632. https://doi.org/10.1002/jaal.1150

Skloot, R. (2010). *The immortal life of Henrietta lacks*. Crown Publishers.

Snapp, S. D., Burdge, H., Licona, A. C., Moody, R. L., & Russell, S. T. (2015). Students' perspectives on LGBTQ-inclusive curriculum. *Equity & Excellence in Education, 48*(2), 249–265. https://doi.org/10.1080/10665684.2015.1025614

Snyder, V. L., & Broadway, F. S. (2004). Queering high school biology textbooks. *Journal of Research in Science Teaching, 41*(6).

Terry, J. (1999). *An American obsession: Science, medicine, and homosexuality in modern society.* University of Chicago Press.

Thoreson, R. (2020, June 15). *US Supreme Court ruling a victory for LGBT workers.* Human Rights Watch. Retrieved June 6, 2023 from https://www.hrw.org/news/2020/06/15/us-supreme-court-ruling-victory-lgbt-workers?psafe_param=1&gclid=Cj0KCQjwn9CgBhDjARIsAD15h0Cy4BfCEgucz1ufWNHV8M9mFT6sttzbBOiDtWFbeoBGbPfPs4L4dT0aAmaXEALw_wcB

Toumey, C. P. (1996). *Conjuring science: Scientific symbols and cultural meanings in American life.* Rutgers University Press.

Townsend, K. L. (2011). The medicalization of "homosexuality". https://surface.syr.edu/honors_capstone/292

Trevor Project, T. (2022). *2022 national survey on LGBTQ youth mental health.* https://www.thetrevorproject.org/survey-2022/assets/static/trevor01_2022survey_final.pdf

Wolfe, G., & Fogwell, N. (2022). DSM discrimination and the LGBT community: Using the history of diagnostic discrimination against sexual minorities to contextualize current issues in transgender and gender diverse mental healthcare. *Psychology from the Margins, 4.* https://ideaexchange.uakron.edu/psychologyfromthemargins/vol4/iss1/2

Yurcaba, J. (2022, October 29). Florida medical board votes to ban gender-affirming care for transgender minors. *NBC News.* Retrieved July 6, 2023 from https://www.nbcnews.com/nbc-out/out-news/florida-medical-board-votes-ban-gender-affirming-care-transgender-mino-rcnarcna54632

Matthew Weinstein is a professor of science education at the University of Washington-Tacoma. His work looks at ethics, ideology, and political economy of science education and schooling. He is the author of Robot World, which examines science education and popular culture, and Bodies Out of Control, which explores the ideologies of science texts. He is also the creator of TAMS Analyzer an open source qualitative coding and analysis tool for Apple computers.

Alysa M. Schafer is a trans-nonbinary secondary science educator teaching chemistry and marine biology at a public urban high school. They obtained a Bachelor of Science degree in biology from the University of Puget Sound and completed their Master of Education at the University of Washington-Tacoma. Alysa's area of focus in research is centered around LGBTQIA+ activism in science. They are also the advisor for their school's gay-straight alliance club.

Chapter 8
When Science Museums Re-imagine Their Social Roles: Responding to the COVID-19 Pandemic

Ana Maria Navas Iannini and Erminia Pedretti

8.1 Introduction

Over the past few years, information (and misinformation) about the coronavirus, prevention strategies, and more recently, vaccination and vaccine guidelines has inundated television, newspapers, and social media (Scheufele et al., 2020, 2021). People worldwide observed the need for responsible decision-making and agency both at individual and societal levels. In the science museum landscape, the COVID-19 pandemic was also experienced as a difficult and contemporary issue impacting the work of these institutions, while simultaneously creating opportunities to generate novel educational programming, exhibitions, and visitor engagement efforts (Burke et al., 2020; Kahn, 2020; Lee et al., 2023; Kim et al., 2022; Schultz et al., 2021; Tan & Tan, 2021). In these troubled and challenging times, science museums have been positioned as important voices and players in promoting social transformation (Cameron et al., 2014; Janes, 2011; Janes & Sandell, 2019; Laurent, 2012; Ng et al., 2017) and social justice (Dawson, 2014; Feinstein, 2017).

In this chapter, we examine how science museums faced collapse and responded to the COVID-19 pandemic through online initiatives that sought public engagement with this complex issue. We acknowledge that many institutions began to revisit and transform their practices through creative strategies and adjustments. For instance, we identified virtual exhibits about coronavirus (and co-related social, political, economic, and environmental issues), online forums on COVID-19

A. M. N. Iannini (✉)
Faculty of Education, Simon Fraser University, Burnaby, BC, Canada
e-mail: anavasia@sfu.ca

E. Pedretti
Ontario Institute for Studies in Education, University of Toronto, Toronto, ON, Canada

C. B. Moura (ed.), *A Sociopolitical Turn in Science Education*, Contemporary Trends and Issues in Science Education 62,
https://doi.org/10.1007/978-3-031-78586-3_8

vaccines, and live events with communities and experts (see, for example, Burke et al., 2020; Kahn, 2020; Lee et al., 2023; Schultz et al., 2021; Tan & Tan, 2021). What we see is a movement away from dominant approaches based on siloed disciplines, unquestionable science facts, and interactive science phenomena—attributes of first, second, and third generation science museums (Amodio, 2013; Friedman, 2010; McManus, 1992) towards approaches that embrace criticality, social responsibility, and civic agency—features of fourth generation science museums (e.g., Ng et al., 2017; Pedretti & Iannini, 2020a; Tagüeña et al., 2011). Although there are many examples worldwide of science museums' responses to COVID-19, we were drawn to three sites—the *Museu do Amanhã [Museum of Tomorrow]* (Rio de Janeiro, Brazil), the *Museum of Science* (Boston, U.S.) and the *Royal B.C. Museum* (Victoria, Canada) because they captured our imagination and revealed possibilities for reimagining institutional goals and their social roles amidst the COVID-19 (post)pandemic context.

Drawing upon: (1) publicly accessible online initiatives (e.g., virtual exhibitions, community conversations, forums, social media programs, descriptions of exhibits, and position statements); (2) scholarship in the fields of museum studies and informal science education in the context of COVID-19; and (3) our combined research spanning 25 years, we examine three institutional responses to COVID-19. These responses encompass education and communication programs and initiatives created and designed for public consumption and engagement. We interpret this work and practice through the science museum generation framework (Amodio, 2013; Bucchi, 2008; Friedman, 2010; McManus, 1992), with an emphasis on fourth generation science museums (Pedretti & Iannini, 2020a, b). In so doing, we consider how the pandemic created momentum for thinking about the social purposes, values, and impact of these institutions in contemporary societies.

8.2 Generations of Science Museums

The idea of describing and understanding the work of science museums through *generations* that are rooted in historical and sociocultural contexts emerged in the 1990s through the work of McManus (1992). In her iconic paper *Topics in museums and science education*, she highlights the importance of considering the history of these institutions' development to make sense of their current education and communication goals. Consequently, discussions evolved in museum studies literature regarding *generations* of science museums. First generation science museums were identified as focusing on collections of objects with historical and biological value. Originating with cabinets of curiosities in the seventeenth century—and currently represented by natural history museums—this generation is often associated with the idea of displaying collections of specimens, objects, and 'wonders' of nature (Friedman, 2010; McManus, 1992). Under the principle of *look-but-don't-touch*, first generation science museums aim to disseminate research conducted about specific collections and to expand knowledge about them (Amodio, 2013).

Second generation science museums are associated with the Industrial Revolution and are represented by technical and industrial museums willing to make public information about scientific progress (McManus, 1992). Still, with a focus on objects with historical value, this generation looks primarily to foster public understanding of science and create a welcoming climate for scientific advancements (Bucchi, 2008). In contrast, third generation science museums are represented by hands-on and interactive science centres. Compared with the previous generations, the focus here changes from objects (e.g., rocks, taxidermized mammals, skeletons, turbines) to ideas and scientific phenomena. As such, third generation science museums look to actively engage visitors with multisensorial experiences (e.g., about gravity, optics, electricity) and with learning opportunities derived from these experiences (Amodio, 2013).

Critics of these three generations of science museums—and what they entail in terms of sociocultural purposes and educational goals—moved scholarship in the field to imagine what the next generation (science) museums might look like. For example, Tagüeña et al.'s (2011) work brings into the conversation the term *glocal*, which emerges from the articulation of knowledge about *global* themes with the reality of a *local* place. The concept of *glocal* has impacted narratives and possibilities for engagement in science museum practices. The challenge is to engage with current and pressing issues of global interest and, at the same time, attend to local situations and projects developed to address these problems. For Tagüeña et al. (2011), *glocal* museums reflect the potential for generating a sense of belonging and attachment in the communities they attend. As such, *glocal* museums are called to remove language and cultural barriers and incorporate closer/local references (e.g., scientific, environmental, artistic, and cultural). Other research studies and perspectives highlight important directions that also speak to the next generation of science museums. These include, for example, promoting dialogue and participation among museum visitors (Mazda, 2004); embracing socioscientific issues as part of exhibitions and other educational practices (Kollmann et al., 2013); fostering scientific citizenship (Hine & Medvecky, 2015); creating more equitable relationships between visitors and museum professionals (Aguirre Rios, 2014); and working with communities (at local levels) to engage in collective actions that preserve diversity and nurture equity and inclusion (Heumann Gurian, 2015). In our previous work related to fourth generation science museums, we describe commitments to social responsibility, and civic agency as hallmarks of this new landscape (Pedretti & Iannini, 2020a, b). With these perspectives in mind, we refined the six key drivers that represent possibilities for re-imagining science museums' spaces and practices (Table 8.1).

In Table 8.1, the six key drivers reflect ways of understanding contemporary work of science museums in the context of COVID-19. We argue that these drivers can serve as a lens to interpret current practices and changing social identities of these institutions—and capture, for example, commitments to communities and organizations (allyship) or broadening of institutional values (empathy). With these six key drivers in mind, we sought to identify patterns, overlaps, and possibly new directions that could emerge as we examined the practices and initiatives of the

Table 8.1 Descriptors of fourth generation science museums' key drivers

Drivers of fourth generation science museums	Descriptors
Fostering change and transformation	Places for activism; forces for public good; mindful institutions with sustained awareness of global and local issues with relevance for their communities.
Promoting productive struggle	Places for ambiguity, controversy, dissonance, and emotional disequilibrium; advocates of (positive) experiences with difficult issues.
Developing allyship	Places for partnerships and networks with different organizations. Inclusive spaces for social well-being; institutions working collaboratively with local communities and organizations towards equity, inclusion, and preservation of diversity.
Fostering empathy	Places that explore, build, and develop empathy with the diverse communities they serve.
Supporting epistemic democracy	Spaces to disrupt traditional science narratives, honour a diversity of voices and share epistemic authority.
Acting as hybrid third spaces	Places that reorganize and reconcile knowledge, experiences, and perspectives that visitors bring with established theoretical foundations of science.

Adapted from Pedretti and Iannini (2020a, b)

three sites discussed in this chapter. In so doing, and not surprisingly, some of these drivers were more prominent than others. More will be said about this later in the chapter.

8.3 Museums and COVID-19: What Does the Literature Say?

Literature (published since 2020) about museums and the work they engaged in during the pandemic can be, broadly speaking, characterized in two ways. In the first instance, we note a body of scholarship that speaks to larger institutional responses to COVID-19. The pandemic galvanized museums to reflect upon the "what and how" of their work, often resulting in a significant shift to the production of online content and the digitalization of collections, exhibitions, and patrimony. The second body of literature speaks to institutions' attempts to produce content about—and foster engagement directly related to—the topic of the COVID-19 pandemic. Below we discuss and frame the literature around these two ideas.

In revisiting the 'what and how' of their work, we noted institutions (1) taking collections and exhibitions online; (2) developing new digital tours and exhibits presented on platforms such as Google Arts & Culture; (3) generating crowd-sourced museum exhibitions and content (e.g., through the sharing of meaningful pictures, images, memories, stories, descriptions); 4) using online tools to 'bring' visitors to visit 'physical' exhibitions (e.g., live videos recorded from the museum);

(5) enhancing onsite exhibit and experiences through online tools; (6) investing in social media presence; (7) improving accessibility through online resources (e.g., breaking language barriers, attending special needs), and (8) developing/increasing online educational programming—e.g., talks, guest talks, workshops, lectures, videos, performances (see, Burke et al., 2020; Crooke, 2020; Kahn, 2020; Lee et al., 2023; Tan & Tan, 2021). The literature describes a number of initiatives that emerged during COVID-19. For example, Lee et al. (2023) share the ways in which a Korean science museum developed and distributed—through social media—"experiment videos, exhibition tour videos introducing each section of the center, and videos explaining individual exhibits together with relevant scientific concepts and principles" (p. 654). Tan and Tan (2021) provide a snapshot of digital initiatives developed by art museums in Singapore to show the extended role of museums as public health resources. They analyzed the initiative *Slow art* developed by the National Gallery Singapore which sought to promote "emotional wellbeing through guided mindfulness practices" (Tan & Tan, 2021, p. 4). This activity included weekly video conferencing sessions via Zoom where participants/visitors were asked to "explore various observation exercises while encountering a unique artwork from the museum" (p. 4). In a different example, Burke et al. (2020) discuss the initiative *The Getty museum challenge* (by the Getty Center, Los Angeles, CA) through which visitors were mobilized as content creators from their own homes to produce crowdsourced museum exhibitions. The challenge asked participants to recreate the Getty's artwork collection with objects at home and share their masterpieces on social media platforms. As such, these initiatives reflect an intent to communicate and engage with the audiences in different ways and through an extended array of contexts and topics. They also reflect possibilities to generate communal knowledge, co-create content, increase the information available for visitors, reach audiences beyond geographical limits, and provide social connection (while acknowledging social distancing regulations).

The second body of literature that we identify is located at the intersection of museums and COVID-19 and reflects attempts to produce content about—and foster engagement with—this difficult and contemporary topic. We found this research to be most helpful and relevant to our work given that we were interested in how science museums addressed COVID-19 through their exhibition practices and other education and communication programs. This direction is represented by the: (1) emergence of temporary exhibits (online and onsite) about COVID-19 and correlated topics such as viruses and vaccines; (2) development and upload of educational content and information about COVID-19 and the pandemic in museums' websites, social media and YouTube channels; (3) collection and preservation of scientific, medical and sociocultural patrimony related to COVID-19; (4) organization of conversational spaces and forums about COVID-19 and its impacts—for visitors and staff; and (5) establishment of partnerships with different organizations and institutions (e.g., universities, research centers, community organizations, governmental agencies)—to support initiatives (see, for example, Burke et al., 2020; Crooke, 2020; Kim et al., 2022; Sotério et al., 2023; Schultz et al., 2021). To illustrate this direction, consider the initiative *Collecting COVID-19* undertaken by the

Science Museum Group affiliated with the London Science Museum (United Kingdom). On the project's website it is stated that curators were "actively, and sensitively, researching stories and identifying objects to collect on the nation's behalf" (Science Museum Group, n.d., para 3). The goals of this initiative/research included collecting materials that could reflect medical, public health, and scientific responses to COVID-19, and items that could reveal social and personal interactions with the disease. As expressed by the Science Museums Group, it was expected that these items could provide a permanent historical record of responses to the pandemic and its impacts, both for future research and (potentially) public display. As such, the Group offered the following guidelines regarding donations of items (Science Museum Group, n.d., para. 12):

- *Tangible and digital forms of public advice and guidance produced in response to the COVID-19 pandemic, from official sources at national and local levels.*
- *Material associated with assessing, monitoring and testing those suspected of having severe acute respiratory syndrome coronavirus 2 (SARS-CoV-2) or those who may already have had it.*
- *Material relating to medical treatment for SARS-CoV-2, in a pre-vaccine environment as well as the provision made for the expansion of care facilities.*
- *Items associated with the scientific response to the COVID-19 pandemic.*
- *Material related to the impact of COVID-19 on our culture, economy and society. These items may reflect significant changes to whole industries, such as our rail or air transport networks. Other items collected might be highly trivial, moving or outlandish, and may at first seem removed from scientific or medical aspects of the pandemic.*

In another example, the Exploratorium in San Francisco (U.S.) hosted monthly *COVID Conversations* online events, where the audience could dialogue with experts about the virus, prospective vaccines, tests, and treatments (Exploratorium, 2020). The intersection between science museums and COVID-19 reflected in the literature, can also be traced in webinars, bulletins, newsletters, and list-serves from museum associations. In this context, we learned about the Canadian Association of Science Museums' (CASC) *COVID-19 advocacy efforts* (CASC, 2020) and the European Network Museums and Science Centres' (ECSITE) *COVID-19 resources page* (ECSITE, n.d.).

8.4 Tales from the Field

While teaching our online courses about science museums[1] during the pandemic, we were compelled to change the backdrop of our classwork and move ourselves and our students to explore museum websites, and online educational materials and resources (instead of physically visiting these institutions, as we had done in the past). The preparation of our teaching assignments prompted us to identify some

[1]By that time, Erminia was teaching her longstanding graduate course T*eaching and learning about science and technology beyond schools,* while Ana was teaching a newly proposed undergrad course entitled *Museums, society and education*

powerful virtual initiatives and calls for action from science museums and to engage with published research about museums and the pandemic. From this assemblage of research scholarship and our collective experience as informal educators and researchers, we selected three tales to tell– from the *Museu do Amanhã* (Rio de Janeiro, Brazil), the *Museum of Science* (Boston, U.S.) and the *Royal B.C. Museum* (Victoria, Canada). These three sites captured our imagination and hearts, hinting at what is possible while revealing elements potentially associated with fourth generation science museums.

As we noted earlier, we are aware that there are many other powerful and compelling examples of COVID-19 responses from science museums around the world that could reflect this fourth-generation movement and commitments to social responsibility and civic agency. However, we were also deliberate in choosing three examples that were representative of different socio-cultural contexts (Brazil, Canada, and the U.S.). We searched for institutions that: (1) developed programs about COVID-19 involving more than a single initiative; (2) sought to engage visitors in different ways; and (3) tended to position visitors as active participants of those initiatives. Paradoxically, our tales come primarily from online engagement due to closures, and so the notion of *visitor engagement* is not lost on us—as researchers and visitors. To circle back, the three stories that we chose to share demonstrate what is possible in the science museum landscape. Drawing upon publicly accessible information, documents, and artefacts– and visiting sites virtually, we gathered our stories. Below we present our three tales.

8.4.1 Museu do Amanhã (Rio de Janeiro, Brazil)

From 2020 to 2021, the Museu do Amanhã (Museum of Tomorrow) located in Rio de Janeiro, hosted the exhibit *Coronacene: Thoughts in times of pandemic*. This exhibit resulted from a partnership between the Museu do Amanhã, the research institute Fundação Oswaldo Cruz (Fiocruz), and the Brazilian TV networks Globo and GloboNews. As expressed on the museum's website, this exhibit was looking to "promote reflection on the impacts of the [COVID-19] illness in the world and perspectives related to lifestyle changes worldwide" (Museu do Amanhã, 2023a, para. 2).

The temporary exhibit *Coronacene* was designed around six rooms/galleries made also accessible to a wider public through an online tour presented on Google Arts & Culture (Museu do Amanhã, 2023a, b, c). Through portraits projected on the walls, the room *Essential*, honoured the work of first-line professionals, providers and caregivers such as doctors, nurses, pharmacologists, cleaning professionals and drivers. The digital tour of this gallery included testimonials of an Indigenous leader working in a community belonging to the city of Manaus, where efforts to produce masks, collect donations, and promote prevention and social distancing strategies through WhatsApp and door-to-door initiatives were undertaken. The gallery *From virus to pandemic* presented contextualized knowledge and scientific evidence regarding COVID-19; this included representations of contagion, a large-size model of the coronavirus, and explanatory videos about the illness and the virus. According to the museum's website, the room *Transformed societies* looked to (purposively) promote discomfort by presenting how the pandemic impacted societies both in terms of social distancing and social inequity. The digital tour of this gallery reveals images of faces wearing masks and videos of social situations/scenarios worldwide during the pandemic projected on the walls. *Memorial to those who have passed* was dedicated to Brazilians who died from COVID-19; in this gallery, hourglasses placed over sand evoked life's short timespan and the fragility of our lives. Proper names written on the walls and a poem by Brazilian author Machado de Assis—*The two horizons*—referred to the mourning and respect for those who passed. The room *Science is the protagonist* focused on the central role that scientific practice (and knowledge) played during the pandemic; here, visitors could explore scientific lab environments and interact with personal items and equipment used by Brazilian researchers looking, in 2020, for a vaccine against the virus. Finally, the gallery *Culture is the way* reflected a shift that the cultural sector (e.g., theatre, music) underwent during the pandemic while moving from *traditional* to *digital*. In this room, the museum's website explains how digital culture was not only portrayed as a form of survival but also as a form of resilience.

Interestingly, *Coronacene* was accompanied by a compelling *Manifesto* published on the museum's website (Museu do Amanhã, 2023c, para. 3):

> *MANIFESTO*
>
> *We beheld most of 2020 through the windows of our homes.*
> *We talked, we cried, we had fun, we distracted ourselves, we felt fears and longings, we rediscovered ourselves, and we adapted.*
> *We persisted.*
> *We needed to keep ourselves in our private spaces.*
> *There, we were safe.*
> *Meanwhile, essential workers set out to protect us, working in hospitals, supermarkets, and pharmacies; televising news, delivering orders, and researching in labs.*
> *The outside world carried on, albeit differently.*
> *We saw our relationships in a new light.*
> *We realized we can always find time for ourselves, help someone and live together with those we love.*
> *Life needs to go on, and the future will be brighter if we respect the planet's limits. We are all connected.*
> *The pandemic has shown us that it is time we learn from science and reexamine actions affecting biodiversity.*
> *The virus will continue, and it is up to us to work together to overcome it.*
> *When all of this is over, we want a Tomorrow that is more sustainable, less unequal.*

The exhibit was accompanied by a series of online conversations with experts called *Conversations for tomorrow*, which could eventually anticipate questions and concerns the exhibit might have raised about the coronavirus and the pandemic. For instance, online users could attend live and comment (or later access) full talks around provocative questions. These events were disseminated through the Museu do Amanhã's social media channel (Fig. 8.1).

Fig. 8.1 Screenshot of the *Conversations for tomorrow's* webpage (Museu do Amanhã, n.d.). Some of the provocative questions guiding these conversations were: *Why should we trust science?* (left), *Where is culture going in post-pandemic times?* (center), and *How did the pandemic transform us?* (right)

8.4.2 Museum of Science (Boston, USA)

Our first contact with the work of the Museum of Science (Boston) around COVID-19 was through the online exhibit *Ask a virtual expert*. While looking for online museum educational resources about COVID-19 for our courses, we found this exhibition and shared it with our students. Soon, we learned this was part of a broader initiative called *Project vaccine: Our best defence* (Museum of Science, 2023a). Different initiatives comprise this project, among them, an onsite exhibit dedicated to the complex process of developing a COVID-19 vaccine. The museum's website (Museum of Science, 2023b) describes how the exhibit was intended to raise important questions about the pandemic, lessons learned, and consideration about social inequities (exacerbated through/during the pandemic). The exhibit provided contextualized evidence-based scientific knowledge related to different types of vaccines and the spread of viruses and included interactive installations where visitors were invited to take a position about pressing issues related to vaccines. Still part of *Project vaccine* is the interactive online experience *Ask a virtual expert*— referred to earlier— that generates a 'dialogue' opportunity between online visitors and Dr. Ashis K. Jha, the Dean of the School of Public Health at Brown University (Museum of Science, 2023c). In this interactive experience (that has been regularly updated during the last few years), Dr. Jha appears sitting on a chair and online visitors can ask him or type questions they might have about the coronavirus, COVID-19, the pandemic and correlated topics (such as vaccines, vaccine boosts and so on). Dr. Jha answers the questions or asks visitors to reframe them for clarity.

In 2020, the Museum of Science organized the forum/panel *Addressing community inequalities during COVID 19 recovery* (Museum of Science, 2020a); it was a free virtual event transmitted live on the museum's social media platform. This conversation reunited professionals—engaged with food bank initiatives, innovation hubs, health centres, and human services from the city of Boston—and members of the public to discuss the ways that COVID-19 has disproportionately impacted underserved communities (Museum of Science, 2020a). For instance, panelists of the event engaged in discussions about the two simultaneous 'forces' of declining hospitalization cases in Massachusetts and rising economic insecurities. Important issues were raised by the organizers of the event such as the loss of health insurance and jobs, and access to healthy food, childcare, and housing in the state. These were recognized as critical issues with a longstanding impact on COVID-19 patients and their families (Museum of Science, 2020a). Additionally, questions posed by the audience considered mental health and equity issues (and privileges), and crisis experienced by community organizations and food banks willing to support underserved communities. Advice on how to help was also requested by participants. In the same year, the Museum of Science hosted *Community conversations*—a free online forum that focused on the (still) novel coronavirus. Beyond specialists in infectious disease research, public health administrators, and community organizers, the museum invited anyone interested in the following, to attend:

> *Making informed decisions for* [them] *and* [their] *family;*
> *The science behind this outbreak;*
> *Local and state resiliency planning;*
> *Community support resources and organizing* (Museum of Science, 2020b, para.2).

Other initiatives developed under *Project Vaccine* included informative resources that could be freely accessed (Museum of Science, 2023c). Among these, the infographic presented in Fig. 8.2 approaches complex issues about underrepresented groups in vaccine trials and the ways in which COVID-19 vaccine clinical trials were designed to truthfully represent the U.S. population's demographics.

8.4.3 Royal BC Museum (Victoria, Canada)

In visiting different museum websites and spaces these last few years, we identified many initiatives related to collecting meaningful stories, objects, and materials related to COVID-19 and the pandemic. In this, our final tale, we draw on the *Collecting for our time project* located at the Royal BC Museum, Victoria, Canada. This initiative sought to gather relevant digital content and cultural materials related to the pandemic to establish a repository for a variety of future uses. In this context, the museum created the webpage *Tell us your COVID-19 story*, where members of the public were invited to share/submit "perspectives, photos, videos and mementoes that will help tell the story of this moment for future generations" (Royal BB Museum, n.d.-a, para. 2). Royal BC Museum's professionals expressed (through the museums' social media channel) how this project reflected a shifting perspective for the museum and a desire to actively pay attention to what is happening locally and

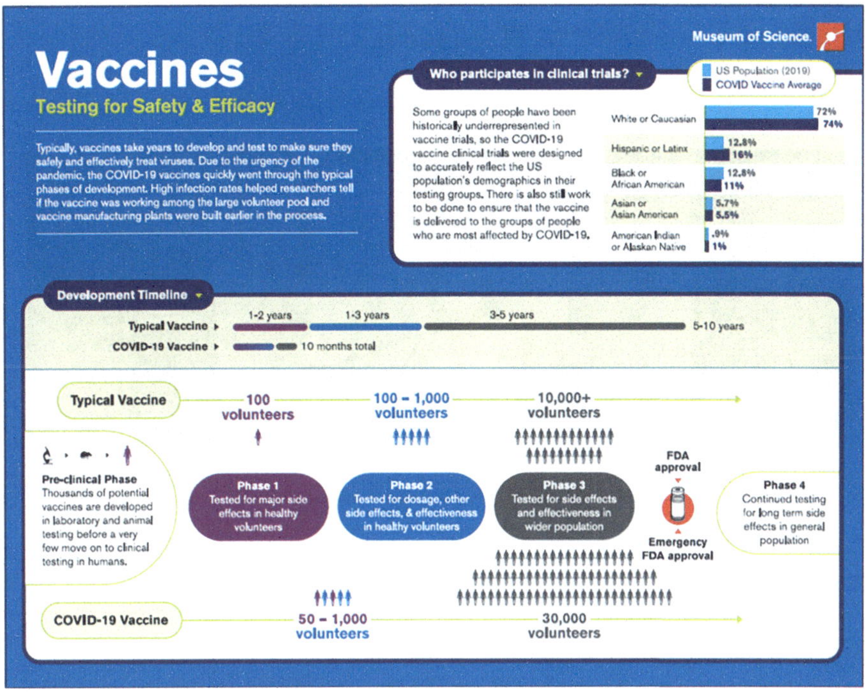

Fig. 8.2 Free downloadable infographic about vaccine testing produced by the Museum of Science both in English and Spanish versions

globally (Royal BC Museum, 2020). As such, the public was invited to share with the institution how they were making sense of the pandemic, for example, through drawings, journals, and pictures. The picture below (Fig. 8.3)—submitted to *Tell us your COVID-19 story*—was accompanied by the following caption on the museum's website:

> [This picture] *captures a moment from Sara's own wedding, during the pandemic. Sara says, 'I don't 100% recommend having a wedding during a pandemic, but I have to say it's gonna be one heck of a story to tell and looking back now I wouldn't change a thing. It made our special day so unique. And at the end of the day the show must go on. Love always wins'*". (Royal BC Museum, n.d.-a)

According to the museum's social media channel, by September 2020, the initiative received over 200 submissions from the province of British Columbia (Royal BC Museum, 2020).

Related to the theme of collecting/presenting COVID-19 stories, was the exhibit *Between us: Adad Hannah's social distancing portraits*, hosted by the Royal BC Museum in 2022. In this exhibit, the American artist Adad Hannah presented video portraits of Vancouverites during the pandemic. These portraits were done in the form of a video of individuals 'posing' as if a picture was being taken by the artist. While being 'frozen in time' (for the video portraits) and in the lockdown as well,

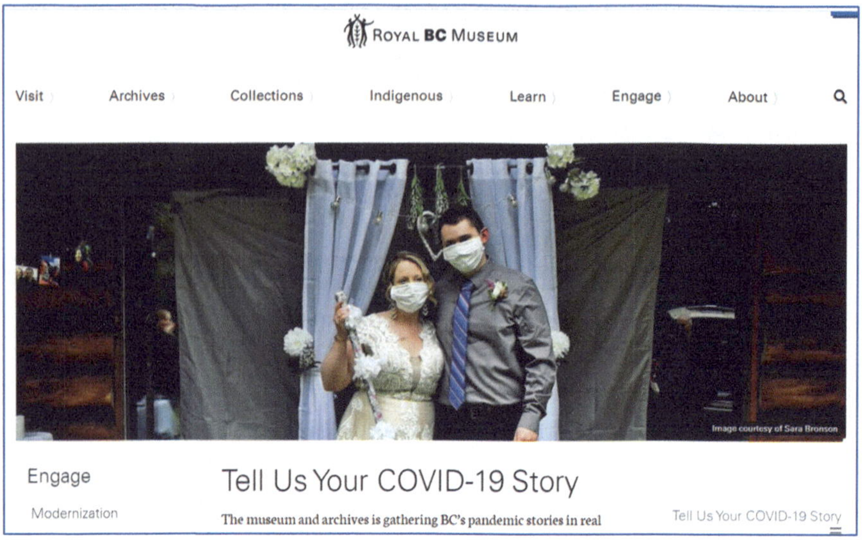

Fig. 8.3 Screenshot of the *Tell Us Your COVID-19 Story* webpage

different people shared their stories of struggle and resilience as they faced the pandemic. The exhibit's website introduced this gallery with the following "With individual sound compositions that can be uplifting, unsettling, or serene, alongside personal narratives from his sitters, Hannah's series personifies our varied emotional journeys" (Royal BC Museum, n.d.-b, para. 3). At the Royal BC Museum, the exhibit was presented in a small room with a sober atmosphere, where these different images occupied all the walls. In the middle of the room, a loop video revealed personal narratives associated with the emotional journeys behind each video portrait.

8.5 Discussion

8.5.1 Interpreting Our Tales

Janes' (2022, p. 730) contemporary reflections about the role of museums facing social collapse, highlight that these institutions have been "absent … in the wide-ranging initiatives to confront global challenges and remain unrecognized as key civic and intellectual resources". Going further, the author states that (academic and professional) discussions about these matters have prompted constructive narratives and surface-level activism but, ultimately, no fundamental change. Agreeing with Janes (2022), we see how *fundamental change* might be viewed as a difficult and evasive endeavour for many institutions. The tales we shared, however, reflect attentive and resilient museum communities that expanded their social roles through renewed institutional values and transformative practices. In interpreting our tales four themes were particularly powerful and prevalent—*empathy*, *epistemic democracy*, the role of *contextualized knowledge*, and *allyship*. Each theme is presented below.

In thinking about institutional values, *empathy* emerged as a key driver of science museums' responses to COVID-19. This is an understandable response, given the pandemic circumstances, the continual lockdowns we lived through, and the often deep isolation people experienced. In this context, empathy can be understood both as a motivator for science museum professionals' work (e.g., as they choose to seek out and try to understand other peoples' accounts and testimonials about the pandemic) and, also, as a desirable goal for their exhibitions (e.g., choosing to curate portraits honouring the work of first-line professionals; collecting and sharing of personal stories related to COVID-19). As we saw in the tales presented (e.g., *Coronacene*, *Tell us your COVID-19 story*, and *Between us*), the sharing of powerful and personal COVID-19 experiences (through different platforms and exhibits) invites visitors to undertake the emotional journey of being in the place of others and acknowledging their lived experiences and feelings during the pandemic.

As we consider what *fundamental change* means in the context of science museums, *epistemic democracy* plays an important role. In all three sites the pandemic moved these institutions to make meaning, alongside visitors, museum professionals, and specialists, about the uncertain event of the coronavirus and its impacts. Information and misinformation about COVID-19 were both at play simultaneously (Scheufele et al., 2020, 2021). Neither museum professionals nor visitors could anticipate what was to come, and so trying to make sense—together—of what was happening (e.g., in the scientific field, public health sector and/or household instances) and what all of us were undergoing (at institutional and personal levels) was a challenging endeavour. As expressed by Kupper (2020, para. 13) "Uncertainty is clearly not a key element of museum concepts". As such, he calls science museums, and other settings engaged in communicating science, to "acknowledge that uncertainty exists… shift the attention to the process of science… focus on the underlying values and perspectives… challenge assumptions and focus on sense-making… and engage in dialogue" (2020, para. 17). Pertinent to our discussion on epistemic democracy—and the possibilities of disrupting traditional science narratives and honouring a diversity of voices—are the dialogic sense-making opportunities created by science museums during the pandemic. Recall the town hall, community conversations, and other dialogic initiatives promoted by the Museum of Science and the Museu do Amanhã. We also see opportunities for visitors to restructure narratives and stories, make sense of them according to their own contexts, knowledge, and experiences, and engage with issues from multiple angles and perspectives (Kupper, 2020).

Another theme that emerges from our tales, as a response to COVID-19, is the idea of *contextualized knowledge* (Klopfer & Aikenhead, 2022; Tagüeña et al., 2011). Humanistic views of science education have advocated for the need to pay attention to the personal and social contexts of science (Klopfer & Aikenhead,

2022). In the science museum landscape, we noted the role that contextualized knowledge played as these institutions attempted to disseminate the science behind COVID-19 and, at the same time, consider complex (local and global) social, economic, and political contexts. Consider, at this point, the community conversations and infographic developed by the Museum of Science. These initiatives tended to focus, simultaneously, on technoscientific information about COVID-19 and co-related themes (e.g., timelines and phases for the production of vaccines; details of treatment in case of required hospitalization) and local issues and forces directly impacting subsets of the U.S. population (e.g., underrepresentation of groups in vaccine trials; economic and health care insecurities). Both initiatives could be seen as important civic and intellectual resources (Janes, 2022), with the potential to foster social transformation at local levels. Here, we recall Tagüeña et al.'s (2011) ideas around the expression and representation of *glocal* perspectives which, in turn, can invite attachment, commitment, and, potentially, agency.

Finally, we note the role that allyship played in the three tales from the field. We are reminded of commitments and calls raised by international organizations (for instance, the International Council of Museums—ICOM [2019, 2020]) encouraging museums worldwide to serve as active agents to offer solutions in critical areas such as the Sustainable Development Goals, climate change education and, more recently, the COVID-19 pandemic (Byun, 2020) through collaboration. The work of the museums we analyzed reflect relationships and alliances created with universities (e.g., School of Public Health, Brown University), national TV Networks (e.g., Globo), research institutes (e.g., Fundação Oswaldo Cruz), and professionals in diverse fields (e.g., medicine, public health, science, art, digital culture). In the (modest) subset of initiatives we examined, we saw creative moments towards building and sustaining connections with visitors—albeit virtual– the world had no choice then. The partnerships and alliances that were created sought to bring together museum professionals, experts, and visitors through online conversations, panels, and social media strategies. Recall the free virtual event *Addressing community inequalities during COVID-19 recovery* developed by the Museum of Science (and described earlier, in this chapter). Here we see the science museum working to establish networks with local communities and organizations in order to support social well-being. Despite these efforts of creating virtual connections with (and among) visitors and establishing partnerships with experts in different fields and with national and local organizations, we did not identify, in the initiatives we analyzed, collaborative work developed alongside local communities in a sustained and programmatic way. When reflecting on social collapse, Janes (2022) recalled the privileged position museums have in society to mitigate crises. His discussions highlight the social impact museums can gain when they are grounded in their communities and become expressions of their locality. In this sense, and similar to Heumann Gurian (2015), we call upon the need to strengthen this kind of collective action and alliances in the pursuit of equity and social inclusion.

8.5.2 COVID-19 and Fourth Generation Science Museums

Our analysis revealed that the unexpected and uncertain events of the pandemic prompted an (implicit) enactment of particular facets of fourth generation science museums (Pedretti & Iannini, 2020b). The pandemic impacted these institutions internally and externally as they had to look inwards, rethink priorities, consider how to establish renewed relationships with visitors and new ways to carry on practices (Burke et al., 2020; ICOM, 2020; Kim et al., 2022; Lee et al., 2023; Tan & Tan, 2021). Our work pushed us to identify overlaps and new directions when institutions faced collapse and responded, as best they could to the pandemic. The interpretations of our three tales pressed us to think about patterns created between some key drivers of this fourth generation, and the contexts within institutions might be operating. In these troubled times, we saw a clustering of a few key drivers: empathy (being in the place of others, sharing stories, and learning from those stories); epistemic democracy (honouring diverse voices, accounts, and experiences in dialogic spaces); allyship (connecting with experts and community organizations); and the emergence of the role of knowledge in relation to context.

Our interpretations caused us to consider the role *knowledge* can play as a potential seventh key driver of fourth generation science museums. Although the function of disseminating information is explicitly related to previous generations (recall, for instance, the public education purposes of first and second generation science museums [Amodio, 2013; Friedman, 2010; McManus, 1992]) a contemporary spin is reflected through the idea of *contextualized knowledge* (Klopfer & Aikenhead, 2022) with the potential for it to become an intellectual and civic resource (Janes, 2022) for visitors AND museum professionals. Different museum communities from around the world acknowledged the need to provide and make accessible content knowledge about the illness of COVID-19 (and its impacts), the coronavirus, prevention strategies, characteristic and proper use of face masks, technoscientific developments linked to the testing, production, and use of vaccines, and so on. Our work suggests that museums disseminated knowledge that was timely and robust (Pedretti & Iannini, 2020b) and necessary to confront misinformation and create a "common domain of expertise to ground the conversation" (Scheufele et al., 2021, p. 1). To this point, Levinson (2010, p. 80) highlights how such an approach "can empower those participating in deliberation by communicating accurate and robust knowledge in comprehensible terms, particularly when distorted accounts of such knowledge are immediately accessible through the new technologies".

Finally, we would like to open up a space to consider the temporal nature of some of the actions and topics of conversation, enacted by science museums while they responded to COVID-19. As the pandemic unfolded, institutions became mindful of what was happening around them and recognized the need to change, update, and edit existing narratives. For example, exhibits and educational resources about the virus, prospective vaccines, and treatments were constantly updated or replaced by other initiatives that could better speak to current times or simply removed. These considerations align with Kollmann et al.'s (2013) call for science museums to represent and foster public engagement with *current* socioscientific issues. However, the rapid way in which information and misinformation about COVID-19 emerged and changed, made us wonder about the challenges that some science museums might have experienced. On one hand, 'temporal responses' (e.g., preparing a temporary exhibit) are part of the repertoire of attending to current socioscientific issues—a hallmark of fourth generation science museums. This is particularly true when attending to communities' needs, motivations, and interests around complex topics affecting their lives. However, responding to socioscientific issues that are unfolding in real-time and possibly dissipating over time, may be difficult to do for small-scale (science) museums that have limited resources, personnel, and capacity.

8.6 Concluding Thoughts

It was our goal to interpret a few museum initiatives related to COVID-19, using the lens of the generations of science museums. As such, we chose to examine initiatives with the potential to reflect a shifting landscape in the museum world. Our work on fourth generation science museums (Pedretti & Iannini, 2020a; b) and controversial exhibitions delving into pressing socioscientific issues (e.g., Navas-Iannini & Pedretti, 2017; Pedretti & Iannini, 2020a) speak directly to the possibility of reimagining the social roles and impacts of these institutions in contemporary societies. Agreeing with Janes (2011) and Janes and Sandell (2019), we acknowledge the power of (science) museums in becoming forces for public good. However, for this to happen, museums need to consider which institutional values matter the most in our challenging times, and how they can better serve different communities through transformative and social justice orientations and practices.

In this spirit, we recall the work of Crooke (2020) and Tan and Tan (2021) as they acknowledged the ways in which museums were relevant during the pandemic. Some became forces of wellbeing and health, inclusive spaces, and means to connect people socially, and emotionally. The literature at the intersection of museums and COVID-19, and the tales we chose to share in this chapter, also reveal how crises generated urgency to act (Crooke, 2020). In times of crisis, diverse science museums became increasingly mindful (Janes, 2011), compassionate, and caring institutions (Tan & Tan, 2021) willing to support and learn from their visitors. Kidman and Chang (2020) also argue that in troubled times "how information is produced and consumed has an immense impact on society, economy, and education" (p. 107). During the pandemic and post-pandemic contexts, science museums played an important role in offering contextualized, relevant and robust knowledge for visitors to engage in conversations, make sense of events, learn, take decisions, act, and help. Finally, in times of crises, reaching out to different experts became essential for science museums to be resilient and rapidly perform, develop work, rethink practice, and create renewed relationships with visitors.

We acknowledge that the work we present here is really the tip of the iceberg with respect to worldwide institutional responses to COVID-19. There are other innovative and bold initiatives, exhibition practices, programs, and collections that are equally compelling and possibly more progressive. This is only a beginning for us. Our next steps include delving further into what is possible and interviewing science museum staff over the next two years. We believe that important questions need to be asked: *What institutional changes implemented during the pandemic still have an impact on science museums' work? What other changes are needed so to become more socially responsive institutions? How are institutional values such as empathy and compassion—emergent in times of crises—reshaping the current work and practice of these institutions? Have science museums maintained efforts to keep connected the online communities they created and supported during the pandemic, and if so how? How are science museums currently envisioning their role as forces for well-being, care, and health at the local level? How, if at all, are allyship and epistemic democracy informing the work of science museums in post-pandemic times?* With these questions in mind, we invite conversations about the longstanding (or temporal) effect of science museums' responses to COVID-19 on their current and future work, and the ways in which science museums can reinvent and re-imagine themselves as they face, confront, and/or take a stance regarding social and environmental crises.

Acknowledgments We want to thank our research assistant, Stéphanie Dodier, for her help with the literature review. An extended thank you to SFU for funding this research through a New Faculty Start-Up Grant (N001082).

References

Aguirre Rios, C. (2014). Science centers. Which role can they play to participate in a city social reconstruction? *Journal of Science Communication, 13*(2), C04. https://doi.org/10.22323/2.13020304

Amodio, L. (2013). Science communication at glance. In A. M. Bruyas & M. Riccio (Eds.), *Science centres and science events* (pp. 27–46). Springer.

Bucchi, M. (2008). Of deficits, deviations and dialogues: Theories of public communication of science. In M. Bucchi & B. Trench (Eds.), *Handbook of public communication of science and technology* (pp. 71–90). Routledge.

Burke, V., Jørgensen, D., & Jørgensen, F. A. (2020). Museums at home: Digital initiatives in response to COVID-19. *Norsk museumstidsskrift, 6*(2), 117–123. https://doi.org/10.18261/issn.2464-2525-2020-02-05

Byun, J. H. (2020). A study on the resolution of ICOM museum definition: Focusing on predictions and suggestions for the museum management. *Korean Association of Arts Management, 54*, 133–153. https://doi.org/10.52564/jamp.2020.54.133

Cameron, F. R., Hodge, B., & Salazar, J. F. (2014). Conclusion: Climate change engagement: A manifesto for museums and science centers. In F. Cameron & B. Neilson (Eds.), *Climate change and museum futures* (pp. 260–280). Routledge.

Canadian Association of Science Museums. (2020, March 27th). *CASC COVID-19 update: Redoubling our efforts to serve our communities*. Retrieved February 20, 2023, from https://www.canadiansciencecentres.ca/News/8924517

Crooke, E. (2020). Communities, change and the COVID-19 crisis. *Museum and Society, 18*(03), 305–310. https://doi.org/10.29311/mas.v18i3.3533

Dawson, E. (2014). "Not designed for us": How science museums and science centers socially exclude low-income, minority ethnic groups. *Science Education, 98*(6), 981–1008. https://doi.org/10.1002/sce.21133

European Network Museums and Science Centres. (n.d.). *COVID-19 resources page.* Retrieved January 23, 2023, from https://www.ecsite.eu/activities-and-services/resources/plans-and-guidelines-re-opening

Exploratorium. (2020, May 6). *New online discussion series from the Exploratorium: COVID Conversations*. Retrieved March 11, 2023, from https://www.exploratorium.edu/press-office/press-releases/new-online-discussion-series-exploratorium-covid-conversations

Feinstein, N. W. (2017). Equity and the meaning of science learning: A defining challenge for science museums. *Science Education, 101*(4), 533–538. https://doi.org/10.1002/sce.21287

Friedman, A. J. (2010). The evolution of the science museum. *Physics Today, 63*(10), 45–51. https://doi.org/10.1063/1.3502548

Heumann Gurian, E. (2015). Nos vamos, nos vemos: Re-imaginar hoy. In F. Typa (Ed.), *Claves para el cambio en los museos de América – El museo reimaginado: Encuentro de profesionales de museos de América* (pp. 92–93). Fundación Typa.

Hine, A., & Medvecky, F. (2015). Unfinished science in museums: A push for critical science literacy. *Journal of Science Communication, 14*(2), A04. https://doi.org/10.22323/2.14020204

International Council of Museums [ICOM]. (2019). *Museums as cultural hubs: The future of tradition.* Report of 25th ICOM General Conference. Retrieved, March 13, 2023, from https://imd.icom.museum/past-editions/2019-museums-as-cultural-hubs-the-future-of-tradition/#

International Council of Museums [ICOM]. (2020). *Museums, museum professionals and COVID-19.* Retrieved March 24, 2023, from https://icom.museum/en/newsmuseums-museum-professionals-andCOVID-19-survey-results

Janes, R. R. (2011). Commentary. *Museums and Social Issues, 6*(2), 137–146. https://doi.org/10.1179/msi.2011.6.2.137

Janes, R. R. (2022). The value of museums in averting societal collapse. *Curator: The Museum Journal, 65*(4), 729–745. https://doi.org/10.1111/cura.12503

Janes, R. R., & Sandell, R. (2019). *Museum activism.* Taylor & Francis.

Kahn, R. (2020). Locked down not locked out–assessing the digital response of museums to COVID-19. *Impact of Social Sciences Blog.* Retrieved, January 8, 2023, from https://blogs.lse.ac.uk/impactofsocialsciences/2020/05/08/locked-down-not-locked-out-assessing-the-digital-response-of-museums-to-COVID-19/

Kidman, G., & Chang, C. H. (2020). What does "crisis" education look like? *International Research in Geographical and Environmental Education, 29*(2), 107–111. https://doi.org/10.1080/10382046.2020.1730095

Kim, M. J., Kang, D. Y., & Martin, S. N. (2022). Exploring informal science education responses to COVID-19 global pandemic: Learning from the case of the Gwacheon National Science Museum in Korea. *Cultural Studies of Science Education, 17*(2), 341–354. https://doi.org/10.1007/s11422-021-10073-5

Klopfer, L. E., & Aikenhead, G. S. (2022). Humanistic science education: The history of science and other relevant contexts. *Science Education, 106*(3), 490–504. https://doi.org/10.1002/sce.21700

Kollmann, E. K., Reich, C., Bell, L., & Goss, J. (2013). Tackling tough topics: Using socio-scientific issues to help museum visitors participate in democratic dialogue and increase their understandings of current science and technology. *Journal of Museum Education, 38*(2), 174–186. https://doi.org/10.1080/10598650.2013.11510768

Kupper, F. (2020, September). The uncertainty of it all. Spokes, 66. https://www.ecsite.eu/activities-and-services/news-and-publications/digital-spokes/issue-66

Laurent, B. (2012). Science museums as political places. Representing nanotechnology in European science museums. *Journal of Science Communication, 11*(04), 1–6. https://doi.org/10.22323/2.11040302

Lee, H., Kang, D. Y., & Kim, M. J. (2023). Navigating into the future of science museum education: Focus on educators' adaptation during COVID-19. *Cultural Studies of Science Education, 18*, 647–667. https://doi.org/10.1007/s11422-022-10142-3

Levinson, R. (2010). Science education and democratic participation: An uneasy congruence? *Studies in Science Education, 46*(1), 69–119. https://doi.org/10.1080/03057260903562433

Mazda, X. (2004). Dangerous ground? Public engagement with scientific controversy. In D. Chittenden, G. Farmelo, & B. Lewenstein (Eds.), *Creating connections: Museums and the public understanding of current research* (pp. 127–144). Altamira Press.

McManus, P. (1992). Topics in museums and science education. *Studies in Science Education, 20*(1), 157–182. https://doi.org/10.1080/03057269208560007

Museu do Amanhã. (2023a). *Coronaceno: Apresentação*. Retrieved, December 1, 2022, from https://coronaceno.museudoamanha.org.br/

Museu do Amanhã. (2023b). *Exposição temporária: Coronaceno – Reflexões em tempos de pandemia*. Retrieved, December 1, 2022, from https://museudoamanha.org.br/pt-br/exposicao-temporaria-do-museu-do-amanha-coronaceno-reflexoes-em-tempos-de-pandemia

Museu do Amanhã. (2023c). *Coronacene*. Retrieved, December 1, 2022, from https://artsandculture.google.com/story/9gVhS6T56XCs0w

Museu do Amanhã. (n.d.). *Conversas para o Amanhã*. Retrieved, December 1, 2022, from https://coronaceno.museudoamanha.org.br/

Museum of Science. (2020a, July 7). *Addressing community inequalities during COVID 19 recovery*. [video file]. You Tube. https://www.youtube.com/watch?v=f9fK4O9tBCc

Museum of Science. (2020b, March 26). *Coronavirus: A community conversation*. [video file] You Tube. https://www.youtube.com/watch?v=8bR5H8WHl2A on March 1st 2023.

Museum of Science. (2023a). *Project vaccine: Our best defence*. Retrieved, September 8, 2022, from https://www.mos.org/project-vaccine

Museum of Science. (2023b). *Project vaccine: Our best defence- permanent exhibition*. Retrieved, September 8, 2022, from https://www.mos.org/exhibits/project-vaccine-exhibit

Museum of Science. (2023c). *Coronavirus: Information and resources*. Retrieved, September 8, 2022, from https://www.mos.org/coronavirus

Navas-Iannini, A. M., & Pedretti, E. (2017). Preventing youth pregnancy: Dialogue and deliberation in a science museum exhibit. *Canadian Journal of Science, Mathematics and Technology Education, 17*(4), 271–287.

Ng, W., Ware, S. M., & Greenberg, A. (2017). Activating diversity and inclusion: A blueprint for museum educators as allies and change makers. *Journal of Museum Education, 42*(2), 142–154. https://doi.org/10.1080/10598650.2017.1306664

Pedretti, E., & Iannini, A. M. N. (2020a). Towards fourth-generation science museums: Changing goals, changing roles. *Canadian Journal of Science, Mathematics and Technology Education, 20*, 700–714. https://doi.org/10.1007/s42330-020-00128-0

Pedretti, E., & Iannini, A. M. N. (2020b). *Controversy in science museums: Re-imagining exhibition spaces and practice*. Routledge.

Royal BC Museum. (2020, September 22nd). RBCM@Home: COVID-19 collecting for our time [video file]. YouTube. https://www.youtube.com/watch?v=pwjHcsL3W60&t=4s

Royal BC Museum. (n.d.-a). *Tell us your COVID-19 story*. Retrieved, October 10, 2022, from https://royalbcmuseum.bc.ca/support-us/tell-us-your-COVID-19-story#:~:text=The%20museum%20and%20archives%20is,to%20remember%20this%20historic%20time%3F

Royal BC Museum. (n.d.-b). *Between us: Adad Hannah's social distancing portraits*. Retrieved, October 4, 2022, from https://royalbcmuseum.bc.ca/visit/exhibitions/between-us-adad-hannahs-social-distancing-portraits

Scheufele, D. A., Krause, N. M., Freiling, I., & Brossard, D. (2020, April 17). How not to lose the COVID-19 communication war. *Issues in Science and Technology*.

Scheufele, D. A., Krause, N. M., & Freiling, I. (2021). Misinformed about the "infodemic?" Science's ongoing struggle with misinformation. *Proceedings of the National Academy of Sciences, 18*(15), 1–3. https://doi.org/10.1016/j.jarmac.2021.10.009

Schultz, G., Pitts, D., & Schmoll, S. (2021). Planetaria and science museums adapting creatively during the COVID-19 pandemic. *Astronomical Society of the Pacific Conference Series, 531*, 94.

Science Museum Group. (n.d.). *Collecting COVID-19*. Retrieved, October 4, 2022, from https://www.sciencemuseumgroup.org.uk/project/collecting-COVID-19/.

Sotério, C., dos Santos, R., Lamim, A., & Linhares Queiroz, S. (2023). Teaching controversial socio-scientific issues in online exhibits of science museums: COVID-19 on the scene. *Disciplinary and Interdisciplinary Science Education Research, 5*(1), 1–15. https://doi.org/10.1186/s43031-022-00069-8

Tagüeña, J., Sánchez, C., & Reynoso, E. (2011). *El museo glocal y sus audiencias*. Museos Universitarios.

Tan, M. K. B., & Tan, C. M. (2021). Curating wellness during a pandemic in Singapore: COVID-19, museums, and digital imagination. *Public Health, 192*, 68–71. https://doi.org/10.1016/j.puhe.2021.01.006

Ana Maria Navas Iannini (PhD, MA, BSc) is an assistant professor at Simon Fraser University (SFU), in the Faculty of Education. She is a researcher in the fields of science education and science communication. She is particularly interested in the ways in which the informal science education sector fosters relationships between science and society. Ana Maria has published 14 papers and book chapters, 4 books, and several technical reports. She has received internal grants to support her research including a FAPA grant by Los Andes University (Colombia) and an New Faculty Start-up Grant (NFSG) grant by Simon Fraser University (SFU) (Canada). Her current research, funded by SFU, relates to mindfulness and response to contemporary socioscientific issues by science museums.

Erminia Pedretti (PhD, MEd, BEd, BSc) is Professor of Science Education at the Ontario Institute for Studies in Education (OISE) at the University of Toronto. Her research and teaching interests include: science museums; controversy; and science, technology, society and environment (STSE) education. She has published 55 articles, 6 books, and 2 teacher education science textbooks, and received numerous Social Sciences and Humanities Research Council of Canada (SSHRC) grants. Her recent grant explored science museums and controversy with a particular interest in science communication models, visitor engagement, and innovative exhibition practices—resulting in a co-authored book (with Navas Iannini) entitled "Controversy in Science Museums: Re-imagining Exhibition Spaces and Practice" (2020).

Chapter 9
Political Demands in Brazilian Science Education Scholarship

Flavia Rezende, Andreia Guerra, and Fernanda Ostermann

9.1 Introduction

In the last two decades, Brazil has experienced significant political changes. In 2002, Luiz Inácio Lula da Silva, from the Workers' Party, was elected president, representing the first center-left government in Brazil. In 2003, the government carried out reforms that responded to neoliberal aspirations, such as the pension reform. At the same time, the government implemented public policies that took Brazil off the hunger map and increased funding for education and scientific institutions. As a result, Lula succeeded with the election of Dilma Rousseff as president in 2010, which made possible the articulation of hegemony in the same direction.

This political orientation was strongly strained in 2013 by the phenomenon of popular demonstrations (also called Marches of June) against increases in public transport fares and its poor quality. Political analysis considers this movement as a cycle of protests composed by distinct and sometimes opposing agendas (Alonso, 2023), from better living conditions to the end of corruption, the privatization of companies and of education systems.

In 2014, Dilma Roussef assumed the presidency for her second term. In March 2016, the National Congress, with the support of a portion of the Brazilian population, decides for her impeachment. Then, the center-right Vice-President Michel Temer took over until 2018, followed by four years of the far-right Bolsonaro's government. Concomitantly with the political insecurity, the coronavirus epidemic in our country reached large proportions in 2020, due to the disregard for the acquisition and distribution of vaccines by the president in office. In 2022, the traversing

F. Rezende (✉) · F. Ostermann
Universidade Federal do Rio Grande do Sul, Porto Alegre, Brazil

A. Guerra
Centro Federal de Educação Tecnológica Celso Suckow da Fonseca, Rio de Janeiro, Brazil

C. B. Moura (ed.), *A Sociopolitical Turn in Science Education*, Contemporary Trends and Issues in Science Education 62,
https://doi.org/10.1007/978-3-031-78586-3_9

power arc from the center-left to the far right ends up, with the election of Luiz Inácio Lula da Silva.

Brazilian recent history, where a denialist government was behind the death of 700,000 people from COVID-19, made us more aware of the relevance of promoting political reflection in education, and in our case, in science education. In this sense, the theoretical conception of the social and articulatory practices given by the theory of discourse (Laclau & Mouffe, 1987) seemed adequate to investigate how Brazilian science education has been leading with this issue. This approach made it possible to elect as the objective of the present study to identify and understand contingent discursive formations, which have been articulated and gained adherence in science education scholarship due to political events which occurred over the long period from 2005 to 2022. Special attention will be given to research produced from 2021 on, which could address discourses on the pandemic and on COVID-19 vaccination in the context of the disregard for the health management of the population by the extreme right-wing government.

9.2 Theoretical Framework: The Social, Hegemony and Articulatory Practices

The post-structural perspectives, as the proposal by Laclau and Mouffe (1987) that we will try to summarize in this section, distances themselves from structuralism and critical perspectives inspired by Marxism, being a possibility of supplanting them, concerning the fixed structures such as social classes and economic or institutional processes such as class struggles.

In the theoretical construction of Laclau and Mouffe (1987), society thought of as a totality that delimits an exteriority with other configurations is abandoned, as there is no single underlying principle that makes homogeneous the set of differences that are part of this totality. Therefore, a constant tension between interiority/exteriority appears, a condition of all social practice, since there is no possibility of total interiority or exteriority. Giving emphasis to discursive processes, the post-critical framework ends up making the essentialist approach to social relations impossible, which reaffirms the precarious nature of identities.

In this scenario, Laclau and Mouffe (1987) give a fundamental role to the concept of articulation, which they understand as "every practice that establishes a relationship between elements such that their identity is modified as a result of this practice" (p. 176). The articulatory practice produces a structured totality—which the authors call *discourse*—which, however, is never fully realized. The discourse, or discursive formation, a concept brought from Foucault and revised by the authors, presents a regularity in dispersion and can be thought of as a set of differential positions. In this type of articulated totality, every identity is relational and subject positions appear dispersed.

The practice of articulation as a fixation of a system of differences is not seen only as a linguistic phenomenon, but as a practice that "must cross the entire material thickness of institutions, rituals, practices of different orders, through which a discursive formation is structured" (Laclau & Mouffe, 1987, p. 185), even though its relational logic is incomplete and permeated by contingency, a terrain where social identities never manage to fully constitute themselves. Neither absolute fixation of identity nor absolute non-fixation is possible, but there are partial fixations around privileged discursive points that the author calls nodal points. In a system with these contours, the elements of discourse are fluctuating signifiers and this fluctuating character penetrates every social identity. Considering the incomplete character of all discursive formation, the ambiguity of the signifier, and its non-fixation to any meaning, the authors understand that all these characteristics can only exist insofar as there is polysemy, which disarticulates the discursive structure. The practice of articulation consists, therefore, in the construction of nodal points that partially fix the meaning; and the partial character of this fixation comes from the opening of the social, resulting in turn from the constant overflow of all discourse (Laclau & Mouffe, 1987, p. 193).

At the same time that closure is articulated and a hegemonic discourse is produced, a relationship of antagonism emerges as a denial of a certain order, and that limits that order only and not a broader totality. However, the limit of the social does not delimit a border signaling something objective and positive that would constitute a new difference, but as something that subverts it, that destroys its aspiration to constitute itself as a full presence. The hegemonic struggle then seeks to "provisionally fix and universalize a particular meaning and simultaneously produce its antagonistic other, expelling it from the hegemonic chain" (Gabriel, 2016, p. 113). The subversion of hegemonic discursive practices takes place in frontier shifts through the production of other universals and antagonisms created by other discursive articulations. It would be possible to consider as an example, in science education scholarship, the articulation of the significant "science", universals such as "technology" and "society" (Aikenhead, 1994; Auler & Bazzo, 2001) to create a "new" curricular approach that would be antagonistic to the hegemonic science curriculum.

As Laclau and Mouffe (1987) point out, the concept of hegemony presupposes the social as an open space, where articulatory practices occur. The authors more specifically define this space as the general field of discursiveness, the same plane where both hegemonizing and hegemonized elements are constituted, which would correspond to different discursive formations. Such formations are not fully constituted, as the exteriority between these formations takes place "between subject positions located within certain discursive formations and elements that lack a precise discursive articulation" (Laclau & Mouffe, 1987, p. 130).

In this theoretical framework, discursive practice involves the articulation of fluctuating signifiers and necessarily takes place through a confrontation of antagonistic articulatory practices whose borders are unstable. The hegemonic relationship resulting from a confrontation can be considered as a type of political relationship, which may constitute, together with other hegemonic nodal points, a social formation. However, no hegemony can constitute the totality of the social at the expense of its own elimination.

With regard to the subject, if the fixed identity is put in check, any educational project becomes impossible, since it is not possible to establish an intersubjective relationship with the other aiming at controlling his identification process. A critical project, for example, that aims at an emancipated and conscious subject is thus destabilized. In the educational context, the educative possibilities are contingent and contextual in a game of unpredictability, uncertainty, and interpretation. However, a post-critical reading of teacher education, as well as the curriculum in general terms and science curriculum in particular, does not promote demobilization; on the contrary, it results in a process of hyperpoliticization, which implies commitment and responsibility for political and contextual actions. The space for political struggle is expanded, since "the other is not the one who holds reason, but who temporarily and contextually holds power" (Lopes, 2013, p. 20).

We propose the following research questions: How are hegemonic discourses constructed in Brazilian science education scholarship? What hegemonic discourses and possible discursive inflections can be perceived due to political and health demands generated in the historical scenario described in the Introduction?

9.3 Methodological Aspects

9.3.1 Outlining the Object of Study

To investigate Brazilian science education scholarship, our first step was to delimit a set of representative journals. We decided, then, to choose the group of journals that have been publishing the most relevant research for the longest period in this field. This group of journals[1] included *Ensaio-Pesquisa em Educação em Ciências, Ciência & Educação, Revista Brasileira de Pesquisa em Educação em Ciências e Investigações em Educação em Ciências.*

We consider the beginning of the period of analysis to be the year 2005 and the end of 2022, understanding that the year 2013 is the middle of this period, as a period of political inflections. The first period, from 2005 to 2013, covers Lula's first two terms and the three years of Dilma's term, that is, a period governed by the center-left. The second period, from 2014 to 2022, is much more politically heterogeneous, as it covers two years of a center-left government, two years of a center-right, and four years of a far-right government.

---

[1]The journals' titles in English could be: Essay-Research in Science Education; Science & Education; Brazilian Journal of Research in Science Education; Investigations in Science Education.

The next step was to search for the signifier "política[2]" in all the articles[3] published in these journals, based on a search for the fragment "politic" in the title, abstract and/or keywords. It is clear that in recent years, the signifier "política" has gained more importance in Brazil, due to the exacerbation of the political polarization. We then proceeded to count the articles published in these two periods, in the selected journals (Table 9.1).

When we compare the data for each journal in the period 2005–2013, C&E is the one which shows a greater role. Regarding the period 2014–2022, the analysis made clear that C&E also presents the most the signifier "política" in the published articles. Considering both periods, C&E presents more articles dedicated to explicitly investigating and taking into account political aspects in their objects of study. This result led us to choose this journal to carry out the research.

Considering the two periods, it was possible to observe the leading role of the C&E in both of them. It is also possible to say that the other journals, whose number of occurrences was low in the first period, started to bring more articles that explicit political issues between 2014–2022, which already gives us a clue about a possible discursive inflection in Brazilian science education scholarship.

9.3.2 *Analysis Procedures*

From a preliminary analysis of the titles, abstracts, and keywords of the 57 articles published in C&E, we identified two groups: 33 articles focused on educational public policies and 24 articles related the signifier "política" to other signifiers.

The next step was to understand the role played by public policy in the group of 33 articles. After reading the abstracts, we divided this group into two subgroups: public policy as an object of study (20 articles) and public policy as an assumption (13 articles). In the subgroup "Public policy as an object of study" the researcher investigates the implementation of a certain public policy. In the subgroup "Public

Table 9.1 Number of occurrences of the fragment "politic" in articles, by journal

Journal	2005–2013	2014–2022
Ensaio–Pesquisa Educação em Ciência (Ensaio)	09	25
Ciência & Educação (C&E)	14	43
Revista Brasileira de Pesquisa em Educação em Ciências (RBPEC)	03	27
Investigações em Educação em Ciências (IEC)	0	38

[2] "Política" is the Portuguese word for "politics".

[3] We do not mean that the absence of this signifier in the article would characterize political neutrality or an apolitical position. We suppose it could be representative of political positions in many different ways, but perhaps more implicit and, therefore, more complex to analyze.

Table 9.2 Articles classified in each category in the three periods

Period	Public policy		"Política"	
	As an object of study	As an assumption	Theoretically constitutive	No relevance
2005–2013	06	03	0	05
2014–2018	10	06	04	05
2019–2022	04	04	06	04

policy as an assumption", the public policy is not investigated, it is taken as a fact or as the research framework. These 13 articles were not considered in the analysis.

In order to interpret the relationship of the signifier "política" with other signifiers in the second group of 24 articles, we analyzed its function in the discourse. We identified a subgroup (10 articles) in which the signifier "política" is theoretically constitutive. In this subgroup, the political dimensions of science education play a central role. In the other subgroup (14 articles) the signifier "política" has no relevance (as in the excerpt "...various contexts of learning and training: social, cultural, *political*, economic…"). These 14 articles were not considered in the analysis.

Therefore, we performed the analysis of 30 articles, distributed in the columns "Public Policy" and "Theoretically constitutive" in Table 9.2. In the period 2005–2013, 06 articles were left to be analyzed. The period 2014–2022 was subdivided into two sets of articles. In the period of 2014–2018, 14 articles were analyzed. In the period of 2019–2022, 10 articles were analyzed. In this period, we also intended to investigate possible relationships between the signifier "política" and the political conditions of the far-right government and tried to map the authors' position on the political context and on the COVID-19 pandemic.

In the following analytical sections, we chose to highlight the categories Public Policy as an object of study and Política as theoretically constitutive because we believe their corresponding articles represent a more relevant contribution to science education scholarship in the sense that theory brings a significant basis to the analysis.

9.4 Political Demands in Brazilian Science Education Scholarship

In this section, we try to summarize, for each period, the articles classified in the categories "Public Policy as an object of study" and "Política as theoretically constitutive", trying to highlight the political demands they express or respond to.

9.4.1 Politics in the Articles from the Period 2005–2013

The analysis of the articles published in this period shows that most of them take a critical perspective to investigate public policies, as for example, curricular policies created before the center-left government.

Of the six articles that take public policy as an object of study, three of them (Lopes, 2005; Busnardo & Lopes, 2010; Carvalho & Rezende, 2013) critically analyze Brazilian curriculum policies. Tozoni-Reis et al. (2013) seek to understand why Environment Education is not present in schools, and based on a Marxist framework, they blame the curricular policies structured by neoliberalism and neoproductivism.

Lopes (2005), Busnardo and Lopes (2010), and Carvalho and Rezende (2013) do not see curricular policies as a fixed structure and determinant of pedagogical practices, either because the very production of the law already hybridizes discourses of both disciplinary groups and government instances or because the processes of recontextualization of the curriculum policy invariably will occur at school. Busnardo and Lopes (2010) go further, assuming curriculum policies as productions of multiple contexts. Supported by Laclau (2005), they analyze the process of hegemonization of meanings in these policies.

Júnior and Sorrentino (2011) analyze the discontinuity of the environmental education policies resulting from the political context and point out bases for their resumption from a Marxist perspective. Based on Bourdieu and Vygotsky, Rezende and Ostermann (2012) analyze the Brazilian Science Olympics as a government initiative that generates unfair competition among culturally pre-selected students.

9.4.2 Politics in the Articles from the Period 2014–2018

Critical studies on public policies continued to play a leading role in this period. Some articles on Environmental Education bring the signifier "política" as a theoretical constitutive but only at the end of the period, we found an article on science education (Watanabe & Kawamura, 2017) that takes "política" in the same way.

Of the 10 articles that take public policy as an object of study, three of them (Ramos & Petrucci-Rosa, 2014; Petrucci-Rosa, 2014; Ferraz & Rezende, 2014) take the National Curricular Parameters for High School as an object of study, in dialogue with the curriculum field, to investigate how curriculum policies are transformed throughout practices in the context of science teaching. Vilanova (2015) uses Fairclough's critical discourse analysis to investigate the meanings that citizenship education acquires in the PNLD.[4] Two articles focus on teacher training policies that aim to solve content "deficits" for practicing teachers. Vieira and Santos (2015) discuss, from a critical perspective, public policies for science and mathe-

[4] National Program of Distribution of Books.

matics teachers' pedagogical training. Rezende and Ostermann (2015) undertake a reflection based on a Marxist framework on the role of professional master's degrees in the pursuit of quality in Brazilian basic education. Supported by socio-historical studies in the field of curriculum, Valla et al. (2014) analyze, based on Goodson's work, curriculum innovations in the discipline of science from 1950 to 1970 in Brazil; and two articles focus on Environment Education: Hofstatter et al. (2016) focus on critical Environment Education and Andrade et al. (2014) propose a model for building Environment Education public policies based on international documents on environmental public policies. Trittini and Lopes[5] (Trittini & Lopes, 2016) take Chilean curriculum policies as an object of study, analyzing this policy in a Laclaunian framework as a process of construction of a hegemonic discourse.

Of the four articles that take the signifier "política" as constitutive of the discourse, three investigate Environmental Education. Two of them take dialectical historical materialism as a framework. Costa and Loureiro (2015) discuss the relationship between interdisciplinarity and critical environment education, from a political-epistemological point of view. Understanding the environmental issue as an interdisciplinary problem, they defend that it should be framed from critical, historical, and complex knowledge. Peneluce et al. (2018) consider that it is possible to develop an emancipatory act with relevant heuristic potential in unveiling the ideological and political-ecological aspects of the socio-environmental crisis by adding dialectical historical materialism to Environmental Education. Barchi (2016) takes Foucault's concept of governmentality to discuss the process of institutionalization of Environmental Education. Watanabe and Kawamura (2017) discuss the role of science popularization in the scientific field based on Bourdieu's sociology.

9.4.3 Politics in the Articles from the Period 2019–2022

In this period, it was possible to observe that the number of articles in which the signifier "política" is theoretically constitutive of the discourse increased. On the other hand, articles in which the signifier "política" refers to public policy as an object of study decreased.

Of the four articles that take public policies as an object of study, Calado and Petrucci-Rosa (2019) analyze the process of curricular restructuring of the degree in Physics of a state public university, seeking to understand the refraction of curricular policies in pedagogical practice. Bandeira and Velozo (2019) and Silva and Loureiro (2020) analyze public policies that were approved after the right-wing Brazilian government took power. Bandeira and Velozo (2019) criticize the curricular policies and the changes made in the PNLD due to the conservative agenda that sought to silence expressions such as "sexual orientation" and "gender identity".

[5] Although this article analyzes a Chilean curriculum policy, it was included because the second author is Brazilian and was the research advisor.

Silva and Loureiro (2020) investigated the voices of teachers in the field of Environmental Education about the BNCC[6] and identified changes that refer to superficiality and reduction of critical content, precariousness, privatization of education, and training aimed at the labor market. Costa and Carmo (2022) analyze the state curriculum document that aims to implement the BNCC approved in 2017, identifying the continuities and ruptures in relation to conservatism and neoliberal aspects present in the legislation.

Of the six articles in which the signifier "política" is constitutive of the discourse, Nascimento et al. (2019) take Bourdieu's theory to discuss how science teaching can contribute to reducing the distance between political culture and school culture. Santos and Auler (2019) defend a Latin American Thought on Science, Technology, and Society to criticize and problematize educational practices that reduce the Society-Technology-Society (STS) perspective to assess the impacts of science and technology on society. De-Carvalho (2021) analyzes examples of neoliberal discourses, in the context of the COVID-19 pandemic, which crossed basic education and which configured an unofficial policy of scientific and social education, opposing the far-right government. This is the first article from all the analyzed periods in which the signifier "política" is related to discussions on gender, class, and race and supported by decolonial references. Lima and Nascimento (2021) present a political-epistemological proposal for science education based on Bruno Latour's symmetrical sociology, referring to the Brazilian political context of the far-right government and the health crisis caused by COVID-19. Souza (2021) presents criticism of government educational projects in the political context of precariousness and devaluation of education workers and political control of teaching by the emerging neoconservative movement in Brazil during the rise of the far-right government. Henning and Ferraro (2022) use Foucault's concept of governmentality to problematize the configuration of Environmental Education aimed at sustainable development.

9.5 Hegemonic Discourses in Brazilian Science Education Scholarship

To analyze the (re)construction processes of hegemonic discourses in Brazilian science education scholarship, we selected two articles from each of the three periods analyzed: one that investigated public policies and another in which the signifier "política" was assumed as theoretically constitutive. Only one article from the period 2005–2013 was selected because there was no other that considered the signifier "política" as theoretical constitutive in this group.

Inspired by Laclau's characteristics of the discourse, we suppose the authors build discourses as in a political process, seeking to respond to a set of demands in

[6] Curricular Common National Basis.

a chain of equivalence around nodal points. We tried to observe the articulations that produced a structured totality (the discourse) in the articles, giving rise to antagonistic relationships.

In the context of an empirical study that aimed to investigate the implementation of Environmental Education, Tozoni-Reis et al. (2013) conclude that the materials used were inappropriate and insufficient, as they did not favor reflection on practice in a broader dimension. The authors state that in the Brazilian educational curriculum parameters of the 1990s, the role of teachers was to develop competencies and skills in order to prepare students to achieve goals established by the assessment systems. In this way, the role of teachers moves away from the role of mediator in the process of human formation. The articulation of demands—for example, the function of Environmental Education as a space of application of activities that guarantee a good performance of skills and competencies—is organized by its antagonism, which considers these activities an impediment to the achievement of good educational indexes that allow international investments in Brazil.

Rezende and Ostermann (2015) undertake a Marxist framework to reflect on the role of professional master's degrees in science teaching in the pursuit of science education quality confronting it to the conditions of secondary education in Brazil. The authors identify multiple demands that are articulated in the constitution of this teacher training policy. Demands such as the best performance of students in national and international exams, a less "theoretical" teacher training closer to school and social reality, and prioritization of continuing training in relation to initial training in order to respond to the preparation of people for a globalized market. All these articulated demands are organized by the antagonism of the lack of quality teacher training. Because teachers are considered as the result of the lack of quality training, it is possible to postulate a solution represented by the professional master's courses for teacher training.

Based on Bourdieu's sociology, Watanabe and Kawamura (2017) analyze the views of a group of Brazilian and European physicists on how the popularization of their research results is understood in the scientific field. The authors argue that disseminator-scientists could have a scientific capital valued in the social game of science, taking the distance from scientists towards the obligations of disseminators as antagonists. Scientists' perception of this activity as a social effort and not as a part of the actions played in the social game of science would be considered as a "negative" capital. However, the authors defend the practice of science popularization as a scientific activity in opposition to the perception of scientific popularization as merely a voluntary activity.

Bandeira and Velozo (2019) build a favorable discourse on textbooks, pointing out several demands that these artifacts could satisfy and focusing on the historical and cultural approach to gender and sexuality relations. Among these demands, the authors articulate discourses on the dichotomy of gender roles, where women's roles are always inferior, including their role in scientific achievements. The authors point out that the demand to discuss cultural and social issues that permeate the human body—including gender and sexuality relations—in textbooks is denied through the biologizing organization of science curriculum. The authors defend positions that liberate gender and sexuality relationships in opposition to the ones that oppress them. They point out that while the 2017's version of the BNCC promotes the discussion of multiple (biological, sociocultural, affective, and ethical) dimensions of human sexuality, a group of experts in the government takes an antagonist discursive maneuver, under the influence of the *Escola sem Partido,*[7] that brings "gender ideology"—a fallacy created by heteronormative categories based on religious fundamentalism—to exclude the terms "sexual orientation" and "gender identity" from the 2018's version.

Henning and Ferraro (2022) criticize the introduction of Education for Sustainable Development (ESD) in Brazil and its repercussions in the field of Environmental Education. The authors point out that the ESD responds to demands from international institutions such as ONU and UNESCO, that aim to define ecologically correct behaviors, such as conscious or 'green' consumption, to guarantee economic and sustainable development. These demands are being articulated in antagonism to an ecologically unfeasible future while blocking reflection on the production process of the capitalist and neoliberal systems. Thus, the criticism of ESD is for taking the environment as a place of alliance between economy and politics, erasing society-nature relations. The authors identify a dispute in the Brazilian academic field in which the ESD intends to silence Environmental Education through the international proposal of sustainable development aligned with neoliberalism and to the standardization of sustainable practices. The authors defend the opposite path: a decolonized perception of sustainability provided by Environmental Education.

9.6 Discursive Inflections

Some discursive inflections have been noticed in the analyzed period. It was possible to identify a significant discursive inflection between the period of 2005–2013—when no articles employed the signifier "política" as a theoretical constitutive—and the period of 2014–2018, when there were four articles in which the signifier "política" was taken as a theoretical constitutive. However, this signifier was used to discuss public policy, which continues to be the priority object of research in this period. In 2015, we found the first article not related to public policies and that incorporates the signifier "política" as a theoretical constitutive. In this article, Costa and Loureiro (2015) claim that Environmental Education has to be discussed from a critical, historical, and political stance and should be related to global environmental issues.

Even though the political climate has been turbulent in Brazil since 2018, it has not impacted the articles published in 2019: of the eight articles published in that year, only one mentions the Brazilian political context. The only article published in

[7] "Escola sem Partido" is a political movement that aims to advance a conservative agenda for Brazilian education, articulated by far-right politicians.

2020 neither mentions the political context. The only three articles that consider politics as theoretical constitutive and refer to the period of the far-right government in Brazil were published in 2021. The analysis of the articles shows little impact of the political reality over Brazilian science education scholarship.

9.7 Concluding Remarks

Taking the analysis as a whole, we can figure out discursive inflections that correspond to Brazilian political demands. In the first period of political stability (2005–2013), the center-left government implemented several education policies which, taken together with increased investment in the sector, created a context of acceptance of government actions by a large portion of Brazilian society. Although most researchers take these public policies as objects of study, questioning them, the signifier "política" is not taken as a theoretical constitutive, whether related to theoretical frameworks or to the political context. This discursive formation, which silences politics could be related to political stability. In the following periods the signifier "política" becomes relevant, even in a short number of articles.

This depoliticization is also revealed by the majority of research papers that reduce political concerns related to science education to current educational public policies, either taking them as object of study or as theoretical framework. However, we recognize that among the articles that deal with public policies as an object of study, there are critical perspectives that contribute to the reflection on the studied policies.

The analysis as a whole reveals the low concern of Brazilian science education scholarship with political and social aspects of reality. This depoliticization is also evident in the absence of articles that discuss the role of science education in facing contemporary societal challenges. In this period, only De-Carvalho (2021) and Lima and Nascimento (2021) refer to the political context and to the COVID-19 pandemic. De-Carvalho (2021) inserts his research in the context of the COVID-19 pandemic and the far-right government. Lima and Nascimento (2021) defend a science education proposal referring to the Brazilian political context and to the health crisis caused by COVID-19.

The low political engagement by Brazilian science education scholarship also became clear when we came across the four articles published in 2022, in which the signifier "política" had no relevance. More articles could have addressed the impact caused by the terrible COVID-2019 pandemic and its consequences, amplified by the lack of health care guidelines and emergency health policies in Bolsonaro's government. In this case, the depolitization of science education scholarship is alarming, considering that the pandemic resulted in the death of 700,000 people in Brazil.

The depolitization of science education is also perceived elsewhere (Tolbert & Bazzul, 2017; Morales-Doyle, 2019). One hypothesis to understand the depoliticization of science education may be its visceral relationship with natural sciences, which are intended to be objective and neutral. This position might also contribute to the depolitization of science education, and therefore, to political actions that has mostly served the interests of neoliberal policies.

The post-pandemic scenario, the tragedies caused by climate change, and threats of the far-right movement in the world led us to strengthen the urgency of bringing politics to the center of science education. We argue in favor of a science education that understands politics not as part of a structured society that aims at certain ends—perception shared by most of the papers analyzed –, but as a contingent process in which acts "undecidibility and decision" (Mouffe, 2017, p 11). In this framework, policy conceived as an intervention project aimed at a common goal in all contexts is eliminated. There is no possibility of a planned and defined action that could represent the end of politics. However, political action as a struggle for the meaning of the different contextual projects carried out by each of us becomes even more important and "it is able to bring a possibility of hope" (Lopes, 2013, p. 21). Thus, our singular, perhaps unique, way of characterizing these antagonistic projects represents opposite pathways that science education might assume in particular contexts in the arena of constant and endless political struggle.

References

Aikenhead, G. (1994). What is STS science teaching? In J. Solomon & G. Aikenhead (Eds.), *STS education: International perspectives on reform*. Teachers College Press.

Alonso, A. (2023). *Treze: a política de rua de Lula à Dilma*. Companhia das Letras.

Andrade, D. F., Luca, A. Q., Castellano, M., Rissato, C. G., & Sorrentino, M. (2014). Da pedagogia à política e da política à pedagogia: uma abordagem sobre a construção de políticas públicas em educação ambiental no Brasil. *Ciência & Educação, 20*(4), 817–832. https://doi.org/10.1590/1516-73132014000400004

Auler, D., & Bazzo, W. A. (2001). Reflexões para a implementação do Movimento CTS no contexto educacional Brasileiro. *Ciência & Educação, 7*(1), 1–13.

Bandeira, A., & Velozo, E. L. (2019). Livro didático como artefato cultural: possibilidades para as abordagens das relações de gênero e sexualidade no Ensino de Ciências. *Ciência & Educação, 25*(4), 1019–1033. https://doi.org/10.1590/1516-731320190040011

Barchi, R. (2016). Educação ambiental e (eco)governamentalidade. *Ciência & Educação, 22*(3), 635–650. https://doi.org/10.1590/1516-731320160030006

Busnardo, F., & Lopes, A. C. (2010). Os discursos da comunidade disciplinar de Ensino de Biologia: circulação em múltiplos contextos. *Ciência & Educação, 16*(1), 87–102. https://doi.org/10.1590/1516-731320190020015

Calado, H. C., & Petrucci-Rosa, M. I. (2019). Formação de professores de Física e interdisciplinaridade: episódios de refração de políticas em narrativas de reforma curricular. *Ciência & Educação, 25*(2), 523–538. https://doi.org/10.1590/1516-731320190020015

Carvalho, R. C., & Rezende, F. (2013). Políticas curriculares e qualidade do ensino de ciências no discurso pedagógico de professores de nível médio. *Ciência & Educação, 19*(3), 555–571. https://doi.org/10.1590/S1516-73132013000300005

Costa, J. G. M., & Carmo, E. M. (2022). Produção do Currículo Bahia e a disciplina escolar Ciências: uma análise centrada nos temas integradores. *Ciência & Educação, 28*, e22057. https://doi.org/10.1590/1516-731320150030011

Costa, C. A. S., & Loureiro, C. F. B. (2015). Interdisciplinaridade e educação ambiental crítica: questões epistemológicas a partir do materialismo histórico-dialético. *Ciência & Educação, 21*(3), 693–708. https://doi.org/10.1590/1516-731320150030011

De-Carvalho, R. (2021). Discursos para Reexistência no Ensino de Ciências: de Baco Exu do Blues a Malcolm X. *Ciência & Educação, 27*, e21031. https://doi.org/10.1590/1516-731320210031

Ferraz, G., & Rezende, F. (2014). Perspectivas de professores de Física sobre as políticas curriculares nacionais para o Ensino Médio. *Ciência & Educação, 20*(2), 497–515. https://doi.org/10.1590/1516-73132014000200015

Gabriel, C. T. (2016). Conhecimento escolar e emancipação: uma leitura pós-fundacional. *Cadernos de Pesquisa, 46*, 104–130. https://doi.org/10.1590/198053143551

Henning, P. C., & Ferraro, J. L. (2022). As lutas políticas da Educação Ambiental nas universidades brasileiras: provocações à governamentalidade neoliberal no âmbito da educação para o desenvolvimento sustentável. *Ciência & Educação, 28*, e22028. https://doi.org/10.1590/1516-731320220028

Hofstatter, L. J. V., Oliveira, H. T., & Souto, F. J. B. (2016). Uma contribuição da educação ambiental crítica para (des)construção do olhar sobre a seca no semiárido baiano. *Ciência & Educação, 22*(3), 615–633. https://doi.org/10.1590/1516-731320160030005

Júnior, L. A. F., & Sorrentino, M. (2011). Imaginário político e colonialidade: desafios à avaliação qualitativa das políticas públicas de Educação Ambiental. *Ciência & Educação, 17*(2), 339–352. https://doi.org/10.1590/S1516-73132011000200006

Laclau, E. (2005). *La razón populista*. Fondo de Cultura Económica.

Laclau, E., & Mouffe, C. (1987). *Hegemonía y estrategia socialista: hacia una radicalización de la democracia*. Siglo XXI.

Lima, N. W., & Nascimento, M. M. (2021). Aterrando no Sul: uma proposta político-epistemológica para a área de educação em ciências do Antropoceno. *Ciência & Educação, 27*, e21041. https://doi.org/10.1590/1516-731320210041

Lopes, A. C. (2005). Discursos curriculares na disciplina escolar Química. *Ciência & Educação, 11*(2), 263–278.

Lopes, A. C. (2013). Teorias pós-críticas, política e currículo. *Educação, Sociedade & Culturas, 39*, 7–23.

Morales-Doyle, D. (2019). There is no equity in a vacuum: On the importance of historical, political, and moral considerations in science education. *Cultural Studies of Science Education, 14*, 485–491. https://doi.org/10.1007/s11422-019-09925-y

Mouffe, C. (2017). Desconstrução, pragmatismo e a política da democracia. In Chantal Mouffe (Org.) *Desconstrução e Pragmatismo*. Mauad Editora Ltda.

Nascimento, M. M., Lima, N. W., Cavalcanti, C. J. H., & Ostermann, F. (2019). Cultura política, desempenho escolar e Educação em Ciências: um estudo empírico à luz de Pierre Bourdieu. *Ciência & Educação, 25*(2), 431–447. https://doi.org/10.1590/1516-731320190020010

Peneluce, M. C., Pinheiro, B. C. S., & Moradillo, E. F. (2018). Possíveis confluências filosóficas e pedagógicas entre a educação ambiental crítica e a Pedagogia Histórico-Crítica. *Ciência & Educação, 24*(1), 157–173. https://doi.org/10.1590/1516-731320180010011

Petrucci-Rosa, M. I. (2014). Políticas curriculares e identidades docentes disciplinares: a área de ciências da natureza e matemática no currículo do ensino médio do estado de São Paulo (2008-2011). *Ciência & Educação, 20*(4), 937–953. https://doi.org/10.1590/1516-73132014000400011

Rezende, F., & Ostermann, F. (2012). Olimpíadas de ciências: uma prática em questão. *Ciência & Educação, 18*(1), 245–256. https://doi.org/10.1590/S1516-73132012000100015

Rezende, F., & Ostermann, F. (2015). O protagonismo controverso dos mestrados profissionais em ensino de ciências. *Ciência & Educação, 21*(3), 543–558. https://doi.org/10.1590/1516-731320150030002

Santos, R. A., & Auler, D. (2019). Práticas educativas CTS: busca de uma participação social para além da avaliação de impactos da Ciência-Tecnologia. *Ciência & Educação, 25*(2), 485–503. https://doi.org/10.1590/1516-731320190020013

Silva, S. N., & Loureiro, C. F. B. (2020). As Vozes de Professores-Pesquisadores do Campo da Educação Ambiental sobre a Base Nacional Comum Curricular (BNCC): Educação Infantil ao Ensino Fundamental. *Ciência & Educação, 26*, e20004. https://doi.org/10.1590/1516-731320200004

Souza, R. D. (2021). O trabalho dos professores de Biologia: a teorização a partir das contribuições dos pesquisadores da área de ensino de Ciências e Biologia. *Ciência & Educação, 27*, e21054. https://doi.org/10.1590/1516-731320210054

Tolbert, S., & Bazzul, J. (2017). Toward the sociopolitical in science education. *Cultural Studies of Science Education, 12*, 321–330. https://doi.org/10.1007/s11422-016-9737-5

Tozoni-Reis, M. F. C., Talamoni, J. L. B., Ruiz, S. S., Neves, J. P., Teixeira, L. A., Cassini, L. F., Festozo, M. B., Janke, N., Maia, J. S. S., Santos, H. M. S., Cruz, L. G., & Munhoz, R. H. (2013). A inserção da Educação Ambiental na educação básica: que fontes de informação os professores utilizam para sua formação? *Ciência & Educação, 19*(2), 359–377.

Trittini, S. A. C., & Lopes, A. C. (2016). Discurso cientificista nas políticas de currículo no Chile (2003–2013): o foco no Programa ECBI – Enseñanza de las Ciências basada en la Indagación. *Ciência & Educação, 22*(2), 279–298. https://doi.org/10.1590/1516-731320160020002

Valla, D. F., Roquette, D. A. G., Gomes, M. M., & Ferreira, M. S. (2014). Disciplina escolar Ciências: inovações curriculares nos anos 1950-1970. *Ciência & Educação, 20*(2), 377–391. https://doi.org/10.1590/1516-73132014000200008

Vieira, K. L. A. S., & Santos, S. S. (2015). Políticas públicas para formação de professores de Ciências e Matemática: complementação pedagógica para bacharéis e tecnólogos. *Ciência & Educação, 21*(3), 575–584. https://doi.org/10.1590/1516-731320150030004

Vilanova, R. (2015). Educação em ciências e cidadania: Mudança discursiva e modos de regulação na política do Programa Nacional do Livro Didático. *Ciência & Educação, 21*(1), 177–187. https://doi.org/10.1590/1516-731320150010012.

Watanabe, G., & Kawamura, R. (2017). A divulgação científica e os físicos de partículas: a construção social de sentidos e objetivos. *Ciência & Educação, 23*(2), 303–320. https://doi.org/10.1590/1516-731320170020002

Flavia Rezende is a professor at the Physics Education Research Program of Universidade Federal do Rio Grande do Sul, Porto Alegre, Brazil, and a researcher of CNPq (National Council for Scientific and Technological Development). Her research interests focus on curricular theories as underpinning to investigate science education practice and scholarship

Andreia Guerra is a professor at CEFET/RJ, Brazil, and coordinates the Research Group on Teaching History of Science and Culture (NIEHCC). Granted as CNPq (Brazilian Research Council) associate researcher, her research interest includes the History of Science for Science Education, Science Studies, and post-colonial theory. She is the past President of IHPST Group (2019–2022) and co-editor of a Brazilian journal on physics teaching (Caderno Brasileiro de Ensino de Física).

Fernanda Ostermann is a professor at the Physics Education Research Program of Universidade Federal do Rio Grande do Sul, Porto Alegre, Brazil, and is a researcher of CNPq (National Council for Scientific and Technological Development). Her research activity covers sociocultural studies of physics teacher education and high school physics.

Chapter 10
Decolonizing Natural Sciences, Environmental and Health Education: Reflections from a Continuous Teacher Training Experience

Ana G. Dumrauf, Silvina Cordero, Verónica Becerro, Luciano Iribarren, and Raúl Esteban Ithuralde

10.1 Introduction

Crises can lead to various outcomes. Their causes, the meanings assigned to new practices that emerge or become widespread, and the reasons behind these changes are all subjects of dispute. Crises inevitably bring about change, which can either lead towards emancipatory social transformation or increased exclusion. These changes involve greater organization and autonomy for popular movements or, conversely, stricter social discipline, which can even worsen the conditions that led to the crisis in the first place. In order to walk along transformative paths, it is crucial to analyze the past to build a better future. This action must inhabit the exception and involve engaging with the current situation through our questions, thoughts, needs, desires, knowledge, and support networks.

A. G. Dumrauf (✉) · V. Becerro
Universidad Nacional de La Plata & CONICET, Buenos Aires, Argentina
e-mail: adumrauf@iflysib.unlp.edu.ar; vbecerro@iflysib.unlp.edu.ar

S. Cordero
Universidad Nacional de La Plata, Buenos Aires, Argentina

L. Iribarren
Dirección General de Cultura y Educación de la Provincia de Buenos Aires, Buenos Aires, Argentina

R. E. Ithuralde
Universidad de Buenos Aires, Universidad Nacional de Santiago del Estero, Universidad Nacional Arturo Jauretche, CONICET & Dirección General de Cultura y Educación de la Provincia de Buenos Aires, Buenos Aires, Argentina
e-mail: esteban@ccpems.exactas.uba.ar

C. B. Moura (ed.), *A Sociopolitical Turn in Science Education*, Contemporary Trends and Issues in Science Education 62,
https://doi.org/10.1007/978-3-031-78586-3_10

In 2020, the educational field grappled with how to address and transform our current life system. This system creates conditions for the appearance of pandemics, generates desensitization to others' suffering, reifies people in various ways, and promotes the predation of territories (Dumrauf, 2020). We recognized that the COVID-19 pandemic was a symptom of a broader civilizational crisis caused by modern rationality's exploitative relationship with nature (Garelli & Dumrauf, 2021).

At the beginning of 2021, Grupo de Didáctica de las Ciencias (GDC), "Sciences Didactics Group", accepted an invitation from the National Institute of Teacher Education and faced the challenge of designing an in-service virtual teacher education course suitable for the complex pandemic conditions. This experience took place from May to November 2021. Our goal was collaborating with thousands of primary school teachers in Argentina seeking Natural Sciences Teacher Education amidst the global socio-sanitary crisis. Despite the critical pandemic context of virtual social bonds, this collective endeavour intended to decolonise and emancipate. To this end, we drew on the theoretical and methodological frameworks and formative experiences we have built to pursue a Natural Sciences, Environmental, and Health Education (NSEHE) grounded in Latin American Popular Education.

In this chapter, we share reflections from our ongoing experiential and theoretical journey, nourished by the GDC's multiple work experiences. We present insights and lessons learned during and after the training course. Finally, by incorporating feedback from students, tutors, and material authors, we explore how some political-pedagogical perspectives from the course were resignified, adapted and integrated into post-pandemic practices. We also examine what factors and elements strengthen the development and implementation of decolonizing proposals in NSEHE. We hope to contribute to the field of NSEHE and invite more teachers to join similar journeys.

10.2 Theoretical-Methodological Framework: Towards the Decolonization of Natural Sciences, Environmental, and Health Education

To reflect in a few paragraphs on the 25 years of collective paths of theoretical and methodological work that we have undertaken as GDC is challenging. To present the perspectives, experiences, and reflections shaping our approach, we will focus on two main areas: initial and ongoing teacher education, socio-territorial environmental and health problems, and conflicts involving peasant and popular movements. Our participatory approach aims to systematize experiences with teachers and social movement members, often integrating both lines of work or creating new ones.

Our journey began in the late 1990s in the field of Natural Sciences Education, which we now see as insufficiently politicized, overly disciplinary, Eurocentric, racialized, and sexist (Hodson, 2010; Bazzul & Sykes, 2011; Rosa, 2018; Dumrauf et al., 2019; Kato et al., 2023; Moura et al., 2023). In response to this context, we have been continually redefining our practices, drawing primarily from Popular Education based on Freirean principles and the lessons learned from collaborating with various academic and non-academic groups (Dumrauf et al., 2016; Dumrauf & Cordero, 2020; Iribarren et al., 2022; Ithuralde, 2020; Mengascini et al., 2024).

We have reconfigured our field of action based on experiential and theoretical fertility emerging from the needs, productions, and knowledge of these others with whom we engaged in a search for dialogic and horizontal construction of knowledge from diverse sources, in addition to scientific knowledge (Cordero & Dumrauf, 2022). In the past decade, we began defining our research area as NSEHE, blending contributions from often disconnected fields and aiming to democratize the relationships in producing academic knowledge.

Critical perspectives from Latin American Popular Education, significantly influenced by Paulo Freire (1993, 2002), have guided our development of political-pedagogical proposals for NSEHE, which aim to sustain a decolonizing and emancipatory intentionality (Dumrauf et al., 2019). We embraced the critical and problematizing view of reality and the proposal for transformative action in the world that Latin American Popular Education presents, foregrounding the political nature of education. Thus, in our Popular and Teacher Education activities, we usually start from the needs and actions of the different groups of participants, aiming to build personal autonomy and community organization for the transformation of reality through dialogic action. We have learned that dialogue must include academic knowledge, lived experiences, and emotions (Merçon et al., 2014). To foster this dialogue, we need to recognize power relationships among participants, pay attention to economic and socio-cultural diversity (generational, ethnic, gender, class, territorial, and so on), and examine the different forms of communication and expression.

This decision led us to explore the existing positions on interculturality in educational contexts (Cucalón Tirado et al., 2017). To address challenges in formal education within contexts of cultural diversity, social difference, and multiple inequalities and to advance in interculturality, we consider it essential to take a stand in the inherent conflicts of social relations. Thus, we align with critical and extended interculturality (Díaz & Rodríguez Anca, 2014), whose actions are not only directed towards marginalized or subalternized groups but mainly towards hegemonic, dominant groups. This perspective includes *white* and *mestizo* individuals as pedagogical subjects. It emphasizes the need to design public policies to deactivate the historical relations of asymmetry and subordination that these groups configured from the most crucial colonial event (the invasion of America) concerning other ethnic groups (Corbetta, 2021). In educational institutions, interculturalization requires cross-cutting different levels: institutional discourse, curriculum, didactic devices and classroom practices (Corbetta, 2021).

We also connect critical and extended interculturality with a decolonizing perspective on NSEHE to challenge the notion of Western knowledge and its claimed universality (Ithuralde et al., 2023). This perspective, which has spread globally since the late fifteenth century, still sustains a dominant position through the coloniality of power and knowledge. Within this framework, the dimensions of race and global ethnic-racial hierarchies permeate all existing social relations, including sexuality, gender, knowledge, class, labour division, epistemology, and spirituality (Grosfoguel, 2017). Rivera Cusicanqui (2018) describes internal colonialism as a structure, ethos and culture reproduced daily in their oppressions and silences, despite the successive attempts for radical transformation advocated by political/intellectual elites, whether in liberal, populist, or Indigenous/Marxistic versions. Coloniality, as both discourse and practice, preaches the natural inferiority of subjects as well as the colonization of nature, marking certain subjects as dispensable and nature as pure raw material for the production of exchangeable goods for the international market (Maldonado-Torres, 2007).

Decolonizing pedagogies, in contrast, embrace utopias and the dialectical relationship between theory and practice. They are based on pedagogical work from the global South, linked to the political sphere, and characterized by integrative and participatory approaches (Guelman & Palumbo, 2015). They appeal to legitimacy, dignity, equality, equity and respect for the self and encourage the creation of other ways of thinking, being, learning, teaching, dreaming, and living (Guelman, 2014). These approaches align with Latin American Environmental Thought in several ways:

1. A critique of Eurocentrism. Such criticism does not seek to deny European thought but rather to displace the Eurocentric view of the world and the way stories are told. This requires a narrative from our own pluralistic visions and versions of the world.
2. A conception of indissoluble linkages between rationality/modernity and coloniality.
3. A view of decolonization as the only way out and the relevance of (decolonized) education as a means to de-Eurocentrize or unravel internal colonialism (Corbetta, 2021).

Guided by the perspectives of Latin American Environmental Thought, critical and extended interculturality, and decolonization, along with an ethic of respect for Others and nature, we propose an education that fosters the construction of territories of life. We integrate NSEHE with feminist perspectives, focusing on political-epistemic recognition of social bonds, lived experiences, and situated relationships (Cabnal, 2010). To achieve this, we seek to create spaces for participation, listening, and addressing the specific needs of teachers and social organizations we work with. Practices of self-care, care for others, and care for territories shape our content and methodologies, aiming to nourish the construction of counter-pedagogies of cruelty that oppose patriarchal order and seek to break free from the hegemonic masculinity mandate. This patriarchal order is intertwined with capitalism and sustains various forms of exploitation and precariousness of life, regulated by a principle of cruelty that entails a decrease in subjects' empathy (Segato, 2018).

By promoting desensitisation to suffering, this order justifies the reification of Others and the predation of territories, as well as the formation of individuals devoid of emotions and feelings whose relationships are primarily based on interests of consumption and utility. In contrast, counter-pedagogies of cruelty value the historical experience of women and sexual dissidents and propose politics linked to territorial and communal issues. They focus more on the process than the product, preserve life by setting limits to its reification and aim to contribute to a historical project of social bonding that encourages reciprocity and fosters community.

Feminisms and theories of care also connect with the Latin American Collective Health approach, which emphasizes the Health/Illness/Care/Attention process rather than a simple Health-Illness binarism. This approach recognizes the subjectivity present in this process and the complexity and multidimensionality of health as a contested field, valuing the living labour in health and the possibility of producing life where we live, existential territories of life (Merhy et al., 2012). It aligns with Latin American Environmental Thought and feminisms by advocating for life in connection with experiences, social struggles, resistances, and claims, thus extending care to all existential territories, not just human life.

10.3 Methodology and Context

Consistent with the perspectives presented, we frame our analysis within the Systematisation of Educational Experiences. This research approach, derived from Popular Education, aligns with our commitment to social transformation (Jara, 2018). Systematization is a participatory process for producing knowledge about educational or social practices. It aims to bring together multiple viewpoints in a comprehensive narrative and combine thick descriptions with interpretations as integral parts of the same process (Messina Raimondi, 2005). Through narrative reconstruction and critical interpretation of the various dimensions, meanings, and rationalities that constitute the experience, it seeks to enhance and contribute to the theorization of connected thematic fields (Torres-Carrillo, 2021). For the reconstruction and analysis of the continuous teacher education course under review, we collected diverse perspectives from participants through virtual forums, recent interviews (conducted two years after the experience), final survey evaluations, written class materials, records of work meetings, and both individual and collective reflections from the team.

To provide context, from March 2020 until the end of 2021, Argentine education was conducted through *Emergency Remote Teaching* (Dussel et al., 2020; Ithuralde et al., 2024), or a combination of in-person and virtual classes, due to the COVID-19 pandemic. The continuous teacher education experience analyzed here was part of a national educational policy aimed at providing education and support for teachers during the pandemic. The goal was to empower in-service teachers by providing new tools, encouraging reflection on their practices, and fostering transformations.

In April 2021, GDC received a proposal to implement a nationwide virtual course and decided to take on this challenge: "especially considering the national scope of the course, the work with teachers (something we miss...), and the possibility of integrating contributions from our different lines of work" (meeting record, 9/4/2021, our translation). In the initial organizational meeting, we formed a group of 16 members, including teachers from disciplines such as Pedagogy, Biology, Chemistry, Anthropology, Physics, Social Communication, and the Arts. We assigned roles, with some members focusing on planning and developing class materials, others on tutoring (which involved daily interactions with participants in forums), and a small group on coordination. We also proposed topics for each of the four classes, considering previously addressed issues such as Environmental Education, Nutrition, Comprehensive Sexuality Education, and Vector-Borne Diseases. We agreed to incorporate cross-cutting strategies that proved effective in educational research during the pandemic, such as strengthening the connection between schools, families, and communities and working with integrated projects that link different subject areas (Ithuralde et al., 2024). We aimed to include perspectives throughout the course, such as critical and extended interculturality. Ultimately, we addressed three proposed topics: Environmental Education, Nutrition, and Health Education with Vector-Borne Diseases, while integrating gender and intercultural perspectives.

Additionally, we included a new lesson on *Diverse Perspectives on Earth and the Cosmos*, a topic we had not previously covered in past courses. This decision explicitly addressed critical and extended interculturality within Natural Sciences Education, an issue we had previously explored in the context of environmental and health conflicts. We organized ourselves into four groups, each developing one class for the course, including selecting relevant content and resources and producing materials for asynchronous work.

10.4 Reflections on the Experience

In analyzing the experience, we summarize the meanings and lessons learned during the design and implementation phases, as reflected by the various participants. This aims to describe elements that may support the adoption of a decolonial approach in the development of NSEHE proposals.

10.4.1 Designing a Proposal for Teacher Education in Natural Sciences, Environment, and Health at the Primary Level from a Decolonial Approach

The course "Teaching in Diverse Scenarios: Natural Sciences, Environment, and Health at Primary Level"[1] proposed a reflective journey aimed at contributing to the design of NSEH-based teaching proposals that were grounded, transformative, situated in territories and adaptable to the evolving pandemic situation. The course,

[1] https://infod.educacion.gob.ar/cursos/603

Table 10.1 Contents and activities of each class

Class	Contents	Activities
Class 0	Introducing the participants and the platform	Forum of introduction: geolocation on a map (padlet); introduction (audio and/or text)
Class 1: Teaching of natural sciences and environmental education	Environmental education, environment and environmental conflicts, Latin American environmental thought. Territories body and land. Humanity-nature	Optional activities: Forums "what is the environment?" and "bodies and territories" Accreditation activity 1: Forum "environmental education, virtuality, face-to-face"
Class 2: Health education—Food	Feeding from complexity and multidimensionality Food and health education	Optional activity: Forum "the route of a dish" Accreditation activity 2: Forum "the route of a dish in primary courses"
Class 3: Education in health and vector-borne diseases	Health education Chagas and dengue as cases to be addressed from a situated, comprehensive and current perspective	Optional activities: Forums "the health education that we dream of" and "the problem of vector-borne diseases in photos of my territory" Accreditation activity 3: Forum: "Rethinking practices in light of what was addressed in class"
Class 4: Different perspectives on the earth and the cosmos	The earth and the universe Interculturality and cultural diversity	Optional activity: Forum "sharing educational experiences" Accreditation activity 4: Forum "socio-cultural context and intercultural practices in primary education"

delivered in a tutored format, consisted of four asynchronous virtual classes and forums and was implemented twice during 2021. Each edition sought to critically and comprehensively analyze regional eco-territorial conflicts and socio-sanitary crises to understand complex local realities within the broader planetary crisis.

The first class (Class 1, in Table 10.1) aimed to establish a dialogue between Natural Sciences Education and Environmental Education from an integral, transversal, and critical perspective linked to Latin American Environmental Thought (Corbetta, 2015). The pandemic context was problematized from two complementary theoretical perspectives in this approach. On the one hand, it was seen as another expression of the planetary environmental crisis (Leff, 2020). On the other hand, it was considered a significant context for transversally addressing Environmental Education, as the initial analysis of Emergency Remote Teaching showed the benefits of integrated work between different curriculum areas facilitated by virtuality. One key aspect of this first class was working with the body-territory concept derived from Latin American Communitarian Feminisms (CMCTF, 2017). Through this theoretical perspective, largely absent from mainstream Environmental Education practices, we aimed to present Environmental Education as an interdisciplinary approach that promotes institutional teamwork, strengthens the pedagogical link with families and addresses community-relevant topics. Additionally, Environmental Education was positioned as a possible starting point for other topics such as nutrition, health, and intercultural approaches to Natural Sciences Education, as addressed by Corbetta (2021).

Class 2 focused on Nutrition and Health Education from a complex and multidimensional perspective. Drawing from Aguirre's analysis (Aguirre, 2019) on diner's taste formation from what "falls on the food plate" -environment, technology, history, economy, rights systems, beliefs, religion, health, and aesthetics- we chose not to discuss one particular way of "healthy eating" but instead explored diverse ways of nourishing based on context, conditions of inequality, cultural practices, and seasonal variations while considering the intertwining of health and nutrition. Thus, from a participative critical approach to Health Education (Fainsod & Busca, 2016), there are no universal forms of healthy eating for all human groups (such as the food pyramid commonly presented in Primary Education). Class 2 focused on Nutrition and Health Education from a complex and multidimensional perspective. Drawing from Aguirre's analysis (Aguirre, 2019) on factors influencing diners' taste formation, from what "falls on the food plate" -environment, technology, history, economy, rights systems, beliefs, religion, health, aesthetics- we chose not to discuss one particular way of "healthy eating" but instead explored diverse ways of nourishing based on context, conditions of inequality, cultural practices, and seasonal variations while considering the intertwining of health and nutrition. Thus, from a participative critical approach to Health Education (Fainsod & Busca, 2016), there are no universal forms of healthy eating for all human groups (such as the food pyramid commonly presented in Primary Education). Addressing the different possible circuits of food production, acquisition, distribution, preparation, and consumption allowed us to discuss, among other aspects, the notion of Food Sovereignty and the subjectivity present in everyday eating practices (Merhy et al., 2012). In this way, in line with feminisms and Latin American collective health, we problematized gender roles in nutrition and the social production of care (Segato, 2018) while introducing the concept of inequality associated with the diversity of nutrition practices. This association would be deepened in class 4.

Class 3 focused on Health Education and vector-borne diseases, aiming to problematize our understanding of health and explore different perspectives and foundations for its comprehension. We positioned ourselves from a participatory and critical approach to Health Education, grounded in a multidimensional, situated, and collective notion of health and within political-pedagogical frameworks that aim for action and transformation within a democratic paradigm (Fainsod & Busca, 2016). We began with critical perspectives on the commodification of health, the medicalization of life, power dynamics in healthcare settings, lack of access, and inequalities influence on diseases, thus problematizing the Hegemonic Medical Model (Menéndez, 2005). As paradigmatic examples of vector-borne diseases, we addressed the issues of Chagas disease and dengue, which are endemic in our country. A comprehensive, multidimensional and situated understanding of these issues requires not only biological knowledge but also the integration of insights and dialogues with other disciplines and diverse forms of popular and ancestral knowledge (Sanmartino et al., 2020; Garelli & Dumrauf, 2022).

Class 4 addressed topics related to the Earth, the Universe, and their changes, as outlined in the National Prioritized Learning Objectives of Natural Sciences. We aimed to reflect on the ideas and cosmologies that teachers and students bring to educational spaces and how these evolve through the teaching process. We sought to integrate the critical and extended interculturality perspective (Díaz & Rodríguez Anca, 2014) and the Experientially Meaningful Teaching model, where individuals experience phenomena under study and situations personally, observing, feeling, and living them as their own, identifying with the process (Camino, 2021). Connecting students with the experiences and curiosity evoked by these phenomena also allowed us, from a decolonial approach (Corbetta, 2021; Guelman, 2014), to problematize the notion of scientific knowledge as supposedly neutral, devoid of sensations, interests, and subjective appreciations, disconnected from the socio-cultural and economic contexts that frame its production, validation, and dissemination. We focus on knowledge constructions from indigenous and peasant communities and popular wisdom that inhabit classrooms but are generally made invisible and unknown. To achieve this objective, we contacted teachers who worked from an intercultural perspective and asked them to send us accounts of their experiences. The class shared these accounts, including voices from indigenous communities that advocated for their rights, expressing their cosmological views on the Earth and the Cosmos. Through the critical and extended interculturality perspective, the class aimed to invite reflection on how diverse cultures and social groups historically understand Natural Sciences phenomena taught and learned in schools and promote intercultural dialogue with scientific knowledge. Following the decolonial approach, we aimed to challenge the idea that explanations of the cosmos are exclusively the domain of scientific knowledge.

The course, reflected in the four classes designed, was a significant outcome of the experience. Throughout the classes, we emphasised the interconnectedness of identity creation in teaching, students' knowledge, and connection with territories through emotions, sensations, and experiences (Cabnal, 2010; Merçon et al., 2014).

We sought to leverage the socio-cultural diversity of the territories to foster a dialogue of knowledge(s), ways of feeling, and ways of living. Scientific, popular, and indigenous knowledge, among others, were invited to exchange within a critical and extended intercultural process. The course also aimed to integrate theory and practice, value the exchange of experiences and teaching knowledge, problematize Natural Sciences, and reconsider our relationship with them. The course also aimed to address NSEH topics characterized by complexity and multidimensionality (analyzing dimensions, dynamics, and conflicts), situate NSEH teaching in the inhabited territory for (re)thinking it, and intersect the prescribed curriculum of Natural Sciences by promoting the development of specific teaching proposals.

10.4.2 *On Collective Work for Designing and Putting into Play a Decolonial Proposal*

We formed a team with diverse disciplinary backgrounds and roles in academic spaces (universities, research institutions, and secondary and higher education teachers). The team members resided in different places in Argentina: the city of Buenos Aires and the provinces of Buenos Aires, Córdoba, Jujuy, Río Negro and Santiago del Estero and also worked in diverse educational environments: rural schools, Popular High Schools, Universities, Institutes of Teaching Education, Primary and Secondary Schools, Young and Adult's Education. The connection with GDC was the point of convergence generated in different contexts: teaching, research and university extension. The diversity of members of the team allowed us to draw materials from a transdisciplinary and situated perspective on different Argentinian territories. The team's diversity allowed us to draw on a transdisciplinary and situated perspective on various Argentinian territories, reflecting different real educational contexts across the country and seeking to dismantle internal colonialism (Rivera Cusicanqui, 2018). This diversity also enabled us to create a contextualized proposal addressing social problems related to science and technology (Massarini & Schneck, 2015), aligning with the content of Teacher Education selected for the course.

In periodic meetings, we exchanged ideas, reflected on produced materials, teaching practices, and our relationships with students, and agreed on didactic strategies and course planning. Virtual collective exchange spaces helped consolidate the team, strengthen tutoring, and support professional growth. Tutors frequently highlighted the importance of collective care within the group. Mutual support was based on shared experiences and knowledge, providing feedback and developing personalized teaching styles through collaboration and standing over the group's shoulders. The collective was thus built as a space for self-care and mutual support for the teaching practice itself, challenging the patriarchal focus on exploitation, lack of empathy, and competition (Segato, 2018).

Collective work also facilitated the development of communication strategies for the virtual environment, differing from those used in face-to-face settings.

> Working as a team and collaboratively implied a learning experience for me. (It was) a lot about learning to communicate, to communicate with each other. Inwardly within the team and outwardly as well, always thinking about the best strategies and ways to communicate through which channels and how to generate a pace of the course without the habits generated by in-person interactions. (TN, our translation)

Through collective work, tutors identified key learnings related to implementing the course, such as transitioning to an entirely virtual format, managing content and activities, and using virtual environments. The formation of the working team around a specific project also fostered a sense of belonging for new GDC members.

Finally, the participant in-service teachers highlighted the importance of collective aspects in an anonymous virtual survey conducted through the platform during the first edition of the course. They valued peer exchange, the opportunity to learn about colleagues' experiences and the feedback and support from tutors. Thus, the communication processes within the teaching team in charge of the course and the strategies built for communication with the participants were reflected in the virtual classrooms.

This pedagogical approach involved participation throughout the course during design, implementation, and evaluation. We promoted a collective teaching practice aimed at democratizing teaching, valuing teachers' and students' knowledge, and fostering transformations in their practices (Freire, 2010). These collective workspaces were designed to promote interpersonal care, as proposed by counter pedagogies of cruelty (Segato, 2018). Teaching NSEH from a decolonial approach requires experiencing the social relationships promoted by this approach. Therefore, the course was designed as a space for decolonial construction through careful and democratic work processes (Guelman, 2014).

10.4.3 On Strategies and Approaches "That Allow Individuals to Be Political Actors in the Construction of Scientific Concepts"

Among the strategies presented in the course, participants found the body-territory approach (Class 1) and the "plate route" (Class 2) to be the most significant ones. Both frameworks address the multidimensionality and complexity of decolonization, environment, and food. They allow for the explicit politicization of teaching by identifying environmental conflicts and their effects and problematizing food production, distribution, consumption, and preparation as human and social activities. These strategies are developed from or inspired by Latin American Communitarian Feminisms (CMCFT, 2017). They also facilitate interdisciplinary dialogue between Natural and Social Sciences and other forms of knowledge:

> I liked it very much body-territory, (and) it served me well. And the food plate, I tell you, was good to incorporate this kind of topics. Not to mention that it widens your gaze.
>
> And being able to relate social aspects to natural ones, because the environment concept opens you up to that possibility, doesn't it? (...) Seeing the food oval and how to analyze the food plate seemed very important to us, seeing how to territorialize the plate. What plate arrives at our table? In other words, where does the food come from if we can identify the origin of that food? (DCS, our translation)

Considering the heterogeneity present in all classrooms, reflecting on the value of diversity of knowledge and, in some cases, an intercultural perspective were among the highlighted aspects of the course's approach. The course challenged the dominant Western knowledge paradigm, contributing to a decolonial perspective (Corbetta, 2021).

Tutors noted that the course was "very formative in theoretical terms... thinking about environmental education, health education, and the intercultural perspective was super, super novel" (TF, our translation). They appreciated that the course was not solely focused on Indigenous peoples, migrants, women, and sexual dissidences but was inclusive of all individuals, especially "white" individuals from hegemonical groups. Tutors valued the opportunity to explore the journey of food from its source to becoming a meal on a plate and the situation of commensality, which allows for connections not only with disciplinary content (and its contextualization) but also with Environmental Education and Comprehensive Sexual Education. The "utopia forum", which encouraged participants to transform their practices and educational spaces based on their feelings and desires, was particularly valued. This activity moved away from the rationalist dominance in Natural Sciences Education, as proposed by communitarian feminist perspectives (Cabnal, 2010). The transversal and situated approaches were especially appreciated:

> Many people, even those who did not share the same perspectives, also affirmed that the transversality of the content was one learning from the pandemic. So, the power of situatedness, right? The possibility of connecting with the community, the school community, as a teaching and learning strategy. (TT, our translation)

The approach of the Pedagogy of Environmental Conflict (Canciani & Telías, 2013; Iribarren et al., 2022) prompted reflections on previous Environmental Education practices:

> Learning about the approach to environmental conflict has helped me problematize our ways of doing things. The ways we had, my colleagues and me, even though we sometimes proposed work to highlight environmental issues, we sometimes fell into the search for solutions that we know are unattainable (...). These classes enabled us to think about ourselves from that complexity. So, deepening that perspective has been crucial for me. (TT, our translation)

Thus, the theoretical frameworks presented in the classes would dialogue with the practices themselves, focusing on their complexity and deepening into relevant perspectives for teachers.

In the forums for each class, participants highlighted several contributions from the course's approach. They noted the importance of:

1. recognizing the value of diversity and social inequality in food practices and incorporating a gender perspective when analyzing them;
2. approaching Health Education from a multidimensional perspective within the Natural Sciences curriculum;
3. practicing interculturality through connections with the territory, Indigenous communities, intergenerational dialogue, and students' lived experiences, approaching a critical and extended interculturality perspective; and
4. promoting collective and transdisciplinary work among colleagues.

Participants referred to the course's novelty in valuing local, ancestral, popular, and student knowledge and teachers' knowledge, alongside perspectives on multidimensionality, complexity, and Comprehensive Sexual Education. According to some participants, the course allowed for self-practice rethinking and enriched it with new didactic strategies and resources through the exchange among colleagues and dialogue with actors in the community, students, and families "since they are sources of knowledge and wisdom, providing diversity and mutual enrichment (...)

in dialogue with scientific knowledge". The multidisciplinary and multidimensional perspectives and the theoretical foundations of the teachers' proposed and implemented approaches were also valued. Finally, the construction of a "new way of approaching Natural Sciences that enables individuals to be political actors in the construction of scientific concepts" was highly appreciated.

According to this analysis, critical elements for building teaching strategies and approaches that politicize scientific knowledge construction include the assessment of various types of links, thoughts and feelings (Merçon et al., 2014), reflection on situations and practices of daily life, recognition and assessment of the complexity of reality and the promotion of creative processes (Dumrauf et al., 2016), elements which support the development of teaching strategies that foster autonomy, movement, life (Iribarren et al., 2021).

10.5 Contributions of the Course to Decolonial Teaching Work in NSEHE in Post-pandemic Era

Some participants noted that the course prompted reflections on their teaching practices, offering insights that extended beyond its duration.

> This axis that I mentioned at the beginning, when I addressed it, although I worked on it and presented it with technological resources, I did not have this intercultural perspective; I approached it from the sciences. And, thanks to the course, I got to know the students better and what they bring, which sometimes we say from a discursive perspective, but sometimes it slips past us. Being able to think about it and incorporate it into lesson planning, considering the different communities from which the students come, which I had not thought about before the course. (TVC, our translation).
>
> It changed my perspective on planning particular content and proposing and evaluating it. It allowed me to reflect on how I was working and the extreme value I gave to scientifically validated content. Now, I realize that all my students have something to teach. (Anonymous survey, our translation).

The openness to new theoretical and methodological perspectives, accompanied by activities and spaces for exchange and debate, stimulated participant practice transformations. These changes included valuing students' knowledge and experience and embracing socio-cultural diversity. Once again, engaging with others proved to be an effective strategy for promoting these individual shifts, which were also reflected in how participants approached content and lesson planning.

Some final projects for course accreditation showed attempts to deepen connections with the communities involved in the schools. However, in most cases, there was no greater territorialization of educational proposals. Strengthening the school-family and community links through territorialization was identified during the pandemic as vital for maintaining the educational bond. From our perspective, this aspect is necessary to foster transformations beyond the school and appeared in several participants' contributions in the "Dream Forum" related to collective health.nities involved in the schools. However, in most cases, there was little advancement in territorializing educational proposals. Strengthening school-family and community links through territorialization was recognized during the pandemic as crucial for maintaining the educational bond. This aspect, highlighted by several participants in the "Dream Forum" related to collective health, is necessary for fostering transformations beyond the school setting.

Interviews revealed reflections on dimensions not directly addressed by the course but inspired by it, such as the relationship between colonialism and Comprehensive Sexual Education (CSE) and the adult-centric approach in school events:

> When we addressed these cultures, it was like: "Tell me what your grandmother said, but now I will tell you what science says ". It was like pretending. Finally, science is what matters. And with CSE [Comprehensive Sexual Education], it is the same, adult-centric. We say with our colleagues, "Yes, yes", but when it comes up in our daily lives, we ignore the child's voice who may not want to do something, not want to dramatize the May 25th event, but we must comply. We continue to think about what they want and their voices. And that [listening to their voices] has to do with guaranteeing rights. (TVC, our translation)

Such reflections, beyond what the course proposed, demonstrate the potential of spaces that encourage the analysis of one's practices through diverse and complementary theoretical perspectives (Latin American Popular Education, Latin American Environmental Thought, communitarian feminisms) to support specific teaching activities and proposals on specific topics made available to teachers.

In various ways, tutors incorporated aspects of the course into their classes in the post-pandemic era. Reflections on planned time, its relation to activities, and timing remained part of their practices, even though they adapted to in-person dynamics. They applied concepts and activities from the course—such as "environmental social representations", "body-territory", "territory-land", "collective mapping", "environmental conflicts", the "plate route" in food education, the "dengue and Chagas booklet," "critical and extended interculturality," and the "kaleidoscopic perspective" on health issues—using materials like stories, videos, and collaborative online tools in their in-person settings after the course ended. These reflections and contributions from shared work continue to promote further transformations in the post-pandemic period. The dialogue with primary school teachers nationwide was also enriching:

> That year, when I was a tutor in the course, I was teaching Citizenship Education in the first year of high school in the city where I live (...) the project that year was about Environmental Education, and we basically implemented many of the proposals presented in the course. Additionally, I could talk about things I had learned through exchanges with other teachers because, suddenly, there were things I did not know about the environmental conflicts that each teacher experienced in their town and city. And I brought them as examples of things happening in the country. (TV, our translation)

For tutors, the exchange and collective reflection spaces were extended and enhanced compared to those for participating in-service teachers. This opportunity led to greater reflection and enrichment, with tutors also recognizing that they learned from the participants. A key aspect noted by both teachers and tutors was that the proposed activities integrated theoretical elements with teaching practice. This integration allowed them to adopt different approaches and reflect on and modify their practices.

10.6 (Re)Thinking/Building Decolonizing Natural Sciences, Environmental and Health Education

In our collective effort to develop decolonizing approaches in NSEHE, we undertook the complex task of implementing the analyzed teacher education course. This article outlines our theoretical and methodological frameworks and details the course and its foundations. We reflected on group work contributions, political-pedagogical approaches, proposed strategies, and activities. Finally, we highlighted the contributions to teaching practices in the post-pandemic period. It was gratifying to see how participants embraced this approach, particularly its political aspect and the value placed on collective work and diversity. This led to transformations in their educational practices.

In previous work (Dumrauf & Cordero, 2020), we identified the need for forming teaching groups focused on "participation and treatment of problems". We emphasized the importance of facilitation roles in building these groups, of decision-making about practices and the testing of "exploratory actions." The challenge in this experience, beyond the virtual format, was to enrich our proposal, making it more complex, conceptually and methodologically. The democratization of teaching practices and teacher education is a crucial aspect of decolonizing/decolonial proposals. The "exploratory action" involved integrating theoretical perspectives from different fields with specific NSEHE content and making this framework available to a considerable number of teachers across the country.

Concerning the elements and aspects that would strengthen the construction of proposals and the implementation of decolonizing practices in the NSEHE, it is worth noting that:

- the construction of spaces and moments of exchange, debate, and collective dialogue is essential and possible even from virtuality;
- this dialogue must critically incorporate first-person Other's and silenced knowledges, recovering diversity and territories in an equal hierarchy concerning scientifically validated knowledge;
- a fundamental step is to problematize the construction of scientific knowledge as a human and social activity in which, among others, there are power relations (with the consequent exercise of violence), corporate and economic interests, gender and ethnic-racial biases, presuppositions of the historical time, emotions;
- shared spaces need to relate self-care, care for others and territories in order to promote trust in relationships and strengthen autonomy;

- didactic proposals with decolonial intentions must necessarily incorporate words, experiences, and emotions of students, families, groups, and communities;
- the pedagogical-didactic proposals that aim to be decolonizing need to recover, as a key element, the transformative and emancipatory character of Latin American Popular Education.

Participant and tutor feedback led us to reflect on issues that need to be addressed or explored in greater depth, which positively challenged us and guided our transformation. Among these issues were the connections between environment and health, which were included in teaching proposals even though they were not explicitly requested. The educational treatment of these connections has been long neglected (Sauvé, 2004). Activities like the Plate Route and Dream Forum proved helpful in exploring this interconnection and reflecting on how to address it in teaching. The Body-Territory activity and its adoption by participants show an interest in integrating feminist communitarian perspectives, prompting us to deepen the connections between Environmental Education, Health Education, and Comprehensive Sexual Education. Systematically exploring these intersections is essential and urgent in the post-pandemic context. One participant highlighted the need to address adult-centric practices in schools, which also affect NSEHE and infringe on the rights of children, adolescents, and youth.

The voices of participants and tutors allowed for reflections on issues that still need to be addressed by us in these intentional decolonizing constructions, or at least not explicitly or with sufficient depth, making us uncomfortable in a positive way that guides our path of transformation. Among these issues were the relationships between environment and health, which were included in the teaching proposals without being explicitly requested. The educational treatment of these relationships has been long neglected (Sauvé, 2004). Activities like the Plate Route and Dream Forum proved helpful in exploring this interconnection and reflecting on how to problematize it in teaching. The Body-Territory activity and its adoption by participants show an interest in integrating feminist communitarian perspectives, prompting us to deepen the connections between Environmental Education, Health Education, and Comprehensive Sexual Education. Systematically exploring these intersections is essential and urgent in the post-pandemic context. One participant highlighted the need to address adult-centric practices in schools, which also affect NSEHE and infringe on the rights of children, adolescents, and youth.

The value of territorialization in education was another important finding of the systematization process. The pandemic context highlighted the significance of collaborative work among teachers, with the communities, and intersectorially. Although these collaborations were already valued in our approach, these findings encourage us to further explore territorializing the promotion of popular organization whenever possible, inviting us to build and strengthen collective processes in an emancipatory sense (Freire, 2010).

Including the arts in didactic proposals developed by teachers encourages us to problematize and expand our perspectives on natural phenomena, environment, and health while exploring the experiential and affective dimensions of NSEHE. While we examined some of these connections through Theatre of the Oppressed (Garelli et al., 2021), extending them to other art forms, such as visual and sound arts, would be fruitful.

The experiential deepening of the critical inclusion of Information and Communication Technologies linked to Popular Education (Mejía, 2013) in Teacher Education has been a significant contribution to our approach (Garelli et al., 2023; Poveda Ducón et al., 2024). Adapting methodologies and content from Popular Education, such as collective mapping or the Plate Route, to virtual formats required creativity, as these were previously developed only in in-person settings. As tutors and coordinators mention, creating classes and organizing pedagogical resources for each topic, utilizing digital applications and different virtual environments to enable remote exchanges and participatory techniques in graphic, sound, and audiovisual languages were novelties that require further experimentation. Building a sense of group cohesion in the virtual realm, supporting the trajectory of participants through the role of tutors, and the collective work among tutors and coordinators during the course were crucial in sustaining and developing their virtual educational experiences. These communication modalities have become integral to our daily lives in our post-pandemic reality.

The course has thus entailed growth and learning for all actors involved through exchanging and appreciating their voices and knowledge: participants, tutors, coordinators, content creators and collectively as GDC. The voices of the different people involved in the course, in particular through their analyses of how they changed their practices, have allowed us to strengthen our theoretical reflections on decolonizing NSEHE by deepening our proposal on some dimensions we already were considering and bringing up others we were neglecting. We hope to continue these dialogues and the production of knowledges in new formative experiences.

References

Aguirre, P. (2019). *¿Hoy los alimentos son buenos para vender, pero no para comer?* [Today, food is good for selling, but not for eating?] dissertation. Congreso Argentino de Cardiología [Argentinian Congress of Cardiology]. https://www.youtube.com/watch?v=JVDbG7_05Kg&t=97s

Bazzul, J., & Sykes, H. (2011). The secret identity of a biology textbook: Straight and naturally sexed. *Cultural Studies of Science Education, 6*, 265–286. https://doi.org/10.1007/s11422-010-9297-z

Cabnal, L. (2010). Acercamiento a la construcción del pensamiento epistémico de las mujeres indígenas feministas comunitarias de Abya Yala [Approach to the construction of epistemic thinking of indigenous feminist community women of Abya Yala]. In ACSUR- Las Segovias (Eds.) *Feminismos diversos: el feminismo comunitario* [Diverse feminisms: community feminism] (pp. 10–26). Asociación para la Cooperación con el Sur. Retrieved Febraury 6 from https://porunavidavivible.files.wordpress.com/2012/09/feminismos-comunitario-lorena-cabnal.pdf

Camino, N. (2021). Diseño de actividades para una Didáctica de la Astronomía vivencialmente significativa [Design of activities for an experientially meaningful didactic of astronomy]. *Góndola, Enseñanza y Aprendizaje de las Ciencias, 16*, 15–37. https://doi.org/10.14483/23464712.16609

Canciani, M. L., & Telías, A. (2013). Aportes teóricos conceptuales para pensar los procesos educativos en escenarios de conflicto ambiental [Conceptual theoretical contributions to think about educational processes in environmental conflict scenarios]. *Revista del Instituto de Investigaciones en Ciencias de la Educación, 34*, 111–122. https://doi.org/10.34096/riice.n34.1446

Colectivo Miradas Críticas del Territorio desde el Feminismo (CMCTF) [Collective Critical Views of the Territory from Feminism]. (2017). *Mapeando el cuerpo-territorio. Guía metodológica para mujeres que defienden sus territorios.* [Mapping the body-territory. Methodological guide for women who defend their territories]. CLACSO.

Corbetta, S. (2015). Pensamiento Ambiental Latinoamericano y Educación Ambiental [Latin American environmental thought and environmental education]. *Voces en el Fénix, 43*, 160–167.

Corbetta, S. (2021). Educación Ambiental y Educación Intercultural: hacia una construcción de puentes desde un pensamiento ambiental y latinoamericano crítico [Environmental Education and Intercultural Education: towards building bridges from critical environmental and Latin American thinking]. *Gestión y ambiente, 24*, 107–130. https://doi.org/10.15446/ga.v24nsupl1.91903

Cordero, S., & Dumrauf, A. (2022). De las prácticas a las teorías y de las teorías a las prácticas: Caminos en la descolonización de la Educación en Ciencias Naturales, Ambiental y en Salud [From practices to theories and from theories to practices: Paths in the decolonization of Education in Natural, Environmental and Health Sciences]. *Escritos sobre la Biología y su Enseñanza*, Número Extraordinario, pp. 3266–3279.

Cucalón Tirado, P., Cordero, S., & Dumrauf, A. (2017). Interculturalidad y educación en ciencias naturales, ambiente y salud en la provincia de Buenos Aires [Interculturality and education in natural sciences, environment and health in the province of Buenos Aires]. In: *Memorias de la XII Reunión de Antropología del Mercosur Experiencias Etnográficas RAM 2017: desafíos y acciones para el Siglo 21 [Memories of the XII Mercosur anthropology meeting ethnographic experiences RAM 2017: Challenges and actions for the 21st century]*. National University of Misiones.

Díaz, R., & Rodríguez Anca, A. (2014). Activismo intercultural: una mirada descolonizadora, crítica e interseccional [Intercultural activism: a decolonizing, critical and intersectional view]. In A. Villa & M. E. Martínez (Eds.), *Relaciones escolares y diferencias culturales: la educación en perspectiva intercultural [School relations and cultural differences: education in an intercultural perspective]* (pp. 171–198). Noveduc.

Dumrauf, A. (2020). ≏Qué sabemos sobre coronavirus? Algunas construcciones y reflexiones en proceso desde quienes habitamos la Educación [What do we know about coronavirus? Some constructions and reflections in process from those of us who inhabit education], https://www.facebook.com/notes/353602372543370/

Dumrauf, A., & Cordero, S. (2020). Un enfoque participativo para la formación docente continua en la Educación en Ciencias Naturales, Ambiental y en Salud [A participatory approach for continuing teacher training in natural, environmental and health sciences education]. *Revista Eureka sobre Enseñanza y Divulgación de las Ciencias, 17*, 160201–160215. https://doi.org/10.25267/Rev_Eureka_ensen_divulg_cienc.2020.v17.i1.1602

Dumrauf, A., Cordero, S., & Mengascini, A. (2016). Experiencias educativas de los movimientos sociales: Contribuciones para la educación científica, ambiental y en salud desde una perspectiva emancipadora [Educational experiences of social movements: Contributions to scientific, environmental and health education from an emancipatory perspective]. *Revista Brasileira de Pesquisa em Educação em Ciências, 16*, 477–497.

Dumrauf, A., Cordero, S., Cucalón Tirado, P., Guerrero Tamayo, K., & Garelli, F. (2019). Hacia nuevos territorios epistémicos: Aportes desde un camino de construcción pedagógica descolonizadora en educación en ciencias naturales, ambiental y en salud [Towards new epistemic territories: Contributions from a path of decolonizing pedagogical construction in education in natural, environmental and health sciences]. In B. Monteiro, D. S. A. Dutra, S. Cassiani, C. Sánchez, & R. V. D. L. Oliveira (Eds.), *Decolonialidades na educação em ciencias [Decolonialities in science education]* (pp. 288–306). Editora Livraria da Física.

Dussel, I., Ferrante, P., & Pulfer, D. (2020). Nuevas ecuaciones entre educación, sociedad, tecnología y Estado [New equations between education, society, technology and the state]. In I. Dussel, P. Ferrante, & D. Pulfer (Eds.), *Pensar la educación en tiempos de pandemia [Thinking about education in times of pandemic]* (pp. 351–364). National Pedagogical University of Colombia.

Fainsod, P. & Busca, M. (2016). *Educación para la salud y género. Escenas del curriculum en acción* [*Education for health and gender. Scenes from the curriculum in action*]. Homo Sapiens Editora.

Freire, P. (1993). *Pedagogía de la Esperanza [Pedagogy of Hope]*. Siglo XXI Editores.

Freire, P. (2002). *Pedagogía del Oprimido [Pedagogy of the oppressed]*. Siglo XXI Editores.

Freire, P. (2010). *La Educación en la ciudad* [*Education in the City*]. Siglo XXI Editores.

Garelli, F. M., & Dumrauf, A. G. (2021). Una mirada al campo de la educación en salud: hegemonía, pandemia y alternativas [A look at the field of health education: Hegemony, pandemic and alternatives]. *Espacio Abierto, 30*(2), 100–116.

Garelli, F., & Dumrauf, A. (2022). Educación en Salud desde la Educación Popular: propuestas para el abordaje del dengue/chikungunya/zika [Health Education from Popular Education: proposals for addressing dengue/chikungunya/zika]. *Pro-Posições, 33*. https://doi.org/10.1590/1980-6248-2020-0127ES

Garelli, F., Mateyca, C., Mordeglia, C., & Cordero, S. (2021). "Pasar las cosas por el cuerpo siempre es distinto a ponerlo en palabras ¿no?": educación en salud, interculturalidad y teatro del oprimido ["Passing things through the body is always different from putting it into words, right?": Health education, interculturality and theater of the oppressed]. *Ciência & Educação, 27*, Article e21045. https://doi.org/10.1590/1516-731320210045

Garelli, F., Di Battista, C., Cochero, J., & Dumrauf, A. (2023). Educación en Salud, Educación Popular y TIC: aprendizajes a partir del desarrollo de la aplicación Caza Mosquitos [Health Education, Popular Education and ICTs: learning from the development of the Caza Mosquitos application]. *Revista de Educación en Biología, 26*(1), 80–94.

Grosfoguel, R. (2017). *¿Qué es el racismo?* [What is racism?] Dissertation, III Doctoral Research Seminar of the Doctoral Program in Migration Studies of the University of Granada. Retrieved February 22 from http://media.ugr.es/index.php/categorias/25-ciencias-sociales-y-juridicas/1212-conferencia-ramon-gosfroguel

Guelman, L. A. (2014). Educación Popular y Pedagogía Descolonizadora en el MOCASE-VC [Popular Education and Decolonizing Pedagogy at MOCASE-VC]. *Intersticios de la Política y la Cultura. Intervenciones Latinoamericanas, 3*(6), 35–49.

Guelman, A., & Palumbo, M. M. (2015). Prácticas pedagógicas descolonizadoras en experiencias productivas autogestionadas de movimientos sociales: el principio formativo del trabajo [Decolonizing pedagogical practices in self-managed productive experiences of social movements: The formative principle of work]. *Revista Interamericana de Educación de Adultos, 37*(2), 48–64.

Hodson, D. (2010). Science education as a call to action. *Canadian Journal of Science, Mathematics and Technology Education, 10*(3), 197–206. https://doi.org/10.1080/14926156.2010.504478

Iribarren, L., Cordero, S., Dumrauf, A. G., Garelli, F., Ithuralde, R. E., & Mengascini, A. (2021). Formación docente desde la educación popular en ciencias naturales, ambiental y en salud: Algunas tensiones y experiencias [Teacher education from Popular Education in Natural Sciences, Environmental and Health: Some tensions and experiences]. *InterCambios: Dilemas y Transiciones de la Educación Superior, 8*(2), 10.

Iribarren, L., Guerrero Tamayo, K., Garelli, F., & Dumrauf, A. (2022). Pedagogías del conflicto ambiental: Aportes desde una experiencia participativa de formación docente en un territorio en disputa [Pedagogies of environmental conflict: Contributions from a participatory experience of teacher training in a disputed territory]. *Praxis Educativa, 26*(1), 1–24. https://doi.org/10.19137/praxiseducativa-2022-260102

Ithuralde, R. E. (2020). Thinking about a popular education in sciences? *Cadernos de Pesquisa, 50*(175), 186–208. https://doi.org/10.1590/198053146644

Ithuralde, R. E., Becerro, V., Cordero, S., Dumrauf, A., & Díaz-Barrios, C. (2023). Compartir historias sobre el cielo y la Tierra en la formación docente continua [Sharing stories about the sky and earth in continuing teacher training]. *Revista de Enseñanza De la Física, 35*, 201–207. https://revistas.unc.edu.ar/index.php/revistaEF/article/view/43310

Ithuralde, R. E., Corbetta, S., Sosa, S., Panosetti, P., Díaz Español, N., Córdoba, F., Taboada, M., Kaliman, R., Banegas, R., Chazarreta, I., Maldonado, J., Bravo, E., Moreno, C., Peralta, T., Soria, M. I., Tavolaro Iacovone, F., Ruiz, E., & Silveti, L. (2024). Una aproximación a la educación remota de emergencia en Santiago del Estero, Argentina, en contexto de pandemia: periodizaciones, estrategias docentes e institucionales y sentires [An approach to emergency remote education in Santiago del Estero, Argentina, in the context of a pandemic: periodisations, teaching and institutional strategies and feelings]. *Cuadernos FhyCS-UNJU, 65*, 99–128. http://revista.fhycs.unju.edu.ar/revistacuadernos/index.php/cuadernos/article/view/969

Jara, O. (2018). *La sistematización de experiencias: teoría y práctica para otros mundos posibles [The systematization of experiences: Theory and practice for other possible worlds]*. CINDE.

Kato, D. S., Galamba, A., & Monteiro, B. A. P. (2023). Decolonial scientifc education to combat 'science for domination'. *Cultural Studies of Science Education, 18*, 217–235. https://doi.org/10.1007/s11422-023-10165-4

Leff, E. (2020). A cada quien su virus. La pregunta por la vida y el porvenir de una democracia viral [To each person their virus. The question about the life and future of a viral democracy]. *HALAC: Historia Ambiental, Latinoamericana y Caribeña, 10*(1), 139–175.

Maldonado-Torres, N. (2007). Sobre la colonialidad del ser: Contribuciones al desarrollo de un concepto [On the coloniality of being: Contributions to the development of a concept]. In S. Castro Gómez & R. Grosfoguel (Eds.), *El giro decolonial. Reflexiones para una diversidad epistémica más allá del capitalismo global [The decolonial turn. Reflections for epistemic diversity beyond global capitalism]* (pp. 127–167). Siglo del Hombre Editores.

Massarini, A., & Schneck, A. (2015). *Ciencia entre todxs: Una producción colectiva desde y para la educación científica [Science among all: A collective production from and for scientific education]*. Paidós.

Mejía, M. R. (2013). La educación popular con y desde las NTIC [Popular education with and from NICT]. In L. Cendales, M. R. Mejía, & J. Muñoz (Eds.), *Entretejidos de la educación popular en Colombia [Interweaves of popular education in Colombia]* (pp. 185–222). CEAAL.

Menéndez, E. L. (2005). El modelo médico y la salud de los trabajadores [The medical model and workers' health]. *Salud Colectiva, 1*(1), 9–32. https://doi.org/10.18294/sc.2005.1

Mengascini, A., Garelli, F., Cordero, S., & Dumrauf, A. (2024). Educación en Salud y Educación Popular: ideas para problematizar, dialogar y transformar [Health Education and Popular Education: ideas for problematisation, dialogue and transformation]. *Paulo Freire. Revista De Pedagogía Crítica, 30*, 125–145. https://doi.org/10.25074/pfr.v0i30.2537

Merçon, J., Camou-Guerrero, A., Núñez Madrazo, C., & Escalona Aguilar, M. A. (2014). ¿Diálogo de saberes? La investigación acción participativa va más allá de lo que sabemos [Dialogue of knowledge? Participatory action research goes beyond what we know]. *Decisio. Saberes para la acción en Educación de Adultos, 38*, 29–33.

Merhy, E. E., Feuerwerker, L. C. M., & Silva, E. (2012). Methodological contributions towards the study of health care production: Lessons from a research study on barriers and access in mental health. *Salud colectiva, 8*(1), 25–34.

Messina Raimondi, G. (2005). La sistematización: acerca de su especificidad [Systematization: about its specificity]. *Revista Interamericana de Educación de Adultos, 27*(2), 163–171.

Moura, C. B., Alsop, S., Camel, T., & Guerra, A. (2023). Science education in a world in crisis: Contributions from the south to a defense of a cultural–historical approach in science teaching. *Cultural Studies of Science Education, 18*, 669–693. https://doi.org/10.1007/s11422-022-10129-0

Poveda Ducón, K., Piegari, E., Boron, I., & Iribarren, L. (2024). Apropiación social de tecnologías libres: Una experiencia de monitoreo ambiental participativo y educación ambiental [Social appropriation of free technologies: An experience of participatory environmental monitoring and environmental education]. *Revista Tecnología, Ciencia y Educación, 28*, 163–190. https://doi.org/10.51302/tce.2024.18667

Rivera Cusicanqui, S. (2018). *Un mundo ch'ixi es posible. Ensayos desde un presente en crisis* [A ch'ixi world is possible. Essays from a present in crisis]. Tinta Limón.

Rosa, K. (2018). Science identity possibilities: A look into Blackness, masculinities, and economic power relations. *Cultural Studies of Science Education, 13*, 1005–1013. https://doi.org/10.1007/s11422-018-9859-z

Sanmartino, M., Mateyca, C., & Pastorino, I. C. (2020). What are we talking about when we talk about education and Chagas? A systematic review of the issue. *Biochimica et Biophysica Acta. Molecular Basis of Disease, 1866*(5), Article 165691. https://doi.org/10.1016/j.bbadis.2020.165691

Sauvé, L. (2004). Perspectivas curriculares para la formación de formadores en educación ambiental [Curricular perspectives for the training of trainers in environmental education]. *Carpeta informativa CENEAM*, 1–12.

Segato, R. (2018). *Contra-pedagogías de la crueldad [Counter-pedagogies of cruelty]*. Prometeo Libros.

Torres-Carrillo, A. (2021). Hacer lo que se sabe, pensar lo que se hace. La sistematización Como modalidad investigativa [Do what you know, think what you do. Systematization as a research modality]. *PRO, 31*, 27–47.

Ana G. Dumrauf is founder member of the Grupo de Didáctica de las Ciencias (IFLYSIB, UNLP-CONICET). She is female professor at Departamento de Ciencias de la Educación, Universidad Nacional de La Plata (UNLP), and researcher at Consejo Nacional de Investigaciones Científicas y Técnicas (CONICET). Her research topics include: Research with participation in Education in Natural Sciences, Environment and Health: theoretical constructions from experiences in different contexts.

Silvina Cordero is founder member of the Grupo de Didáctica de las Ciencias (IFLYSIB, UNLP-CONICET). She works as researcher and professor at the Departamento de Ciencias de la Educación and the Instituto de Investigaciones en Humanidades y Ciencias Sociales, Facultad de Humanidades y Ciencias de la Educación, Universidad Nacional de La Plata (UNLP). Her research topics include: Teacher training in Natural Sciences, Environmental, and Health Education.

Verónica Becerro is a member of the Grupo de Didáctica de las Ciencias (IFLYSIB-CONICET, UNLP) and a PhD candidate in Education Sciences at the Universidad Nacional de La Plata (UNLP). She is also a secondary school teacher and teaches nursing students in courses related to Health Education and Communication at Instituto Superior de Formación Técnica N° 221. Her doctoral research focuses on health education, community participation, and youth engagement in underserved neighborhoods.

Luciano Iribarren is a member of the Grupo de Didáctica de las Ciencias (IFLYSIB, UNLP-CONICET). He works as a professor in continuous teacher training state institutions in Natural Sciences and Environmental Education in Dirección de Formación Docente Permanente, Dirección General de Cultura y Educación, Provincia de Buenos Aires. His research topics include: Research with participation in Natural Sciences, Popular Education, and Environmental Education.

Raúl Esteban Ithuralde is a member of the Grupo de Didáctica de las Ciencias (IFLYSIB UNLP/CONICET). He is a professor at CCPEMS/CEFIEC, Facultad de Ciencias Exactas y Naturales, Universidad de Buenos Aires; a Consejo Nacional de Investigaciones Científicas y Técnicas (CONICET) researcher; and professor of Intecultural Pedagogy at the Educación, Ambiente, Hábitats y Territorios team (Instituto de Estudios para el Desarrollo Social – Facultad de Humanidades, Ciencias Sociales y de la Salud, Universidad Nacional de Santiago del Estero/CONICET). His research topics include: Research with participation in adverse educational management conditions and Teacher Education in Natural Sciences, Environment and Health Education.

Chapter 11
The Role of Intercultural Science Education: Reorienting a Socio-political Agenda for Latin-America

Jairo Robles-Piñeros and Susana Abella

11.1 Introduction

> *Soy parte de quienes alzan la voz para parar la destrucción de los ríos, bosques y páramos, de aquellos que sueñan en que un día los seres humanos vamos a cambiar el modelo económico de muerte, para darnos paso a construir un modelo que garantice la vida*[1].
> — Francia Márquez

Among the most outstanding events of the last decade in terms of global impact and its consequences for the environment and society in general is the pandemic caused by COVID-19 (Robinson, 2020; Waitzkin, 2023). An event that left enormous consequences in all spheres of society, both on a planetary scale in ecological terms, and on a social scale in cultural, economic, and political terms, embracing how complex our world is. This could be classified as a phenomenon resulting from the enormous devastation of ecosystems caused by savage capitalism, i.e. the fact of causing that a virus broke the ecological barrier and caused a tragedy like the one experienced in this pandemic. It is evidence of how devastating capitalism can be; authors such as Moore (2016, 2017) and Arons (2023) affirm that in addition to accepting the concept of Anthropocene referring to the impact of human activity in processes of transformation of the planet, we should be seriously treating the

[1] "*I am part of those who raise their voices to stop the destruction of rivers, forests and moors, of those who dream that one day human beings will change the economic model of death, to give way to building a model that guarantees life*". Words spoken by Francia Marquez, vice president of Colombia, and first Afro woman to hold such an important position in a Latin-American country.

J. Robles-Piñeros (✉)
Universidad Pedagógica Nacional, Bogotá, Colombia

S. Abella
Secretaria de Educación del Distrito (SED-Bogotá), Bogotá, Colombia

C. B. Moura (ed.), *A Sociopolitical Turn in Science Education*, Contemporary Trends and Issues in Science Education 62,
https://doi.org/10.1007/978-3-031-78586-3_11

concept of the *Capitalocene* as the result of the unlimited exploitation of resources to maintain a predatory production system.

According to Moore (2017) and Arons (2023), the concept of Capitalocene not only refers to the devastating consequences of capitalism in recent decades, but on the contrary, it accounts for a process that has been carried out almost since the fifteenth-century European colonial expansion. Thinking about the devastation of the Western world model has resulted in the fact that there are countless times in which by prioritizing a single way of life and a production system, many others have literally been turned off and disappeared, and one of the regions that has historically suffered this process is precisely Latin America. Society needs political, cultural and educational changes. If there were more engineers, doctors, and scientists with political power, empathy towards a situation like the pandemic would prioritize the most vulnerable (Done, 2020); hence, scientific training would be a fundamental part of post-pandemic reflections. Teaching science not only promotes the training of scientific adults, but also that of a literate society for decision-making, as well as the humanities and, above all, ethics.

Latin America is considered a region with enormous biological and cultural diversity; in other words, it is a region with great biocultural diversity (Robles-Piñeros et al., 2024). But, in the same proportion, it's a region with great disparity, inequity and poorness as the result of centuries of exploitation, looting and decimation by the countries of the global north. Since the fifteenth century, Eurocentrism brought with them a single way of seeing the world, a single reality, and a single system of thought that, consequently, produced a process of criminalization, persecution and consequent prohibition of any type of knowledge system that was not within the established canon (García-Canclini, 2004). This posture reduced the enormous cultural diversity to a pale and flat landscape of homogenizing worldviews, establishing a single way of knowing (Burke, 2022; Concheiro, 2023).

Throughout the colonial period, Latin American territories witnessed the devastation not only in biological terms but also in cultural and epistemological ways, due to the imposition of a knowledge system brought by modern Western lifestyles (Gentili, 2009). This phenomenon had a process of officialization through the imposition of educational systems that were controlled by the church, just after the period of the so-called independence struggles, most countries handed over control of education to the church. This process was known, in Colombia for example, as "El Concordato" (Peñaloza & Robles-Piñeros, 2022). Education was allocated to a citizen training program focused on Western standards and models, leaving behind any other form of knowledge and/or worldviews.

Nowadays, as García-Canclini (2004) affirms, we are moving from a multicultural world, where the juxtaposition of ethnicities and human groups predominated, to another world, globalized and intercultural. The distinction is important because it implies two different modes of social production: While multiculturalism only presupposes the acceptance of the heterogeneous, interculturality suggests that the different constitute each other and, thus, become what they are in relationships of negotiation, conflicts, and exchanges. This involves understanding societies where social interactions cause tensions, encounters, and disagreements (Molina-Andrade,

2020; Tsing, 2011). In the field of education, specifically science education, this phenomenon has not been foreign. On the contrary, in the last decades of the twentieth century and those leading up to the twenty-first century, the concern of academics and educators has been marked by the recognition and appreciation of other forms of knowledge to rescue knowledge systems that were decimated, forgotten, and segregated from both the social world and the school space (Molina-Andrade, 2020; El-Hani, 2022).

Intercultural scientific education under the idea of science as a cultural activity has demonstrated the need to build dialogue between science and other forms of knowledge (Baptista, 2018; Robles-Piñeros, 2023), not only on epistemological terms but also ontological ones, committed to the recognition of the people who make up social groups, their realities, needs and value systems. That makes scientific education a process committed not only to the epistemological aspects of science and other knowledge systems but also a process that addresses needs and realities that were "unmasked" after the pandemic (Pérez-García, 2021; Rapanta et al., 2021). For this chapter, we develop the idea that giving the ISE a character of political commitment allows for carrying out a process of training science teachers not only with cultural but also political sensitivity and reorientation of agendas for the Latin American countries.

Based on the above, this reflective essay proposes a discussion, based on philosophical and theoretical-methodological foundations, on the role of Intercultural Science Education (ISE) in the initial and in-service training of science teachers, to develop intercultural attitudes in the classroom closer to the otherness and a culturally and politically sensitivity profile, to dialogue and negotiate with different knowledge systems. While African and Latin American philosophies have extensively addressed intercultural dialogue, social ontology in the Global North still faces difficulties in the absence of well-established traditions of collaborative, intercultural, and interdisciplinary research (Ludwig et al., 2024). Simultaneously, this essay seeks to show that multiple approaches to philosophy are achievable by engaging scholars from different disciplines in dialogue with local populations.

In this way, the chapter goes through five sections: (1) A theoretical framework on Intercultural Science Education; (2) An approach to southern ontologies concept and the philosophical commitments to ISE; (3) A discussion towards a culturally and politically sensitive profile of science teachers in Latin America; (4) A proposal of Culturally and politically engaged biology teaching; and (5) Some considerations about the ISE as a political statement to reorient political agendas to Latin America.

11.2 A Framework for an Intercultural Science Education (ISE)

Nowadays, one of the great challenges for Intercultural Scientific Education (ISE) is facing the constant situations of hierarchies between the knowledge systems (El-Hani, 2022; Robles-Piñeros, 2023) that are present in the classrooms in the global south: curricula, programs, teaching materials, practices and, above all, the discourses that perpetuate themselves in a disparate relationship between Western Modern Science (WMS) and Indigenous Local Knowledge (ILK). In this sense, one of the main concerns in the countries of the Global South is how these knowledge systems are treated when they are considered, and the consequences of those attitudes. Throughout the history of scholarly education, scientific knowledge has been presented in the classroom and disseminated from a view of the superiority of Western science. This reaction is also generated by confusion facing environmental and ethical problems and social inequalities resulting from the dominant economic development model, mainly based on technology and knowledge (Massarini & Schnek, 2015).

This has led various social and cultural groups around the world, specifically in the Latin-American region, to become sceptical and critical about issues related to scientific knowledge and begin to defend the recovery of other forms of knowledge e.g. in Brazil, Netherlands, and South Africa, Ludwig et al. (2023, 2024), Robles-Piñeros et al. (2020), and Robles-Piñeros et al. (2023) in Chile with the work of Quintriqueo-Millán, S. y Quilaqueo-Rapimán (Quintriqueo-Millán et al., 2019) among many others (Gómez Galindo et al., 2014) around the Global South. Considering the intercultural character of societies in which we live, especially in times of hegemonic globalization (Santos, 2010; García-Canclini, 2004), it becomes inevitable to think of education as a space for cultural encounters and disagreements (Ladson-Billings, 1995; Aikenhead & Jegede, 1999; Baptista & El-Hani, 2009; Robles-Piñeros, 2023). It is essential, then, to recognize that, in the interactions of different people, in their diversity of ethnicities, cultures, genders, social groups, etc. (Tsing, 2011), not only conflicts emerge, but also possibilities for dialogue and exchange. In other words, an intercultural society presents several "contact zones," which challenge us to overcome dichotomous understandings and construct a more complex perspective.

According to some authors from the region (El-Hani, 2022; Molina-Andrade, 2020; Pinheiro & Baptista, 2021; Robles-Piñeros et al., 2024), ISE is a relatively new area of study in science education. Its roots can be found in the first part of the 1980s when research on cultural aspects of science education emerged. ISE research focused on the interactions between diverse peoples and cultures, often shaped by historical oppression, inequality, and conflict, and this worry also extends to the educational settings, raising concerns about the need to establish curricula, classroom designs, instructional strategies, in general, to create an institution of higher education that was "sensitive" to the cultural variety that "developed" nations were dealing with mass immigration (Aikenhead & Lima, 2009).

A main aspect of the ISE is its recognition of differences (Molina-Andrade, 2020), characterized for being aligned with a perspective of exchange within disparities and inequalities, seeking a dialogue that allows a collaborative interaction between knowledge systems (Peñaloza et al., 2023); today is recognized that ISE is distinguished by a dynamic interaction between knowledge systems. From the perspective of understand the classroom as an intercultural space (because at least two cultures will always be present, the scientific culture and that of the students), the

ISE seeks the develop of an educational system that is sensitive to acknowledge epistemological differences in terms of the search for a possibility for mutual support and the collective construction of knowledge (El-Hani, 2022; Robles-Piñeros et al., 2020; Ludwig & El-Hani, 2020) This perspective of science education finds its theoretical basis in studies like Aikenhead and Jegede (1999) and Banks (2008) with their proposal of cross-cultural education, among others these authors were pioneers in the field of cultural difference in science teaching, however, the current ISE perspective goes beyond the mere recognition and search for "cultural inclusion" and enters to the field of dialogue process not only in terms of epistemological differences, but also recognizing ontological and value differences, since this will allow us to go beyond a merely inclusive process in curricular terms.

Nowadays, some of the theoretical and methodological proposals that have consolidated the ISE as a strong, prominent, and relevant line of research for science education in the countries of the Global South[2] have been positioned. In the beginning, in the first decade of the 1990s, the ISE was configured as a proposal framed within the need to recognize the influence of context and culture in science's teaching and learning process. Later it was evident the need for structuring a theoretical-methodological framework for the development of the intercultural proposal, with several proposals built in Latin America only to mention a few, such as the conglomerates of relevance (Molina et al., 2009); bridges of intercultural dialogue (Molina & Mojica, 2013); the contextual cognition tables (Baptista, 2018; Baptista & Molina-Andrade, 2023); and more recently the so-called trading zones (Robles-Piñeros et al., 2020) and, Design of intercultural education as knowledge dialogue (El-Hani, 2022).

In the last decade, one of the pillars in the consolidation process of the ISE in Latin America and the global south is related to aspects that the decolonial approach has brought to the table to be discussed. One of those aspects is the issue of epistemological injustices and the hierarchization of knowledge, understanding the need to make visible the communities that, over time, have been exploited and stolen through the knowledge mining carried out by Western countries in the consolidation of Western Modern Science.

[2] The term "Global South" is not geographical, but geopolitical term, the "Global South" thus appears as a label that is both multiple and neutral to designate States historically relegated to the margins of the world order. In 1926, in an essay titled *Alcuni temi della questione meridionale,* the Italian communist and revolutionary theorist Antonio Gramsci used the term to refer to the inequalities between northern and southern Italy. In doing so, according to Garland Mahler, Gramsci applied a political framework that later informed the theorization of the South in the North: "a deterritorialized geography of the externalities of capitalism... to account for subjugated peoples within the borders of the richer countries, so that there is South in the geographical North and North in the geographical South».

11.3 Southern Ontologies and the Philosophical Commitments of ISE

The intercultural project develops its proposal based on Epistemological Pluralism (De Sousa-Santos, 2010), Scientific Pluralism (Ludwig & Ruphy, 2021), and Transdisciplinary Philosophy of Science (Ludwig, El-Hani, et al., 2023), recognizing that knowledge is produced by heterogeneous epistemic communities inside and outside academia, making it possible for other communities to exist, those other knowledge systems have their place located in their sociocultural contexts of origin (Chilisa, 2012). In this way, a proposed scientific education project prides itself on being sensitive to this epistemological diversity as a fundamental value of its epistemic position (Wiredu, 1998), but not only in epistemological terms, also engaged with ontological commitments of participants of the process, and their value systems, proposing emphasizes how knowledge systems are related since, depending on how the relationship between knowledge occurs, an intercultural position can be configured (Rist & Dahdouh-Guebas, 2006; Monceri, 2014).

Intercultural Science Education is based on two important foundations. The first is the teacher's development of an intercultural attitude to foster a dialogue process between knowledge systems. This can be seen from the development of a culturally sensitive profile when teaching science (Robles-Piñeros et al., 2023), capable of recognizing the different contact zones, zones of dialogue between systems and as a relationship between world visions or, in Galison's (2010) terms, negotiation zones or trading zones. The second point, derived from the first, accounts for a process of scientific training with an intercultural character, this means a teacher who understand the epistemic and ontological issues of science and its relationship with differentiated knowledge systems (Ludwig et al., 2024). This latter topic accounts for a treatment perspective aimed at developing an education that, in Freirean terms, addresses plurality and respect for difference (Freire, 2005; Peñaloza et al., 2023).

Regarding the ontological issue as a challenge for the intercultural project, it is necessary to recognize that any process of intercultural relationship will be framed by friction. To this meeting point between cultures, Tsing (2011) emphasizes the importance of recognizing that over time the processes of intercultural contact have been crossed by friction between cultures. This friction is the starting point for intercultural relations, which are framed by epistemic injustices, recognizing that, more than an epistemic challenge (Fricker, 2007), there is an ontological one observed from the perspective of the meeting of beliefs, realities, and different metaphysical framings (Bricker, 2016). It characterizes the need for a differentiated treatment and the creation of contacts, or negotiation zones (Galison, 2010), that allow finding meeting points between disparate knowledge systems from an ontological perspective (Ludwig & El-Hani, 2020; Ludwig et al., 2024).

Reflections on ontological and epistemological commitments are relevant to this educational-political project, especially to understand the epistemic framework or epistemological situations that guide the construction processes of individual and collective knowledge, particularly in educational communities. According to

Ludwig and collaborators (Ludwig et al., 2024), it is necessary in the process of knowledge production, to go beyond a utilitarian perspective of the ILK at the service of academic knowledge, or in the worst of situations, subordinated to Western knowledge. Therefore, when reference is made to intercultural dialogue (Boogaard & van Norren, 2021), it is necessary to understand that cultures do not dialogue itself, those who dialogue are the subjects that inhabit the cultures, and these subjects bring with them their epistemological commitments, their reality (ontologies) and the values systems that are the result of the interaction between the knowledge systems and the ontologies of each cultural group.

Therefore, science teachers must recognize other subjects from different cultures to expand their practices, worldviews, and epistemological frameworks without attributing epistemic or even ontological superiorities to one knowledge system. For this reason, ISE commits to epistemologies, ontologies, and values from the different communities, these communities with which the teacher will dialogue in the countries of the region, those communities that have been forgotten, oppressed, and left behind historically and, for this reason, ISE has a political commitment too.

11.4 Towards a Culturally and Politically Sensitive Profile in Science Teacher Education

When we refer to the science teacher's ability to recognize, value, and dialogue with knowledge systems other than scientific ones, understand their specificities, and promote fruitful exchanges for both knowledge systems. We are in accordance with Robles-Piñeros and collaborators (Robles-Piñeros et al., 2023) mentioned as a Culturally Sensitive Profile (CSP) of the science teacher. In this process, researchers realize how it is possible to identify a "teacher profile" sensitive to cultural diversity by analyzing their epistemological, ontological, and ethical commitments facing the cultural, epistemological, and ontological diversity of the communities. This becomes a key factor for the science teaching process that concerns interculturality.

As already mentioned, the ontological aspects and values of knowledge systems must also be considered, not only to forge a dialogue in terms of merely epistemological differences. Therefore, teacher training (pre-service and in-service) is a key piece in building intercultural scientific education (Baptista & Molina-Andrade, 2023), because, the science teacher must be aware not only of the epistemological positioning that it must have in the teaching process but must also be sensitive to ontological diversity since, based on the commitments it acquires about knowledge, reality, the world, what it should and should not be (Bazzul, 2016). It will develop a critical position regarding epistemological ethnocentrisms and guide activities in intercultural scientific education (Robles-Piñeros et al., 2023).

However, referring only to the development of a Culturally Sensitive Profile (CSP) is not enough when we know that there are issues that go beyond the merely educational field, and when we talk about intercultural scientific education, we refer to an educational process that is on the side of the oppressed (Freire, 2005; Peñaloza et al., 2023), the forgotten, the nobodies, those communities that have historically suffered all kind of violence, as a result of epistemological and ontological violence produced by models that have legitimized epistemological and ontological injustices throughout the countries of the global south (Parra-Romero, 2023). All these tensions emerge as a need to think about how the educational process addresses these disparities that the education model itself historically perpetuated (Giesinger, 2023).

For this reason, in this chapter, we propose to go beyond the development of a Culturally Sensitive Profile for science teachers, and we argue the need to develop a Politically Sensitive Profile (PSP) too, understanding that the educational process, being a cultural act is a political act (Stojanov, 2023). The idea of the politically sensitive profile is based on the assumption that, just as in the culturally sensitive profile, the science teacher must recognize and dialogue not only with the epistemological aspects (convergent and divergent), but also with the ontological and value aspects; of these latter topics are framed within a purely political field, because subjects, communities, and countries of the global south have always been decimated by international and national policies that have been imposed on them, and that have resulted in countless testimonial injustices, and for this reason, there is a commitment with the development of democratic education (Elgin, 2011; Culp et al., 2023).

A teacher's identity as an educational professional is consolidated to the extent that he recognizes the part of his history that leads to problematizing the professional and social "self" to configure a locus of enunciation. In addition, it is necessary to appropriate the different historical events that give rise to the complex teaching current that is nourished by culture, politics, and the economy. For this reason, political training and recognition of the struggles that have triggered the current state of education in every Latin American country are needed to vindicate it. For example, being a teacher in a country like Colombia has always involved struggles for social and labour recognition, in the nineteenth century the teacher was not recognized as an intellectual, but rather a subject with exemplary and impeccable conduct who must have the necessary training to teach. This brings a responsibility because he/she dedicates his life to service others. As Gallego (1991) mentions, the teacher is a subject monitored by all those involved in the educational community, such as pairs, principals, parents, society in general, and at a macro level by the state.

On the other hand, another mechanism of control is the evaluation, and textbooks also serve as tools that monitor the processes carried out by the teacher. Being that they are in the public domain to "verify" compliance with what the law expects to do curricularly; then the teacher is at the service of the whims of the government, which, of course, moves under interests evidenced by the different decrees and laws imposed that disadvantage teachers. It can be thought that being a socially exposed subject for a long time reduces the general interest in being teachers, the consequence can be shown by crises in multiple countries related to teaching shortages.

This further justifies the need for a Politically Sensitive Profile (PSP). To build status in the teaching profession, it is considered a priority that teachers have initiatives to update their knowledge from different fields and that this leads the teacher to be a researcher in their field to strengthen education at various levels (from schools to universities), impeding the advance of the sloppy idea that any professional can be a teacher without pre-service training (Culp et al., 2023). This occurs, for example, in Colombia with Decree 272 of 1998, which is repealed by the government of Álvaro Uribe, and gives way for Decree 2566, which deprofessionalizes the profession allowing any professional to teach, causing the loss of the teachers' identities and overshadowing their status. Being a professional in education involves recognizing the historical conjunctures that shape current education and understanding them from the epistemological, ontological, ethical, and political aspects.

In Latin America, there are a lot of social groups with economic vulnerability, which increases difficulties in accessing a balanced diet. In Colombia, reports since 2015 by the National Survey of Nutritional Situation (ENSIN) of households facing food shortage due to the low incomings. It is a social and political concern because food security depends on governments, but every family is also responsible when deciding the products they access. Low consumption of sugary drinks, sausages, and packaged foods is expected for a healthy diet. These products were of low priority during the COVID-19 pandemic, which increased the purchase of first-order necessities such as fruits and vegetables (Bejarano-Roncancio et al., 2020), strengthening agricultural production.

Based on the above situation, it is evident the importance of a perspective in intercultural scientific education for biology education, for example, that legitimizes the field and crops as a curricular dimension for all Latin America in rural and urban education by including I-P-A-R perspective: (1) Interculturality and citizen participation, (2) Pedagogies for peace, (3) Agriculture and, (4) Responsible production projects, aiming at food security (FS) and sovereignty. Suppose the science curriculum is crossed from early childhood, with these 4 dimensions, in that case, the application of scientific knowledge for making nutritional and political decisions becomes tangible. It even contributes to SA from home or communities with citizen participation in formal and non-formal biology education.

It is worth mentioning that this proposal is not about including more aspects to the science curriculum, it is about turning the curriculum towards an intercultural turn, involving the families, the children, who are not only farmers, or not all of them are professionals, but they are people with experiences and cultural and biological (biocultural) knowledge that enrich learning. Including the countryside in the traditional curriculum, without political reflections such as the privatization of food, will be empty content without significance for students. The intercultural curriculum recognizes the voices and knowledge of others and creates opportunities to reflect the amount of loss in terms of biodiversity and food security, in favour of the "capitalocene" as the axis of the current biospheric crisis (Moore, 2016).

11.4.1 On a Proposal of Culturally and Politically Engaged Biology Teaching

We propose a curriculum for Biology Education based on teachers' sociopolitical commitments, knowing that their students' decision-making must be based on informed positions about the destiny of nature and society. The scientific knowledge that does not recognize intercultural constructions without real applications and transforming the relationship of human beings with the entire environment in a responsible way results in reductionist concepts, promoting a contrary vision of a scientific education approach with political views of peace between social individuals and biodiversity.

Achieving peace with biodiversity implies that, from the history of every individual, their perception of nature and biodiversity has improved. According to Gardner and Stern (2008), every process must include psychological, evolutionary, and socio-ecological processes as direct influencers of how people think about the environment and how they react to cross-border situations. When an individual experiences some environmental change in their close or family environment, this will have a greater impact on them than if they only recognize an external event somewhere else in the world.

One way to be at peace with the environment could be from the interactional theory, where people are more aware when they have seen some environmental change because they will reproduce this information, in addition, "he worries about trying to do something to help when there is an attachment with a specific place" (Gifford et al., 2011). Then, emotional bonding with specific places can be generated from childhood, because when people have passed this stage without "attachments to nature", they lose more than an individual relationship with the landscape, "a cultural element is lost, thus, the result is the decrease of collective memory" (Wertsch, 2009).

One reason why people do not help ethically and socially to take care of nature may be that they believe that climate change is not real, or perhaps they are already normalized knowing many contaminated ecosystems, considering water scarcity as normal in the "underdeveloped countries", or that many species must be in extinction danger because "everything changes". "Cognitive dissonance" is the ability to ignore evidence contrary to existence (Wicklund & Brehm, 2013). This innate optimism has allowed humanity to remain calm, without the intention of fostering a change in the face of global environmental problems that are deteriorating all ecosystems (Beckford & Nahdee, 2011). The educational field implies curricular transitions, understanding that today there are curricular standards that must be modified towards science education from ethical and political relationships with nature by promoting attachments.

Now, we will explain the four dimensions we propose for the different levels of scientific education. We take as a starting point the Basic Standards of Curricular Competencies from Colombia (as an example) from the Ministry of National Education (MEN, 2004), which converge with the curricular comparative studies of what is taught in science for Latin America according to Furman (2020). Most countries align their vision with "international standards of good practices and with the principal objective of scientific literacy as a fundamental preparation for a world in which scientific knowledge and the development of thinking skills are fundamental to developing full citizenship" (Furman, 2020).

It is important to start from the document so that the proposal of a curriculum for science teaching, with 4 dimensions from a cultural-political profile, is easily suitable in different educational spaces, without fear of "deviating" from the political-educational demands (Fig. 11.1).

Interculturality and citizen participation—it is essential to dialogue through the recognition, valuing, and interaction with worldviews of subjects owned from formal, informal, and Indigenous, Traditional, or Local knowledge. Furthermore, it expands the vision about the knowledge of some cultures, which is patented or scientifically disseminated. This vindicates the role of the citizen by being an active participant in the decision-making and actions in his territory, also understanding the natural phenomena implicit in daily life.

Pedagogies for peace—it requires alterations in human behaviour because the topic is not just about being at peace with other humans, but about reconciling the relationship with nature, moving forward seeing it more than a resource, and

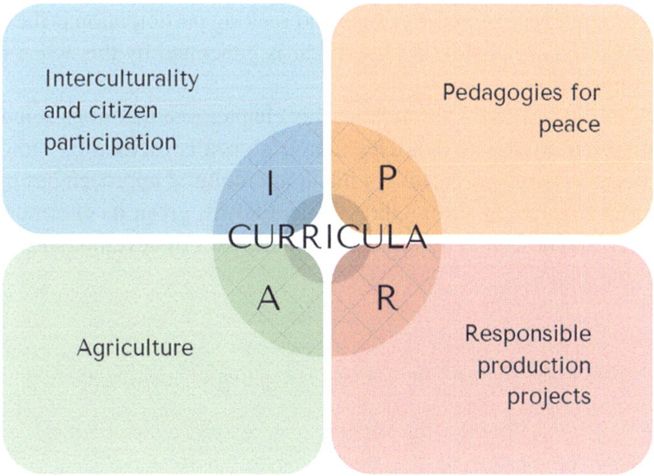

Fig. 11.1 The four dimensions (I-P-A-R) for an Intercultural Science Education Curriculum (ISEC) with Socio-political Engagement (SPE)

understanding ourselves as part of it. Consolidating the political position leads to reflections on the importance of having equal conditions for a dignified life, not just for covering primary needs, but embracing differences.

Agriculture—addresses the need to take the discussion seriously by our countries about food sovereignty and sovereignty over our crops, it is a reality that it is increasingly difficult to approach agriculture without considering the enormous impact of GMOs, patented seeds, and the privatization of the genetic code and, therefore, the privatization of life itself.

Responsible production projects—gather learning in natural sciences from the four dimensions, which strengthen research processes and try to solve real problems. Sauvé (1999) points out that environmental issues require education due to the lack of democracy, which reproduces the same obstacles in different parts of the world. Therefore, civil society must commit to the eco-social, there must be critical surveillance of the public and proposals regarding the management of resources.

In society, learning must foster an eco-social action reflectively, since environmental knowledge emerges from all contexts, so scientific education is also important because it educates about eco-citizenships, in this way democracy is the centre of the proposal. Consequently, eco-citizenship moves to a critical, ethical, political, and ontological dimension (Robles-Piñeros & Abella, 2021). Then the relationship with the teaching process emerges, it is responsible for teaching how to live with values and that experiences are a consequence of them.

By listening to what students say regarding the environmental problems surrounding them, they become involved with actions and ideas. It is better to show students what is happening by making approaches that may not lead to social change but are worth taking the risk if it helps them be critical and reflective (Sauvé, 1999). One of the consequences of being critical and seeking participation is the possibility of falling into activism; however, a great fear is generated by this when it is worse to fall into immobility.

Although in general terms throughout this chapter one mentions science education, it is important to denote that the example shown is focused on biology teaching; this is because Biology teaching with an intercultural approach has much more development in theoretical and methodological terms, given its enormous conceptual framework and how directly related it is to the axes of development of the IPAR proposal (Table 11.1).

Table 11.1 A proposal for Biology Intercultural Education with cultural and political commitment for Latin America

<table>
<tr><th colspan="5">Biology teaching curriculum for ise (example)</th></tr>
<tr><th rowspan="2">Transversal dimensions (I-P-A-R)</th><th rowspan="2">Six axis for a socio-political profile</th><th colspan="3">Main natural phenomena (community among 13–15 years)</th></tr>
<tr><th>Characteristics and importance of genetics information</th><th>Energy transference</th><th>Genetic manipulation</th></tr>
<tr><td>Interculturality and citizen participation</td><td rowspan="4">1. Valuation of the public and the common
2. Recognition of diverse ways of life and knowledge
3. Building consensus and dissent
4. Valuation of different ways of participation and knowledge production
5. Appropriation of participatory mechanisms
6. Collective construction of proposals and alternatives and co-production of knowledge</td><td>Hereditary diseases and treatments from different views (including indigenous) (2)
Filial relationships, ways to establish common ancestors (2,3)</td><td>A historical journey about the different types of energy used by humans, from fire to nuclear energy (6).
Who in my territory still does not have power and why? (1,5).</td><td>Genetic manipulation, different positions (3,4). Processes for obtaining drinks based on corn and rice: Fermentation (4).</td></tr>
<tr><td>Pedagogies for peace</td><td>Biological, sociocultural, and sexual diversity as part of difference and otherness. (2,3)</td><td>Let's generate a thermodynamic system to solve a community problem (1,6).</td><td>The importance of accessing medical, technological, and educational services on equal terms (1,6).</td></tr>
<tr><td>Agriculture</td><td>Soil, humidity, and the environment: Criteria for selection processes in agricultural production (5)</td><td>The physical and chemical properties of chemical compounds used in agro-industrial processes and their influence on food security and sovereignty (3,6)</td><td>Microbiology, and indigenous and peasant knowledge in agriculture (2). Transgenics for the organism and ecosystems (4).</td></tr>
<tr><td>Responsible production projects</td><td colspan="3">Identify needs such as: gardens, worm farms, composters, that improve responsible production processes for the community. Having these spaces, identifying problems, and formulating projects with possible solutions to perpetuate them.</td></tr>
</table>

11.5 Final Remarks: The ISE as a Political Statement

We position ourselves and defend scientific education in and for indigenous, local, peasant, and Afro communities that recognize and respect epistemic diversity in classrooms. This process must be intercultural, problematizing the interactions between scientific knowledge and practices and indigenous-local knowledge and practices about nature. We defend a scientific education that aims to train scientifically literate citizens, in the sense of expanding their cultural knowledge with scientific knowledge, being increasingly capable of acting critically and autonomously in the face of the complexities of their sociocultural environments and those who are scientifically and technologically influenced, without, however, being compelled to deny the knowledge and practices that integrate their cultural identities, due to a supposed epistemic superiority of Western science.

Understanding the educational process as a space for dialogue that is not only inter-epistemic, but also inter-ontological, and with a series of ethical characteristics in promoting an intercultural attitude in teaching practice, is, without a doubt, a huge step in the process of consolidating research in intercultural scientific education. It is necessary to consider the epistemological and ontological beliefs not only of students when proposing a pedagogical action but also of teachers, who are key players in negotiating meanings in the teaching process.

This type of approach allows how teachers relate to knowledge and its nature to become visible; the need to problematize questions about how cultural diversity (epistemic, ontological and ethical) is faced in the classroom and how it is possible to characterize a culturally sensitive profile in science teaching, a vital issue in Latin American countries, in which each classroom is an intercultural space. An intercultural scientific education is based on two important points: the first concerns the development of an intercultural attitude on the part of the teacher, a process of dialogue between systems of knowledge, seen from the development of a culturally sensitive profile in teaching capable of recognizing different contact zones, zones of dialogue between systems and as a relationship between worldviews; and second, a politically sensitive profile, that accounts for the agency of the science teacher with political awareness and the ability to read the socio-political contexts in which he or she operates professionally, this as a vital point.

The latter, in a way, responds to a perspective of approach aimed at developing an education that, in Freirean terms, attends to plurality and respect for difference that allows meeting points between disparate knowledge systems from an epistemic and ontological, but which always proposes paths of dialogue and coexistence to seek socio-political solutions in different cultural contexts. And finally, for this reason, it is proposed that the ISE is not only a relevant research perspective for Latin America, but it is a political declaration as a process of vindication and defence of the biocultural heritage of our region.

References

Aikenhead, G. S., & Jegede, O. J. (1999). Cross-cultural science education: A cognitive explanation of a cultural phenomenon. *Journal of Research in Science Teaching, 36*, 269–287. https://doi.org/10.1002/(SICI)1098-2736(199903)36:3<269:AID-TEA3>3.0.CO;2-T

Aikenhead, G., & Lima, K. E. C. (2009). Science, culture, and citizenship: Cross-cultural science education. *Revista Brasileira de Pesquisa em Educação em Ciências, 9*(3), 1–15.

Arons, W. (2023). We should be talking about the Capitalocene. *TDR: The Drama Review, 67*(1), 35–40. https://doi.org/10.1017/S1054204322000697

Banks, J. (2008). *An introduction to multicultural education* (5th ed.). University of Washington.

Baptista, G. C. S. (2018). Tables of contextual cognition: A proposal for intercultural research in science education. *Cultural Studies of Science Education, 13*, 845–863. https://doi.org/10.1007/s11422-017-9807-3

Baptista, G. S. C., & El-Hani, C. N. (2009). The contribution of ethnobiology to the construction of a dialogue between ways of knowing: A case study in a Brazilian public high school. *Science & Education, 18*, 1–18.

Baptista, G. C. S., & Molina-Andrade, A. (2023). Science teachers' conceptions about the importance of teaching and how to teach Western science to students from traditional communities. *Hu Arenas, 6*, 704–731. https://doi.org/10.1007/s42087-021-00257-4

Bazzul, J. (2016). *Ethics and science education: How subjectivity matters*. Springer Briefs in Education Vienna.

Beckford, C., & Nahdee, R. (2011, September). Teaching for ecological sustainability incorporating indigenous philosophies and practices. What work? In *Research into practice research monograph #36*. Ontario Association of Deans of Education.

Bejarano-Roncancio, J. J., Samacá-Murcia, L., Morales-Salcedo, I. S., Pava-Cárdenas, A., Cáceres-Jeréz, M. L., & Durán-Agüero, S. (2020). Caracterización de la seguridad alimentaria en familias colombianas durante el confinamiento por COVID-19. *Revista Española de Nutrición Comunitaria, 4*, 235–241.

Boogaard, B., & van Norren, D. (2021). Development' perspectives from the global south: Learning from Ubuntu and Buen Vivir philosophies. In D. Ludwig, B. Boogaard, P. Machnaghten, & C. Leeuwis (Eds.), *The politics of knowledge in inclusive development and innovation*. Routledge.

Bricker, P. (2016). Ontological commitment. In E. Zalta (Ed.), *The Stanford encyclopedia of philosophy*. https://plato.stanford.edu/archives/win2016/entries/ontological-commitment/

Burke, L. E. C. A. (2022). Foregrounding intersectionality in considerations of diversity: Confronting discrimination in science teacher education. *Research in Science Education, 52*, 1157–1170. https://doi.org/10.1007/s11165-021-10001-1

Chilisa, B. (2012). *Indigenous research methodologies* (306 p.). SAGE Publications.

Concheiro, L. (2023). *Marxismo neotropical: variaciones latinoamericanas sobre los modos de producción*. Unidad de Investigación sobre Representaciones Culturales y Sociales. Universidad Nacional Autonoma de México. ISBN. 978-607-30-8129-0.

Culp, J., Drerup, J., & Yacek, D. (2023). *The Cambridge handbook of democratic education*. Cambridge University Press.

De Sousa-Santos, B. (2010). Para além do pensamento abissal: Das linhas globais a uma ecologia de saberes. In: B. de Sousa-Santos & M. P. Meneses (eds.), *Epistemologias do sul* (2.ª ed.) (pp. 23–71). Almedina/CES.

Done, W. I. T. B. (2020). Post-pandemic science and education. *American Journal of Physics, 88*, 518.

Elgin, C. (2011). Science, ethics, and education. *Theory and Research in Education, 9*(3), 251–263. https://doi.org/10.1177/1477878511419559

El-Hani, C. N. (2022). Bases teórico-filosóficas para o design de educação intercultural como diálogo de saberes. *Investigações Em Ensino De Ciências, 27*(1), 01–38. https://doi.org/10.22600/1518-8795.ienci2022v27n1p01

Freire, P. (2005). *Pedagogia do Oprimido* (8th ed.). Paz e Terra.

Fricker, M. (2007). *Epistemic injustice: Power and the ethics of knowing*. Oxford University Press.

Furman, M. (2020). *Aprender ciencias en las escuelas primarias de América Latina*. UNESCO: Montevideo. https://unesdoc.unesco.org/ark:/48223/pf0000375199consultado, 19(10), 2021.

Galison, P. (2010). Trading with the enemy. In M. Gorman (Ed.), *Trading zones and international expertise: Creating new kinds of collaboration* (pp. 25–52). The MIT Press.

Gallego, A. A. (1991). El maestro: historia de un oficio. *Revista Colombiana de educación*, 22–23.

García-Canclini, N. (2004). *Diferentes, desiguales y desconectados. Mapas de la Interculturalidad*. Gedisa.

Gardner, G. T., & Stern, P. C. (2008). The short list: The most effective actions US households can take to curb climate change. *Environment: Science and Policy for Sustainable Development, 50*(5), 12–25.

Gentili, P. (2009). Marchas y contramarchas. El derecho a la educación y las dinámicas de exclusión incluyente en América Latina (a sesenta años de la Declaración Universal de los Derechos Humanos). *Revista Iberoamericana De Educación, 49*, 19–57.

Giesinger, J. (2023). The COVID-19 pandemic and democratic education. In J. Culp, J. Drerup, & D. Yacek (Eds.), *The Cambridge handbook of democratic education (Cambridge handbooks in education)* (pp. 594–606). Cambridge University Press. https://doi.org/10.1017/9781009071536.039
Gifford, R., Kormos, C., & McIntyre, A. (2011). Behavioral dimensions of climate change: Drivers, responses, barriers, and interventions. *Wiley Interdisciplinary Reviews: Climate Change, 2*(6), 801–827.
Gómez Galindo, A. A., García Franco, A., & González Galli, L. (2014). Enseñanza de la evolución desde una perspectiva intercultural dialógica. *Revista CITECSA, 5*(8), 91–103.
Ladson-Billings, G. (1995). Toward a theory of culturally relevant pedagogy. *American Educational Research Journal, 32*(3), 465–491. https://doi.org/10.3102/00028312032003465
Ludwig, D., & El-Hani, C. N. (2020). Philosophy of ethnobiology: Understanding knowledge integration and its limitations. *Journal of Ethnobiology, 40*(1), 3–20. https://doi.org/10.2993/0278-0771-40.1.3
Ludwig, D., & Ruphy, S. (2021). Scientific Pluralism. *The Stanford Encyclopedia of Philosophy* (Winter 2021 Edition), Edward N. Zalta (ed.), https://plato.stanford.edu/archives/win2021/entries/scientific-pluralism/
Ludwig, D., El-Hani, C., Gatti, F., Kendig, C., Kramm, M., Neco, L., Nieves-Delgado, A., Poliseli, L., Renck, V., Ressiore C., Reyes-Galindo, L., Rickard, T., De La Rosa, G., Turska, J., Vergara-Silva, F., & Wilson, R. (2023). Transdisciplinary philosophy of science: Meeting the challenge of indigenous expertise. *Philosophy of Science*.
Ludwig, D., Banuoku, D. F., Boogaard, B., El-Hani, C., Guri, B. Y., Kramm, M., Renck, V., Ressiore, C. A., Robles-Piñeros, J., & Turska, J. (2024). Southern ontologies. Reorienting agendas in social ontology. *Journal of Social Ontology, 10*(2). https://doi.org/10.25365/jso-2024-7691
Massarini, A., & Schnek, A. (2015). *Ciencia entre todxs: Tecnociencia en contexto social: Una propuesta de enseñanza*. Ciudad Autónoma de Buenos Aires, Argentina.
MEN. (2004). Estándares Básicos de Competencias en Ciencias Naturales. *Formar en ciencias, el desafío*. Recuperado de. https://www.mineducacion.gov.co/1780/articles-81033_archivo_pdf.pdf
Molina, A., & Mojica, L. (2013). Enseñanza como puente entre conocimientos científicos escolares y conocimientos ecológicos tradicionales. *Magis, 6*, 37–53. https://doi.org/10.11144/Javeriana.m6-12.ecpe
Molina, A., Martínez Rivera, C. A., Mosquera Suárez, C. J., & Mojica Rios, L. (2009). Diversidad cultural e implicaciones en la enseñanza de las ciencias: Reflexiones y avances. *Revista Colombiana De Educación*, (56). https://doi.org/10.17227/01203916.7582
Molina-Andrade, A. (2020). Science education research in South America, social cohesion and cultural diversity. In C. El-Hani, M. Pietrocola, E. Mortimer, & M. Otero (Eds.), *Science education research in Latin America*. Brill.
Monceri, F. (2014). Taking diversity seriously: On the notion of intercultural philosophy. In W. Sweet (Ed.), *What is intercultural philosophy?* (pp. 81–94). The Council for Research in Values and Philosophy.
Moore, J. W. (2016). Anthropocene or Capitalocene? Nature, history, and the crisis of capitalism. *Sociology Faculty Scholarship* https://orb.binghamton.edu/sociology_fac/1
Moore, J. W. (2017). The Capitalocene, part I: On the nature and origins of our ecological crisis. *The Journal of Peasant Studies, 44*(3), 594–630. https://doi.org/10.1080/03066150.2016.1235036
Parra-Romero, A. (2023). *Knowing from conflict: Interculturality as a space of interaction for the production of knowledges*. Latin American Science, Technology and Society. https://doi.org/10.1080/25729861.2023.2207866
Peñaloza, G., & Robles-Piñeros, J. (2022). How the relation between religion, politics, and science (R-P-S) gave rise to school biology: The circulation of Darwinism in Colombian textbooks (1870–1977). *SN Social Science, 2*, 137. https://doi.org/10.1007/s43545-022-00390-z
Peñaloza, G., Robles-Piñeros, J., & Baptista, G. C. S. (2023). Science education and cultural diversity: Freire's concept of dialogue as theoretical lens to study the classroom discourse of

science teachers. *Cultural Studies of Science Education, 18*, 95–114. https://doi.org/10.1007/s11422-023-10158-3

Pérez-García, Á. (2021). Retos y desafíos de la educación post pandémica. *Aula De Encuentro, 23*(1), 1–4. https://doi.org/10.17561/ae.v23n1.6246

Pinheiro, P. C., & Baptista, G. C. S. (2021). Em busca de referenciais culturais para a educação científica. In G. C. S. Baptista, P. C. Pinheiro, & L. M. S. Farias (Eds.), *Educação científica por meio da interculturalidade de saberes e práticas*. EDUFBA.

Quintriqueo-Millán, S., Arias-Ortega, K., & Maheux, G. (2019). Contenidos disciplinarios del currículum escolar en el marco de un pluralismo epistemológico-intercultural en contexto indígena. In S. Quintriqueo-Millán, & D. Quilaqueo-Rapimán (Org.), *Educación e Interculturalidad: Aproximación crítica y Decolonial en contexto indígena*. Ediciones Universidad Católica de Temuco.

Rapanta, C., Botturi, L., Goodyear, P., et al. (2021). Balancing technology, pedagogy and the new normal: Post-pandemic challenges for higher education. *Postdigital Science and Education, 3*, 715–742. https://doi.org/10.1007/s42438-021-00249-1

Rist, S., & Dahdouh-Guebas, F. (2006). Ethnosciences: A step towards the integration of scientific and indigenous forms of knowledge in the management of natural resources for the future. *Environment, Development and Sustainability, 8*(4), 467–493. https://doi.org/10.1007/s10668-006-9050-7

Robinson, W. (2020). Global capitalism post-pandemic. *Race & Class, 62*(2), 3–13. https://doi.org/10.1177/0306396820951999%20journals.sagepub.com/home/rac

Robles-Piñeros, J. (2023). Hacia la consolidación de una Educación Científica Intercultural: Una propuesta de investigación necesaria para América Latina. Ricardo, C. (Dir.). Cano, J. (Dir.). Ruiz, A. (Dir.). *Educación intercultural y las mediaciones tic: un abordaje desde la interdisciplinariedad*. Editorial Universidad del Norte. https://editorial.uninorte.edu.co/gpd-educacion-intercultural-y-las-mediaciones-tic-un-abordaje-desde-la-interdisciplinariedad-9789587895285-652070a9c5c46.html

Robles-Piñeros, J., & Abella, S. (2021). Sobre la pertinencia de una didáctica de la ecología: Una perspectiva para América Latina. In E. F. Amortegui & J. Mosquera (Eds.), *Didáctica de las Ciencias Naturales: perspectivas latinoamericanas: aportes a la formación del profesorado y la educación científica* (1st ed.). Universidad Nacional de Tierra del Fuego.

Robles-Piñeros, J., Ludwig, D., Baptista, G. C. S., & Molina-Andrade, A. (2020). Intercultural science education as a trading zone between traditional and academic knowledge. *Studies in History and Philosophy of Science Part C: Studies in History and Philosophy of Biological and Biomedical Sciences, 84*, 11337. https://doi.org/10.1016/j.shpsc.2020.101337

Robles-Piñeros, J., Baptista, G. C. S., & Molina-Andrade, A. (2023). Caracterizando um Perfil Culturalmente Sensível (PCS) no ensino de biologia: Subsídios na formação continuada de professoras de ciências para uma educação científica intercultural (ECI). *Investigações Em Ensino De Ciências, 28*(1), 39–55. https://doi.org/10.22600/1518-8795.ienci2023v28n1p39

Robles-Piñeros, J., Molina-Andrade, A., Baptista, G., Amador-Rodriguez, R., & Peñaloza, G. (2024). Approaches with an intercultural focus for biology education in Latin America. In A. Marzabal & C. Merino (Eds.), *Rethinking science education in Latin-America. Contemporary trends and issues in science education* (Vol. 59). Springer. https://doi.org/10.1007/978-3-031-52830-9_6

Santos, B. S. (2010). *Descolonizar el saber, Reinventar el Poder*. Ediciones Trilce, Montevideo.

Sauvé, L. (1999). La educación ambiental entre la modernidad y la posmodernidad: en busca de un marco educativo de referencia integrador. *Tópicos, 1*(2), 7–27.

Stojanov, K. (2023). Global justice and democratic education. In J. Culp, J. Drerup, & D. Yacek (Eds.), *The Cambridge handbook of democratic education*. Cambridge University Press.

Tsing, A. L. (2011). *Friction*. Princeton University Press. ISBN: 9781400830596.

Waitzkin, H. (2023). "Post"-pandemic capitalism: Reform or transform? Comment on "ensuring Global Health equity in a post-pandemic economy". *International Journal of Health Policy and Management, 12*, 7936.

Wertsch, J. V. (2009). Collective memory. In *Memory in mind and culture* (pp. 117–137). Cambridge University Press.
Wicklund, R. A., & Brehm, J. W. (2013). *Perspectives on cognitive dissonance*. Psychology Press.
Wiredu, K. (1998). Can philosophy be intercultural? An African Viewpoint. *Diogenes, 46*(184), 147–167.

Jairo Robles-Piñeros graduated in Biology at the National Pedagogical University (Colombia) and holds master's degree in History, Philosophy and Science Teaching from the Federal University of Bahia (Brazil), PhD in History, Philosophy and Science Teaching from the Federal University of Bahia, and PhD in Education from the Interinstitutional Doctorate in Education from the Francisco José de Caldas District University. He is a professor of the master's program in Contemporary Studies in the Biology Teaching and associated teacher at biology department at National Pedagogical University (Colombia). He is a member of INTERCITEC (Interculturality, Science and Technology) research group. He has research experience in the fields of science education, biology education, ethnobiology and conservation, Intercultural science education, and philosophy and history of biology.

Susana Abella graduated in Biology, holds a master's degree in Education in Natural Sciences and technology, and is PhD candidate in Education from Interinstitutional Doctorate in Education (DIE) at the Francisco José de Caldas District University. She currently works as a Natural Sciences teacher linked to the District Education Secretariat (SED) at the Aquileo Parra School. She has worked at the university level in academic spaces such as Special Educational Needs (NEEs) from accessibility, ICT, and inclusive didactical design for pre-service teachers at the District University. She belongs to the Experimental Science Teaching Research Group (GREECE); she has research in modeling for teaching biology and ecology through Information and communication technology (ICT) (video games and Geographic Information Systems).

Chapter 12
Rethinking Scientific Literacy Based on Covid-19 Experiences from One Caribbean Country

Joanne Nazir

12.1 Introduction

For Trinidad and Tobago (T&T), the intense period of the Covid-19 pandemic occurred between March 2020–December 2022. For almost three years, the country, like most of the world, was gripped in the peculiar situation that threatened to change human lives in unforeseeable ways. During that time, I was living on the island of Trinidad, working as a science education lecturer at a local university. For those three years, I lived mostly in an unattached dwelling house in a rural part of the island, did most of my professional work online, did no international travel and ventured out sparingly mostly for basic needs like food, pharmaceuticals, and financial errands. Since the island had relatively strong digital communication capacity, I was able to keep well-informed of what was happening in the world, despite my physical isolation, through the internet, social media, cable television, and local media outlets (radio, television and newspapers). Indeed, a noteworthy characteristic of T&T during that time was that information was never disrupted or in short supply. From the very beginning of the pandemic, citizens were inundated with information from government sources, non-governmental agencies, businesses, colleagues, friends and family; via the media modes listed above. Many opportunities were also provided for the public to share their thoughts, ideas and questions via call in radio and local television programs.

During the first months of the pandemic science educators worldwide had begun assessing what was happening and thinking about the implications for science education (e.g. Bencze & Alsop, 2020; Verma et al., 2020). I also began paying attention to reactions of the T&T public to the pandemic, actively analysing what I was

J. Nazir (✉)
University of the West Indies, St. Augustine, Trinidad and Tobago
e-mail: Joanne.Nazir@sta.uwi.edu

C. B. Moura (ed.), *A Sociopolitical Turn in Science Education*, Contemporary Trends and Issues in Science Education 62,
https://doi.org/10.1007/978-3-031-78586-3_12

witnessing for insights that could inform local science education endeavors. This work resulted in a published paper, Nazir (2021), in which I discussed emerging trends I observed in the early part of the pandemic and their implications for local science education. The paper encouraged local science education stakeholders to consider additional options for teaching science including explicit teaching of the nature of science, adopting a culturally relevant science-in-context approach and urgent development of appropriate digital pedagogies.

It is now August 2024, and the declaration of the end of the pandemic by the World Health Organisation is over a year old (United Nations, 2023). Life on the islands is still recovering. Workplaces have long returned to face-to-face, international travel has restarted and socioeconomic life is apace again. In public, people have re-established greeting each other with hugs, masks are rarely worn and while public wash sinks are still visible everywhere, very few persons seem to be using them. In the international academic world, much has been written on the impact of the pandemic on science education (e.g. Pietrocola et al., 2021; Ray & Srivastava, 2020; Reiss, 2020). Major lessons coming out include: the need for science teachers to be flexible and adaptable in their teaching, the limitations and opportunities of using online platforms to teach science, and the emotional aspects of science learning. While reading these publications, I have continued to think about the T&T Covid experience and compare it to the international experience. Many of the conclusions drawn by international scholars resonate in my local context. But there are aspects of the T&T experience that are not fully reflected in this published literature. In this paper, I focus on three peculiar aspects of the T&T Covid-19 experience and speculate on the meaning of these for scientific literacy internationally. These are the T&T public's: (i) misconceptions of the nature of science, (ii) naïve perceptions of the relationships between science and society; and (iii) a preference for alternative non-scientific ideas in leading their lives.

12.2 Science Education in the T&T Context

Foreigners always asking me why in Trinidad people so happy. What to do? Man, I have to tell them the truth. (Lord Funny, 1966)

Trinidad and Tobago is the southernmost country in the Caribbean island chain. It is made up of two separate islands that exist in close proximity, but exhibit differing geographic and sociocultural features. Trinidad, the larger of the two islands, is rocky as it was formed when it broke away from the South American mainland millennia ago. Tobago, on the other hand, is a coral island on a volcanic base that grew out of the ocean. Both islands, like most Caribbean territories, have a long colonial past, having been settled and fought over by various European powers during the 15–19th centuries. For much of this period, the islands had separate histories because they were separately administered, but that changed in 1889 when the British incorporated both, into a single political entity. During the colonial period, the islands were used as plantations to supply the world market with sugar, cocoa, coffee and sea island cotton. Initially, Amerindian natives were used as slave labor on these plantations, but that population was quickly decimated and replaced by large numbers of African slaves and later indentured labourers from India. For various sociopolitical reasons, the population has continued to be supplemented over the years by immigrants from China, the Middle East and South America. After a long period of colonial rule, the twin-island country gained independence from Britain in 1962 (Williams, 1964). Since then, it has steered its own course in governing and developing itself as an autonomous nation in the world.

Currently, the country has a population of 1.4 million (Government of Trinidad and Tobago, Central Statistical Office, 2017). This population is mixed, multireligious, multiethnic and culturally tolerant. Indeed, the twin traits of religiosity and tolerance are traits enshrined in the national anthem, which proudly describes T&T as a place where "every creed and race find an equal place" (Castagne, 1962). At the same time, citizens also carry the scars of the many years of slavery and oppression the country suffered in the past. Today, this latter legacy continues to be manifested in the persistent existence of high levels of crime, violence and corruption in the society (McVeigh, 2023). Yet despite some negative traits and its small size, T&T is relatively well-developed in the region in terms of healthcare, education, infrastructure and digital technologies. This is because the country is well-endowed with petroleum resources which, since the early 1980s, have made the country one of the richest Caribbean territories. Other than its mineral wealth, T&T is known worldwide for being the birthplace of artforms like *soca* and *chutney* music, steelpan and a unique carnival festival; cuisine that are fusion mixes reflecting its varied population; and its own local sociocultural norms, dialect and humor. The islands have also produced several internationally known athletes. Today, the country is acknowledged as a unique place and *Trinis*, the fond term for citizens of the islands, a unique people. Trini traits are perhaps best illustrated in the words of local soca lyrics:

> *As crazy as we might seem to be*
> *We still fight to be a family*
> *Indian, African or a Chinee*
> *Syrian, French-Creole and Portuguese*
> *We vex with a spirit fiery*
> *Some people say God is a Trini...*(Rudder, 2003)

Extending from its colonial past, the local education system follows the British style. There are four levels of education: early childhood, primary, secondary and tertiary. Education is compulsory and has been provided freely for over 40 years at the primary and secondary levels. The first three levels of education are administered by a local Ministry of Education. This government body is responsible, among other things, for hiring teachers, writing curricula, and keeping the school system in good working order. At the tertiary level there are several well-established institutions, offering a range of qualifications for students to continue their educational development. These include: The University of the West Indies; The University of Trinidad and Tobago; and the College of Science, Technology and Applied Arts of Trinidad and Tobago. In a bid to ensure equity and educational opportunity, tertiary education continues to be funded in various ways by the T&T government. In 2015 it was determined that 95% of all children were enrolled in primary and secondary schools and the overall basic literacy rate of the population was 99% (Oxford Business Group, 2020).

Science is a compulsory subject taught throughout the elementary and secondary levels of education. Integrated science is taught in primary schools and the first three years of secondary school. Every student is required to read at least one science subject in the two terminal years of secondary school. They can choose integrated science, a pure science or an applied science. Official curriculum guides produced or endorsed by the Ministry of Education determine what is taught in school science classes. A quick look at any of these guides (e.g. Government of Trinidad and Tobago, 2014; Caribbean Examinations Council, 2009) demonstrates the overarching philosophy guiding local science education efforts. Established science knowledge is taught in a spiraling fashion over the years in a bid to inculcate students into core science ideas, comparable to what is found in developed countries. There are few quality studies on the effectiveness of local science education efforts. However, the latest PISA (Program for International Student Assessment) results show T&T in third place in science in Latin America and the Caribbean, and overall, approximately 2 years behind the average science performance of students in OECD (Organisation for Economic Cooperation and Development) countries (Soledad Bos et al., 2016). These statistics resonate with my experience of over 30 years, as a science educator at the elementary, secondary and tertiary on these islands. The science education system seems relatively well-developed and formal student uptake of the discipline is good.

12.3 Trinidad and Tobago's Covid-19 Experience

The Covid-19 phenomenon unfolded in the country in phases. The main events of these phases relevant to this article are summarised in this section. The chronology presented here can be verified through newspaper articles and other publicly available documents (e.g. De Souza, 2022; Rodriguez, 2020; Trinidad Express, 2022).

Special Lockdown When the Covid-19 pandemic was declared in March 2020 by the World Health Organisation, the Trinidad and Tobago government put the islands under special lockdown. In practical terms this meant the immediate closure of all schools, non-essential workplaces, services and recreational facilities, and the country's borders. Citizens were required to stay at home and only go out for essential requirements like food and emergency medical attention. A strict curfew was instated that prevented citizens from leaving their homes at night. Social gatherings were limited to five persons. With schools being closed, many teachers tried to sustain teaching and learning through digital means, similar to what was happening internationally. During these early days this occurred sporadically depending on the resources and skill of individual teachers and schools. Testing of suspected Covid-19 cases began and identified cases were isolated to a parallel health care system consisting of separate primary care, stepdown and quarantine facilities. During this period the country recorded just below 120 infections, which was considered very low, so that the measures implemented during this phase were widely hailed to be effective by international analysts (Hunte et al., 2020). Indeed, the hope was that local spread had been effectively curtailed.

Curbing Local Spread On the heels of this initial success, a partial reopening of the country began in early June 2020. Most workplaces and businesses reopened. However, entertainment, sporting and recreational activities remained locked down. Persons were required to wash hands, wear masks and limit social gatherings to no more than 25 persons. A reopening of schools occurred to allow face to face classes to restart for students in terminal examination levels of the education system. By this time, the local ministry of education had launched its own remote and online education systems, which consisted of a daily schedule of classes aired on local television and radio stations for various levels of pupils, and a moodle-type online learning platform outfitted with conferencing software. The latter allowed every teacher in the school system to create subject classrooms and hold synchronous classes with their own students. Students who had not returned to schools were expected to continue their schooling through these means. However, the situation changed in July 2020 when the number of Covid cases began rising again and the Ministry of Health deemed that local cluster spread had begun. Schools were immediately reclosed when contact tracing exercises to isolate new infections led to the identification of several schools as being sites of active transmission. Remote distance and online education efforts by this time were better organised. Teachers were gaining confidence in the use of the new digital options. Schooling of all students was required to occur through these means. In the larger public sphere, some measures were reinforced to curb the control of this new wave of infection including the implementation of shift systems to limit the numbers of workers in any given workplace, limiting of public gatherings to 10 persons, and tightening of rules for social distancing in public spaces and transportation systems. In addition, all public spaces and private businesses were required to regularly sanitise high touch surfaces, conduct temperature checks at entrances, and install handwashing stations. During this period, regulations regarding non-compliance with mask wearing, numbers in social gatherings, and curfew breaking were enforced, in response to an uptick in persons breaking the Covid 19 public health ordinances. Starting from this time to the end of the pandemic, a number of persons, including youth, were fined for breaking these regulations. By September 2020 the number of recorded cases of Covid 19 was 4463 and 72 deaths had been recorded.

Managing the New Normal By October 2020 it was clear that the situation was not ending soon. Community spread was entrenched. No-one could say for certainty when normalcy would return. Efforts to keep the economy and the essential systems of the country working continued using the health measures described in the previous phase. Schools remained closed. During all of 2020 the social life of citizens was severely disrupted. Activities for religious festivals/holidays (extremely important to Trinis) were curtailed, with families limited to celebrating in small home-based pods. Carnival 2021 occurred in an anemic way, without the usual public parades, parties and social gatherings. Every evening the local news outlets continued reporting statistics of the number of new and existing cases of the disease in the country, and the performance of the parallel health care system. Despite a steady rise in the statistics, the mantra of the government had shifted. Preventing local spread was replaced by the slogan of managing the new normal, further explained as saving lives while keeping the business of the country running. A new hope was growing backed by international media reports that multiple vaccines for the virus were on the horizon. This hope became a reality for T&T when the Indian and Chinese governments donated sizeable batches of vaccines to the country in March 2021.

The Mass Vaccination Drive Vaccine donations enabled the government to kick-start a vaccination drive in April 2021 which continued into March 2022. Later the initial supply of vaccines would be supplemented with further donations, purchases and partnerships made with large non-governmental health agencies, international vaccine suppliers and the governments of several countries. Throughout the drive vaccines were never in short supply. The types of vaccines initially available were AstraZeneca and Sinopharm but in the last quarter of 2021 the Janssen and Pfizer vaccines also became available. The campaign to encourage the uptake of vaccinations was well planned and persistent. The role out was based on a needs basis. Essential services like medical personnel were the first to be scheduled. Next, the elderly and immunocompromised were given the opportunity to be vaccinated. By July 2021 the vaccine was available for all adults. To make the uptake easy and efficient many sites across both islands were created to dispense vaccines. These included district health centers, local community centers, mass vaccination sites, pop-up sites in malls and government buildings, and vaccination buses to service rural areas. As part of the vaccination drive, a mass communication campaign was carried out to encourage the populace to take the vaccine. The communication blitz included a variety of traditional and social media avenues and incorporated the voices of local entertainers and celebrities. Despite all of this effort, the public seemed peculiarly vaccine hesitant. As time passed, and fully equipped vaccination sites became emptier, other measures to encourage vaccination were implemented. Most notably, employees were mandated by some employers to get vaccinated to enter workplaces, several essential services were only accessible if one presented a vaccine card, and leisure/entertainment offerings could only be patronised by the vaccinated. In September 2021, to further incentivize parents to get themselves and their children vaccinated, schools were reopened partially for face-to-face classes with the promise of full reopening once the vaccine levels reached appropriate levels. Remote and online education measures were continued, but the emphasis was on the goal of a full return to face to face classes. By the end of 2021 the situation became more urgent by the arrival of the highly infectious delta and omicron virus strains to the islands, and a consequent spike in the number of Covid 19 infections being reported. Indeed, the largest number of daily infections and deaths for the entire period of the pandemic was reported in December 2021. At one point the parallel health care system caring for Covid patients was almost overwhelmed with an occupancy rate of over 90%. Still, the problem of vaccine hesitancy could not be overcome. By the end of year only 47% of the population was fully vaccinated.

Living with the New Normal Early in 2022, the position of the government shifted again. It acknowledged that the Covid-19 virus was now a part of human existence and that citizens would have to learn to live with this new phenomenon. Fortunately, the new strains of the virus were not as deadly as feared and Covid-19 cases began dropping. The mass vaccination drive was winding up. The country's borders were reopened. The nightly curfew was lifted. Carnival was held in a restrained way, with vaccination proof being necessary for participation in events and parties. In March 2022 the prime minister made a parliamentary speech summarising the government's response to the pandemic. He lamented the dumping of over a quarter of a million doses of unused vaccines. He ended his speech by announcing that over the next few months all special public health ordinances would be slowly rolled back to return life activities to pre-pandemic normalcy. As the months went by, this was indeed the case. In April 2022 schools reopened fully for face-to-face classes and online schooling ceased. With time, all activities leisure/entertainment, businesses, religious resumed. While mask wearing, social distancing, and hand washing were recommended they were no longer requisite or legally punishable. By Christmas 2022 most things seemed normal. Finally, on the 31st of July 2023, the last public health ordinance, the wearing of masks in medical facilities, was discontinued (Daily Express, 2023). For Trinidad and Tobago, the Covid-19 event was officially over. In the end, the country had recorded 191, 496 infections, 4370 deaths and achieved 47.7% full vaccination (PAHO, 2022).

12.4 Relationships with Science Education Revealed

From the very beginning of the Covid 19 situation, the T&T government committed itself to taking a scientific approach to dealing with the pandemic. Given the controversial nature of this term (Moura et al., 2021) the meaning of taking a scientific approach in this context is clarified by the words of the country's prime minister, Dr. Keith Rowley, in a newspaper story from August 2020, as "...the use of the scientific data in risk management" in a bid to give "all people the best chance to get the best available results in a dangerous situation" (Daily Express News Extra, 2020). As a science educator this response caught my attention. I began keeping a journal of how this approach was unfolding and trying to interpret what this meant for science education. I continued journaling my experiences and observations until the end of December 2022. Analysis of my journal revealed a persistent resistance of a sizable segment of the public towards the scientific approach the government used in dealing with the pandemic. This was manifested in news reports, stories and incidents I summarised in my journal of significant numbers of persons willfully breaking and/or finding creative ways to evade public health measures. For example, persons evaded home quarantine orders by conveniently not answering cell phone calls that they knew health officials were using to check in with them. Covert Covid-19 parties became a trend for people to socialise in numbers that nullified social distancing mandates. Always carrying around a snack/drink as an excuse for wearing of masks below the chin, when in public, under the pretense of essential food consumption, became a common practice. But perhaps the most telling evidence of the public resistance to the pandemic response was vaccine hesitancy. As described in the section above, despite a comprehensive effort, only 47.7% of persons in the country were fully vaccinated. When questioned about vaccine hesitancy many of the unvaccinated demurred by saying things like they were "still doing their own research" or they were "weighing their options"; while others stated religious reasons, for example, being warned by religious leaders that the vaccine represented the mark of evil; conspiracy theories for example, the vaccine contained a tracking microchip; and long-term health fears, for example, the vaccine would harm sexual health. Still others claimed they did not want to be experimented on or that they had their own less invasive remedies to prevent being infected. While there are many possible explanations for these reactions, in the rest of this section, I focus on three, that have direct implications for science education.

12.4.1 Misconceptions of the Nature of Science (NOS)

According to many science education scholars understanding the nature of science is an essential aspect of scientific literacy (e.g. Akerson et al., 2011; Hodson, 2008; Jenkins, 1990). Understanding NOS requires knowledge of how the scientific enterprise works and what scientists do in real life. This is important because it is only through a thorough understanding of NOS that individuals can truly gain conviction about the laws and theories of science to make decisions requisite for living in a world that is increasingly mediated by science and technology. The state of public understanding of the NOS has long been a question that science education researchers worldwide have pursued (e.g. Miller et al., 1997; Irwin & Wynn, 2003). According to this body of research, many people, despite evidence to the contrary, continue to believe that: science is a relatively straightforward and quick process; science produces facts and sure knowledge that once discovered changes little over time; science is the idealistic quest for knowledge; and science has explained most things already so that new problems only require application of known ideas.

The pandemic provided opportunities for the T&T public to expose their true perceptions about the nature of science. What was revealed is how deeply ingrained these common misconceptions about the NOS continue to be. In the beginning, many seemed to think Covid-19 would be a relatively short-lived phenomenon, which the medical fraternity would soon end through the dispensation of effective treatments and a vaccine. The public seemed totally unaware of the vaccine development process. Indeed, the need for trials and extensive testing of experimental vaccines was met with incredulity. People questioned why the process should take so long when vaccines are an established technology. Many also expressed confusion about the existence of many different teams of scientists working, often secretly and competitively, in separate locations across the world. This latter simply did not fit with the idealistic view that scientists work selflessly for the good of humanity, unimpeded by sociopolitical interests. Another issue that confirmed other public misconceptions of the true NOS was the call for mask-wearing to provide personal protection and prevent the spread of the disease. Throughout the pandemic, claims, often backed by reasonable sounding evidence, seemed to be coming on a regular basis from various expert groups about this practice. That science could provide no sure knowledge, because evidence could be interpreted in various ways, was a source of confusion of the public. Overall, strongly held misconceptions of the Trinidad public about the NOS, acted as a motivator for skepticism regarding public health measures the government was trying to implement.

The healthy existence of misconceptions demonstrated in this context reinforces the need for the explicit teaching of NOS throughout the school years, that many scholars have been advocating for over the past two decades (e.g. Lederman, 2007; McComas, 1998). Further, it also shows which aspects of the NOS may be essential, if a main goal of scientific literacy is to encourage the public to choose scientifically based actions in making life decisions. Based on what has been discussed in this paper, necessary ideas to teach about the NOS should include: a realistic understanding of the relationships between science, inquiry and empirical evidence; the idea that scientific ideas are tentative and only become established over time; and that scientists work within a sociocultural milieu. Recently, Osborne and Pimentel (2022) pointed out that most of the science taught in schools consists mostly of well-established laws and theories. Perhaps what the pandemic experience shows is that this practice only serves to reinforce some serious misconceptions about the NOS. It also suggests the urgent need for school science to teach some examples of messy science in the making, so as to inculcate a realistic understanding of what science is and how it occurs.

12.4.2 Naïve Ideas of Relationships Between Science and Society

It is well established that science has had far reaching effects on how society works (Cassirer, 1951). The methodologies and knowledge of science have dominated societal development over the past 200 years (Becker, 1994). Science produces knowledge which has been used to generate technologies and artefacts that impact everyday life in unimaginable ways. Most of the changes in health, communication, energy use, transport, sport, agriculture, construction, and recreation that have taken place since the eighteenth century are results of science, technology, and engineering working in tandem. For the foreseeable future, science is expected to continue to be a major mediator of the way we live. However, while science is an entity of power, the extent of that power has been criticized since its inception. Some critics have argued that science methods and knowledge are limited to the physical world and thus provide incomplete information about all aspects of reality (Midgley, 1992). Others cite numerous examples of science having been used for negative purposes—for example, as a hegemonic tool to denigrate various groups of peoples and cultures (Gould, 1981). For these critics, science lacks the epistemological, ethical, and emotional bases required to be the sole guidance for human decisions on complex issues.

The pandemic experience demonstrated that the T&T public continue to hold naïve understandings of the relationships between science and society. As noted earlier, vaccine hesitancy was a major problem in the country during the Covid-19 event. Two exemplar fears that members of the public expressed during the country's mass vaccination drive were that there was a tracking microchip in the vaccine, and that the vaccine was an attempt to experiment on the local population. On the surface these fears may seem unreasonable. But when one recalls the colonial past of these islands and the atrocities that occurred (Williams, 1964), the fear of the T&T public, that those in the first world could have been using the pandemic as an opportunity to exploit the developing world becomes more understandable. While any form of public understanding of the relationships between science and society may be positively interpreted, a closer examination of the related incidents that occurred during the pandemic also showed that public understandings of this important science-society relationship are still underdeveloped. During the pandemic many people were able to express their fears, but unable to critically interrogate those fears so as to debunk unhelpful conspiracy theories and verify valid concerns that could act as sources of healthy personal agency. Indeed, so potent were these naïve ideas that they could not be assuaged by any type of reassurance presented by the local health authorities. In fact, some attempts to interrogate these types of naïve ideas seemed to further alienate and frustrate those who held them. A second phenomenon that demonstrated the naivete of the T&T public regarding the science-society dynamic had to do with the relationships between the public, media and science. As noted earlier, at the beginning of the pandemic T&T had a relatively well-developed digital media culture. Many people were already using popular social media platforms as a source of communication. As the pandemic progressed and people became physically isolated, they became more reliant on these media platforms not only as a source of connection but as a source of information. As time passed, the information being spread on these platforms increased greatly. Furthermore, much of this information proved contradictory on almost every aspect of the pandemic. Being a bit too trusting of newer forms of media along with only nascent understandings that science could be used by bad actors for negative purposes, combined with not knowing how to discern credible from non-credible information to make the public increasingly confused and dismayed. As time passed rather than questioning the quality of the media sources they were relying on, people began doubting the validity of the science-based response to the pandemic. Public trust in science itself seem to diminish.

Recognising the complexity of the science-society dynamic, and the need for it to be actively addressed in science classrooms, a growing fraternity of science educators have been suggesting that science should be taught in contextually relevant ways that honour the complexity of the lived world. Several pedagogical approaches have been suggested, such as teaching through socially acute questioning (SAQ), socioscientific issues (SSI), and science technology societies and environments (STSE) (Bencze et al., 2020). Pedretti and Nazir (2011), in particular, suggest that science teaching must examine the interfaces between science and the social world. What occurred in T&T during the pandemic, further justifies the adoption of a science-in-context approach to science education to improve scientific literacy for all. Science educators need to be aware of context-specific factors that may be acting as barriers scientific literacy, for example, the possibility that the T&T fear of science being used to exploit humans is perhaps an artefact of historical wounds from a dark colonial past and provide opportunities to interrogate these issues in their classrooms. Recently, in response to the explosion in communication technologies and the evolution of fake news, Hottecke and Allchin (2020) have further suggested that the process of science communication is a phenomenon that school science needs to address ungently. This is borne out by the T&T experience during the pandemic where poor understanding about the sociopolitical relationships between science and media proved a major problem.

12.4.3 Preference for Alternative Non-scientific Knowledges

The previous section established science as an entity of power that continues to shape almost every aspect of living in the world today. However, before science became dominant, people had other knowledge systems that have persisted in modern society. The relationships between science and alternative ideas are complex. For many, science has epistemological characteristics that makes it incompatible with other forms of knowledge. Stephen Gould (2002) is clear that science and religious knowledge represent non-overlapping magisteria with different methodologies and domains of interest. Earlier in this paper the colonial past of T&T was described. The current population is diverse because of the way the islands were settled. Settlers coming from far flung parts of the globe, brought their traditional and cultural knowledges with them. Because of the oppressive colonial society they entered, for many years many of these knowledges were orally and secretly preserved. Over time through cross-fertilisation something new evolved, what June George (1999) calls *local indigenous knowledge*. While there is still no account that fully describes this indigenous knowledge, before the pandemic most Trinis seemed to be living their daily lives shifting between it and western modern knowledge seemingly seamlessly to suit the circumstances they encountered. Generally, science was seen as part of western modern knowledge also commonly called *book knowledge*. The tacit understanding was that while book knowledge may be correct, it does not fully explain all phenomena and may not be the best approach to deal with all situations.

However, during the pandemic, the existence of these various knowledge systems became problematic. Aspects of T&T indigenous knowledge emerged as major competitors to science-based responses being implemented by the government. Aspects of indigenous knowledge that were particularly contradictory to science knowledge included: alternative health theories and magical religious thinking. T&T indigenous knowledge contains a variety of non-scientific ideas about the causes and treatment of ill-heath. For example, the notions that infectious diseases are not caused by germs solely but by various factors being out of balance in the body like heat and cold, and toxins accumulating in the organs and blood. Consequently, a number of alternative health treatments exist, that are practiced by local people. For example, sweating may be encouraged to counteract certain diseases, and herbal remedies are taken to assist in recovery or protect against illness. During the pandemic people supplemented or replaced social distancing and mask wearing, with practices like avoiding sudden temperature changes, eating spicy foods, and consuming various concoctions to clean and strengthen their bodies in the hope that these would cure or prevent the disease. Another aspect of T&T indigenous knowledge is the strong existence of magical and/or religious thinking. Supernatural forces in the form of karma, gods, demons, spirits and the evil eye are believed to exist. These are believed to be important actors in the physical world that can affect human lives. Diseases can be a manifestation of these forces. Practices like praying, and magical rituals are necessary to appease or ward off supernatural effects. During the pandemic, certain religious groups insisted that the pandemic was the result of some type of supernatural influence. When places of worship were shut down to limit social congregation, some religious leaders advised their members that the faithful could not get sick through congregational worship and persisted in communing. Later some of these same groups, resisted the vaccination because it represented collusion with evil forces.

The relationship between alternative knowledges and science learning has been considered by several science education researchers over the years (e.g. Aikenhead, 2001; Atwater, 2000; McKinley, 1996; Ogawa, 1995). One interpretation from this body of research is that alternative knowledges may serve as a barrier to scientific literacy by acting as a source of cognitive and identity conflict for learners. Two widely differing ways of dealing with alternative knowledge in the science classroom have been proposed. Posner et al. (1982) propose teachers provide carefully crafted experiences to demonstrate the erroneousness of alternative ideas students may hold in order to create cognitive dissonance which the teacher can then use to indoctrinate students into more reasonable scientific knowledge. In recognition of the intertwining complexity of culture, knowledge and identity, some scholars have criticized this type of conceptual change pedagogy as a kind of violence being perpetuated on persons in societies where modern western worldviews are not dominant especially indigenous, non-white populations (Aikenhead, 2001). They have suggested a different type of pedagogical approach called border crossing. This approach asks educators to present both scientific theories and alternative knowledge alongside each other. The teacher's job is to ensure that students understand scientific theories while respecting alternative ideas learners may hold. Students are free to choose the type of knowledge they ultimately will use in their daily lives. Before the pandemic, for places like T&T, border crossing would have been advocated for by those concerned with the sociocultural concerns described above (Herbert, 2008). However, during the pandemic, the T&T public demonstrated a strong preference for alternative knowledges, and a tendency to eschew proffered scientific knowledge. Furthermore, different groups were acting from different aspects of alternative knowledge. This led to a situation where it was often very difficult to get consensus on actions to cope with the pandemic. The latter suggests that further work may be needed on a border crossing approach to science education, one that respects peoples' cultures, identities and indigenous knowledge systems, but also develops learners' capacities to accommodate other knowledge systems and deal with competing ideas in flexible but sustainable ways that act for the well-being of all in a society.

12.5 Conclusion

Based on existing evidence before the pandemic (e.g. analysis of existing curricula and existing and statistics), Trinidad and Tobago seemed to have a relatively well-developed formal science education system that was producing scientifically literate graduates. The pandemic experience challenged that notion. In this paper, I identified three ways it did so, that is by revealing widespread public misconceptions of the nature of science, naïve ideas about the relationships between science and society, and the public's preference for using alternative ideas (other than science) in leading their lives. Hodson (2008) identifies three aspects to scientific literacy that school science needs to focus on. These are learning science content, learning about science and doing science. Working with this understanding of scientific literacy and specifically focusing on learning about science, within the paper I suggested ways in which educators can mitigate the challenges the T&T experience with the recent pandemic revealed. The need for the explicit teaching of nature of science and the relationships between science and society are reinforced as integral parts of science education, along with the identification of certain critical areas within these aspects of science education that need to be addressed. With respect to the NOS, learners need to be provided with opportunities to understand the messy and tentative work of scientists in the real world as opposed to the idealistic images that persist in the public imagination. With respect to science-society dynamics, learners need to develop more critical understandings of how science can be used by various actors including politicians, media and economists to shape lives and impact societies.

I began writing this paper in July 2023. However, rereading it in 2024, further thoughts have arisen in my mind. Throughout the pandemic, one overarching thought kept recurring in my mind, "How could a population which should have a good level of scientific literacy, become so easily confused and deliberately choose to act in non-scientific ways during a crisis?" In addition to the mitigation measures I already discussed, it now seems to me the T&T pandemic experience suggests that educators may also need to rethink what scientific literacy means and the complexity of achieving it. Some important questions that may be considered are: Can one be scientifically literate and still mistrust science? Is it part of the teacher's job to build trust in science? How can science teachers provide realistic images of science and still maintain hope in its idealistic potential? How can science education change to respectfully accommodate people whose identities are rooted in experiences that may be at odds with the basic values and assumptions of science?

Science educators have long realised that science teaching is intertwined with an underlying philosophy of science. This philosophy can be abstract and idealistic. However, in ending this paper, I would like to add two Trini adages that may provide food for thought:

Science is strange but people are stranger.
A teacher should be teaching people not the curriculum.

Perhaps it is time for the science education fraternity to focus less on the abstract concept of achieving scientific literacy and more on developing *scientifically literate people*. In other words, focus more on human nature and the human aspect of teaching science by fully considering the sociology and psychology of diverse complex learners, in a bid to inspire people globally to be innovative, resilient, adaptable problem solvers in the face of unexpected challenges.

References

Aikenhead, G. (2001). Integrating western and aboriginal sciences: Cross-cultural science teaching. *Research in Science Education, 31*(3), 337–355.

Akerson, V. L., Buck, G. A., Donnelly, L. A., Nargund-Joshi, V., & Weiland, I. S. (2011). The importance of teaching and learning nature of science in the early childhood years. *Journal of Science Education and Technology, 20*, 537–549. https://doi.org/10.1007/s10956-011-9312-5

Atwater, M. M. (2000). Multicultural science education. *The Science Teacher, 67*(1), 48–49.

Becker, G. S. (1994). *Human capital: A theoretical and empirical analysis, with special reference to education* (3rd ed.). University of Chicago Press.

Bencze, L., & Alsop, S. (2020). Reimagining science and technology education in the Covid-19 portal. *Journal for Activist Science & Technology Education, 11*(2), i–vii.

Bencze, L., Pouliot, C., Pedretti, E., Simmoneaux, L., Simmoneaux, J., & Zeidler, D. (2020). SAQ, SSI and STSE education: Defending and extending "science-in-context". *Cultural Studies of Science Education, 15*, 825–851. https://doi.org/10.1007/s11422-019-09962-7

Caribbean Examinations Council. (2009). *Integrated science syllabus*. Caribbean Examinations Council.

Cassirer, E. (1951). *Philosophy of the enlightenment*. Princeton University Press.

Castagne, P. (1962). *National Anthem of Trinidad and Tobago*. https://www.youtube.com/watch?v=loUs9DcIjiE

Daily Express. (2023, July 31). *Masks no longer a must.* Trinidad Express Newspaper https://trinidadexpress.com/news/local/masks-no-longer-a-must/article_5124e9da-2f51-11ee-8703-63836053e235.html

Daily Express, News Extra. (2020, August 18). *How being a volcanologist helped me deal with pandemic.* Trinidad Express Newspaper https://trinidadexpress.com/news-extra/how-being-a-volcanologist-helped-me-deal-with-pandemic/article_09a06558-e13e-11ea-9aa7-275860f8134f. html

De Souza, J. (2022). *From lockdown to freedom: A look back as Covid19 restrictions fall.* The Newsday Newspaper. https://newsday.co.tt/2022/07/16/from-lockdown-to-freedom-a-look-back-as-covid19-restrictions-fall/

George, J. (1999). Indigenous knowledge as a component of the school curriculum. In L. M. Semali & J. Kincheloe (Eds.), *What is indigenous knowledge? Voices from the academy* (pp. 79–94). Falmer Press.

Gould, S. J. (1981). *The mismeasure of man.* W. W. Norton & Company.

Gould, S. (2002). *Rocks of ages: Science and religion in the fullness of life.* Ballantine Books.

Herbert, S. (2008). Collateral learning in science: Students' responses to a cross-cultural unit of work. *International Journal of Science Education, 30*(7), 979–993.

Hodson, D. (2008). *Towards scientific literacy: A teacher's guide to the history, philosophy and sociology of science.* Sense Publishers.

Hottecke, D., & Allchin, D. (2020). Reconceptualizing nature-of-science education in the age of social media. *Science Education, 104*, 641–666. https://doi.org/10.1002/sce.21575

Hunte, S., Pierre, K., St. Rose, R., & Simeon, D. T. (2020). Health systems' resilience: Covid-19 response in Trinidad and Tobago. *The American Journal of Tropical Medicine and Hygiene, 103*(2), 590–592. https://doi.org/10.4269/ajtmh.20-0561

Irwin, A., & Wynn, B. (2003). *Misunderstanding science? The public reconstruction of science and technology.* Cambridge University Press.

Jenkins, E. (1990). Scientific literacy and school science education. *School Science Review, 71*(256), 43–51.

Lederman, N. (2007). Nature of science: Past present and future. In S. K. Abell & N. Lederman (Eds.), *Handbook of research on science education* (pp. 831–880). Routledge.

Lord Funny. (1966). *Sweet Trinidad.* Retrieved from https://www.youtube.com/watch?v=ttGmcn64Pro

McComas, W. (1998). *The nature of science in science education: Rationales and strategies.* Kluwer Academic Publishers.

McKinley, E. (1996). Towards an indigenous science curriculum. *Research in Science Education, 26*(2), 155–167.

McVeigh, T. (2023). *The Caribbean Island of Trinidad grapples with a 'national crisis' of violence.* The Guardian. https://www.theguardian.com/global-development/2023/jun/16/the-caribbean-island-of-trinidad-grapples-with-a-national-crisis-of-violence

Midgley, M. (1992). *Science as salvation: A modern myth and its meaning.* Routledge.

Miller, J. D., Pardo, R., & Niva, F. (1997). *Public perceptions of science and technology: A comparative study of the European Union, the United States and Japan.* Fundacion BBV.

Moura, C. B., Nascimento, M. M., & Lima, N. W. (2021). Epistemic and political confrontations around the public policies to fight Covid-19 pandemic. *Science & Education, 30*, 501–525. https://doi.org/10.1007/s11191-021-00193-3

Nazir, J. (2021). Reviewing science education in Trinidad and Tobago through the lens of the Covid-19 pandemic. *Journal of Education and Development in the Caribbean, 19*(2), 145–171. https://doi.org/10.46425/j519027625

Ogawa, M. (1995). Science education in a multiscience perspective. *Science Education, 79*, 583–593.

Osborne, J. F., & Pimentel, D. (2022). Science, misinformation, and the role of education. *Science, 378*(6617), 246–248. https://doi.org/10.1126/science.abq8093

Oxford Business Group. (2020). The Report. Trinidad and Tobago 2015. https://oxfordbusiness-group.com/trinidad-tobago-2015.
PAHO, Pan-American Health Organisation. (2022). *Annual Report 2021. Trinidad and Tobago: Responding to Covid-19 and preparing for the future.* PAHO/TTO/22-0001.
Pedretti, E., & Nazir, J. (2011). Currents in STSE education: Mapping a complex field, forty years on. *Science Education, 95*(4), 601–626.
Pietrocola, M., Rodrigues, E., Bercot, F., & Schnorr, S. (2021). Risk society and science education: Lessons from the Covid-19 pandemic. *Science & Education, 30*(2), 209–233.
Posner, G. J., Strike, K. A., Hewson, P. W., & Gertzog, W. A. (1982). Accommodation of a scientific conception: Toward a theory of conceptual change. *Science Education, 66*(2), 211–227.
Ray, S., & Srivastava, S. (2020). Virtualization of science education: A lesson from the Covid-19 pandemic. *Journal of Proteins and Proteomics, 11*, 77–80.
Reiss, M. J. (2020). Science education in the light of Covid-19: The contribution of history, philosophy and sociology of science. *Science & Education, 29*(4), 1079–1092.
Rodriguez, K. (2020, April 23). *Timeline: Chronology of Covid-19 in T&T.* Trinidad Express https://trinidadexpress.com/coronavirus/timeline/chronology-of-covid-19-in-t-t/collection_01c6118a-85c1-11ea-acc2-0b8cb81e5c7a.html
Rudder, D. (2003). *Trini to de bone*. Retrieved from https://youtu.be/JVdlpZ4M-Hw
Soledad Bos, M., Elias, A., Vegas, E., & Zoido, P. (2016). *PISA Latin America and the Caribbean.* Inter-American Development Bank.
The Government of Trinidad and Tobago, Central Statistical Office. (2017). *T&T population reaches 1.4 million.* https://cso.gov.tt/news/tt-population-reaches-1-4-million/
The Government of Trinidad and Tobago, Ministry of Education. (2014). *Secondary school teachers' guide, Science.* Curriculum Development Division.
Trinidad Express. (2022). *PM announces Covid restrictions rollback.* Trinidad Express https://trinidadexpress.com/newsextra/pm-announces-covid-restrictions-rollback/article_96005586-9bef-11ec-8c7a-77e92df117ef.html
United Nations (2023). *WHO chief declares end to Covid-19 as a global health emergency.* UN News https://news.un.org/en/story/2023/05/1136367
Verma, G., Campbell, T., Melville, W., & Park, B. (2020). Science teacher education in the times of the Covid-19 pandemic. *Journal of Science Teacher Education, 31*(5), 483–490. https://doi.org/10.1080/1046560X.2020.1771514
Williams, E. (1964). *History of the people of Trinidad and Tobago.* Andre Deutsch.

Joanne Nazir is a lecturer at the School of Education, St. Augustine Campus. She holds degrees in Science and Education from The University of the West Indies; and a doctorate in Curriculum Studies and Teacher Development from the University of Toronto. Her research interests are in the areas of science education, environmental sustainability education, and teacher educationg.

Chapter 13
Can there Be a Science of the Sacred? Thinking with Students' Feminist and Pluriversal Worldmaking Practices

Sara Tolbert and Mahdis Azarmandi

13.1 Introduction

We start this chapter with the experiences of two students across very different science classrooms contexts: the first one takes place in an introductory high school science classroom in the southwestern United States, and the second takes place in a kura kaupapa Māori (Māori culture/language immersion primary school) in Aotearoa New Zealand. We begin with the experiences of students to foreground the important contributions they make to a new critical imaginary for science education (Torres Olave et al., 2023). After sharing these provocations, we take up the students' concerns/actions and bring them into conversation with feminist and decolonial scholars to ask, can there be a science of the sacred? We consider how the disembodying and desensitising practices of school science are ones that must be abandoned, if we are truly committed to 'ongoingness', that is, (the troubles of) living together respons-ably within in our (post)pandemic world (Haraway, 2016).

S. Tolbert (✉)
University of Canterbury, Christchurch, Aotearoa, New Zealand

Monash University, Melbourne, New Zealand
e-mail: sara.tolbert@canterbury.ac.nz; sara.tolbert@monash.edu

M. Azarmandi
University of Canterbury, Christchurch, Aotearoa, New Zealand

C. B. Moura (ed.), *A Sociopolitical Turn in Science Education*, Contemporary Trends and Issues in Science Education 62, https://doi.org/10.1007/978-3-031-78586-3_13

13.2 Talking Back to Exploitative Technoscientific Practice

13.2.1 "I Wanted to Bury Its Heart, At Least" (Adapted from Tolbert, 2019; See Also Krishnamoorthy & Tolbert, 2022)

The first provocation is from a high school dissection activity. The teacher had used a crowdfunding site to raise funds for students in her introductory science course to participate in a cat dissection. As part of a larger research project, Sara met periodically with a small group of student researchers in the class. During one of these meetings, the subject of the cat dissection came up as something that was particularly disturbing to them (2 of 3 students). What was notable about their concerns was that they were not opposed to the dissection, per se, but rather to the way in which the remains of the cats were disposed of:

NICOLE: Well, I didn't do any dissecting. I was watching. It's so sad.
SARA: It was sad?
NICOLE: Yea.
SARA: What about it made you feel sad?
NICOLE: I don't know. I just was stuck with like, we're just going to throw it away at the end of it all. That made me sad. I don't think we should throw stuff away like that.
SNOOPY: Did you want to just bury it?
NICOLE: I wanted to bury its heart at least. [Ms.] Bell cut open its heart and I was sad, like that's sad. It's really horrible.

...

NICOLE: It also taught me, you have to cut off your emotions for this situation, because I want to cry. And I think I did cry. I teared up, and I was like, oh, my god.

...

NICOLE: My grandfather, back in the day, he lived in Mexico. They used to hang the pigs from a tree and they would do it [slaughter the pig] there...a couple of years ago when I was eleven, I was in Mexico because we went down there for a party. I think it was for a quinceañera [celebration of a girl's 15th birthday]. They bought a pig from the butcher and then they slaughtered it themselves. They cut off the head and the whole thing. I was ok with that because they're going to eat it.

13.2.2 "Why Do We Need to Kill Them"?

This second provocation is generated from an online resource on Ahi Pepe. Ahi Pepe is a citizen science project in which students create a moth reference collection as part of their efforts to identify, learn about, and monitor local moth species in their community. The resource includes a series of publicly available videos about the project, its background, resources for teachers (such as how to create a moth reference collection and support students to use the Participatory Science Platform to upload identification data). There is also a module on how one school negotiated culturally situated ethical conflict in their participation in the activity. This online resource is one Sara uses with preservice teachers to help trouble notions of relationality and ethics in science and science education (Krishnamoorthy & Tolbert, 2022). In this example, at the beginning of the citizen science project, a 5-year-old student asked the question, "but why do we have to kill [the moths]?" According to their teacher, the child's question "slowed the project until we could all as a whānau [classroom family], students, teachers and the wider whānau have a conversation about how that [euthanizing moths] could happen...the students were trying to find a way forward. They want to be the scientists and use this equipment and gain these skills but they want to minimize the impact on their cultural beliefs." These concerns from students led to the co-development of tikanga [protocols], in collaboration with scientists supporting the project, for working with the moths which included traditional practices.

The video transcripts are also published online (Science Learning Hub, n.d.). The excerpts below are taken from the videos and transcripts.

TIAHUIA KAWE-SMALL(Kaiako [teacher]): My cultural lens, which is a Māori lens, placed over science has been interesting. We've had some cultural clashes in terms of Western ideology and science and how we teach our tamariki about kaitiakitanga. Everything in the world has a mauri, it has a life force, and therefore we should respect that. And with the science, it was a tikanga challenge when we had to euthanise the moths. We're supposed to be cherishing these creatures in our school environment, and now we have to think about putting them to sleep. And this was a challenge for myself and my team of teachers and also for the students, so much so we had one student, which was quite profound, "Why do we have to kill them?"

That slowed the project until we could all as a whānau—students, teachers and the wider whānau—have a conversation about how that could happen.

DR BARBARA ANDERSON *(Scientist facilitating the citizen science project): One of the reasons that it's really important to sacrifice them is in order to make a reference collection because it makes identification easier. From the science point of view, reference collections and museum collections are really important because they are a bank of information that is curated, and it's kept safe for future generations.*

It was like his second week at school—and so this is this little 5-year-old kid—and we're sitting in the circle and, you know, at the end of every discussion, we say, "Any questions?" And Te Apanui, 5 years old, says, "Why do we need to kill them?"

We ask the questions some time back when we first start being a scientist but then we kind of just keep going. And it's good sometimes to pause and think, "OK, so why am I actually killing these moths? Do I need to do this? What's my reasoning?" And yet this little group of kids, they were saying, "No, we don't want to kill them."...So the solution to this was to take that on board and take that learning into the project. And so we've developed a protocol that we're all comfortable with—from the science point of view, from the kura point of view and from the Kāi Tahu point of view and also from the Orokonui Ecosanctuary point of view. Whaea Tiahuia has made a nice karakia [prayer/incantation], *which we say beforehand.*

TIAHUIA KAWE-SMALL: Pao is an important element of appreciation in tikanga Māori. The pao for the pepe tuna was to give thanks to Tāne Mahuta o Rēhua for gifting them to the Earth, and we have just now used them for our science and we'll respect them. So the students use the pao to acknowledge the higher sources who've gifted these pepe to us to learn from and to be respectful of so that we can continue on with kaitiakitanga and the sustainability across all the elements that we're learning here at kura. It's part of our tikanga.

13.3 Feminist Science Studies: Embodying and Sensitisizing in the Sciences

> *Those who contemplate the beauty of the earth find reserves of strength that will endure as long as life lasts. There is something infinitely healing in the repeated refrains of nature – the assurance that dawn comes after night, and spring after winter.*—Rachel Carson, 1962.

We take this opportunity to sit with these students' critiques of school science, and the ruptures in normative school science that their critiques produce. We find their critiques have been incredibly generative for us. We have wondered at how the students' critiques of science positioned them as more than just 'learners' but rather critical interlocuters of science (Tolbert, 2019). As science studies scholars have pointed out, the doing of science is inextricable from the socio-political and historical dimensions of the doing (Barad, 2007). When students are critical interlocuters of science, they learn not only about scientific facts and theories, but they also learn about affordances and limitations of science, the intersections of science with heteropatriarchy, colonialism, and racism, and the inevitable muckiness of doing science (Krishnamoorthy & Tolbert, 2022; Liboiron, 2021). As critical interlocuters, they learn to do stay with the trouble of science as part of what it means to do more respons-able science, thereby reshaping what is possible and doable and perceptible in (school) science (Haraway, 2016; Tolbert, 2019; Tolbert & Bazzul, 2020).

Part of the work of staying with the trouble of science and science education is both coming to terms with how the 'disembodying experience' of science education came to be—and reclaiming the body (bodies), and its (their) spiritual-affective experiences. Both provocations at the beginning of this chapter highlight how children and young people offer thoughtful and creative ways to reclaim the spiritual-affective dimensions of doing science. Haraway (2016) writes about the task of our times as being one in which we must.

> make kin in lines of inventive connection as a practice of learning to live and die well together in a thick present....In urgent times, many of us are tempted to address trouble in terms of making an imagined future safe, of stopping something form happening that looms in the future, of clearing away the present and the past in order to make futures for coming generations. Staying with the trouble does not require such a relationship to times called the future. (p. 1).

What we focus on in the students' provocations is how they were trying to find ways to resolve the tensions they were experiencing so that they could more fully engage with the science. What we notice about their concerns is that they were not articulating a refusal of science,[1] or advocating for 'stopping something from happening.' Rather, they were advocating for attention to the sacred—that is, a more embodied and spiritual acknowledgement of the "Other" (the cat, the moth, respectively), a way to 'make kin in lines of inventive connection as a practice of learning to live and

[1] We want to make it clear here that though students in this case were not articulating refusal, we view refusal as an important tool for justice-oriented research, science, education, and activism (see Azarmandi & Tolbert, 2024; Tolbert et al., 2018; Tolbert, 2019).

die well together' (Haraway, 2016, p. 1). We turn our attention to how the students in both provocations encourage us to contemplate the ways in which 'bracketing' our spiritual-emotional, or sacred, connections with each other and within the more-than-human world has limited our possibilities for 'ongoingness.'

Feminist science and technology studies—"while they include almost every conceivable methodological and disciplinary approach" are "profoundly committed to the possibility of making better and more reliable representations of nature informed by new values" (Rose, 1994, p. 231). Feminist science studies have for several decades challenged the idea that doing science requires a separation between mind and body, or emotion and reason. For example, Hilary Rose (1994) argued that "The faculties [of reason and emotion], which have been constructed as separate and arranged in hierarchal dualities, need bringing together in a way which is both non hierarchical and anti-foundationalist. All the faculties need developing, for knowing requires them all" (p. 50).

Feminist science scholars have pointed to the ways in which scientific practice is in essence an embodied practice, but the process of establishing science as 'objective truth' was a disembodying process. For example, as Sara (and Marc Higgins) have written:

> Following [the] purchase and analysis [of Isaac Newton's unpublished papers], John Maynard Keynes would go on to state that Newton was 'not the first of the age of reason' but 'the last of the magicians' [Stengers, 2001, p. 160]. Stated otherwise, his approach to science was not devoid of spirituality, but rather the norms of becoming-scientist at the time made it such that he could not openly discuss the multilogical approach that informed and fueled his thinking (as part of the practice of modest witnessing) (Higgins & Tolbert, 2018, p. 283).

Microbiologist Caitlyn Allen (2001) commented how, for her, doing science is an embodied and spiritual experience, and how in her work as a scientist, she finds that "...the natural world sings with meaning, and that the study of bacteria can be intensely and spiritually satisfying" (p. 26). In her classic account of geneticist Barbara McClintock's work, *A Feeling for the Organism*, Evelyn Fox Keller revealed how McClintock experienced a kinship with, and a love for, the 'objects' of her she study. Citing McClintock, Fox Keller (1984) wrote:

> I was down there. I was part of the system. I was right there with them, and everything got big. I even was able to see the internal parts of the chromosomes—actually everything was there. It surprised me because I actually felt as if I were right down there and these were my friends ... the main thing is that you forget yourself (p. 117).

Similarly, Nicole and Te Apanui (students) in both provocations pointed out that the faculties of emotion and perception, and a love for the 'specimen,' are, for them, required for doing, and learning, science. But they are also asking for something beyond the faculties of emotion and reason, something transcendental.

13.4 What of the Spiritual in Science?

> *Today...science and spirituality have the potential to be closer than ever, and to embark upon a collaborative endeavor that has far-reaching potential to help humanity meet the challenges before us.* (His Holiness the Dalai Lama, 2005, pp. 208–209)

As an expression of their love, students in these provocations identified ritual practices to support "learning to live and die well together" (Haraway, 2016, p. 1). For Nicole, burying the heart of the cat as a form of consecration would have helped her more fully engage some of the tensions—and acknowledge the grief—she experienced during the dissection. She wanted to participate, but she could not come to terms with the absence of ritual practices that could have honoured the cat's sacrifice. For Te Apanui, Māori tikanga [cultural protocols] such as waiata [traditional song] and karakia [incantations] helped him address the ethical dilemma he experienced around killing the moths in order to protect the moths. But these students' critiques beg the question, can there a place for the sacred in school science?

Sandra Harding (2019) points out that modern sciences "positioned themselves against the 'enchanted world'" for a variety of philosophical and strategic reasons (p. 127). She asserts that a "secular stance has been a founding feature of the legacy of what is regarded as enlightened Western thought" (p. 128). A secular science was critical to the resisting the hegemony of the Catholic Church in Europe. Harding also points out how a secularist positioning galvanized philosophers of science in the Vienna Circle, in the context of rising fascism and antisemitism. A secularist stance forced many to come to terms with the idea that no supernatural power would save them from what was to come; therefore, humans must rise up to resisting the growing fascist regime. Secularism in the sciences continues to be used as an important political and epistemological tool. For example, secular Euro-Western rationalisms have been used in the context of India as anti-oppressive counterpoints to Hinduvta and religious-based caste oppression (Krishnamoorthy & Tolbert, 2022; Nanda, 2003). Secularist positionings in this regard have facilitated key political and epistemological developments.

Staying with the trouble of secularism, we also observe how a strident secularism in the sciences has been used to render 'less than' those knowledge systems, such as many Indigenous knowledge systems, in which spirituality is entangled—with detrimental effects on those knowledge systems and the societies in which they are practiced (Harding, 2019). A refusal to entertain any notion of the spiritual or sacred dimensions of knowledge production has also produced racist consequences (Sullivan, 2010; West, 1989). Harding points out, however, how the particular brand of secularism in the sciences has Protestant undertones, and "misreads the nature, practices, and benefits of religious and spiritual commitments for indigenous and other cultures" (p. 129). She further argues that "the secularism versus religion binary is neither as obvious nor useful as had been supposed" and manifests "a symptom of the lack of critical self-analysis regarding what appear as clearly religious and cultural values and interests that also infuse Western sciences and their philosophies" (p. 131). It is for these and other compelling reasons that rethinking secularism has become a focus of recent intellectual and political debate (Harding, 2019).

John Heron, in his book, *Sacred Science: Person-centred Inquiry in the Spiritual and the Subtle* (1998), highlights how qualitative methodologists Lincoln and Denzin have traced the "journey to join science and the sacred," from Durkheim to Mary Douglas (p. 8). Citing Lincoln and Denzin, Heron comments how qualitative research, in a sense, has largely been about returning 'concerns of the spirit' to the disciplines (p. 8). Both Heron (1998) and Lincoln and Denzin (1994) draw from the work of Peter Reason (1993), who argues that a sacred science is a science of enchantment with nature, thoughtful reflection, and, borrowing from Buddhism, 'right action':

> the purpose of right action in the present time must be for healing: for healing the rift between persons and the experience, between persons and others, between groups and societies, between our human existence and the requirements of the ecology of the planet, and between ourselves and the realm of the spirit, however one may wish to speak of that. (Reason, 1993, p. 282)

Reason (1993) points to the development of scientific consciousness as an "alienated consciousness" in its total separation from nature. He, like feminist science studies scholars, links this alienated consciousness to our current social, economic, and ecological crisis. Citing Morris Berman's (1981) book, *The Reenchantment of the World,* Reason argues that the crisis is also an epistemological one:

> The view of nature which predominated in the West down to the eve of the Scientific Revolution was that of an enchanted world. Rocks, trees, rivers, and clouds were all seen as wondrous, alive, and human beings felt at home in this environment. The cosmos, in short, was a place of belonging. A member of this cosmos was not an alienated observer of it but a direct participant in its drama. (Berman, cited in Reason, 1993, p. 275).

Both Reason and Heron argue that a secular science fails to address the crises of our current times, crises which necessitate deep connections with the sacred dimensions of ourselves and our world.

13.5 Exploring Socio-political Dimensions of a Sacred Science

> *A sacred science, I believe, is grounded in this immediate present experience of a world that is sacramental. The world is known as an epiphany, that is, an embodiment, a manifestation, an integrated revelation of the divine* (Heron, 1998, p. 8).

Attending to students' desires for sacred science is a move toward the sociopolitical in science education: sociopolitical acts in science education must include a refusal of hegemonic positionings regarding "what counts as 'legitimate' knowledge about the natural world and/or what counts as 'science learning'" (Tolbert & Bazzul, 2017, p. 326). Furthermore, interrogating how science and education make some things visible and some things invisible is a necessary sociopolitical act—in fact, "the very substrate of politics" (Tolbert & Bazzul, 2020, p. 83). Part of that work requires shifting not only our epistemological but also our ontological orientations—not only recognizing and knowing our interconnectedness with other human/more-than-human beings, land, and water, but 'becoming/being' in relation. The students' critiques about the treatment of the moths and the cats make the sacred/spiritual dimensions of science experience—and being-in-relation—visible, whereas in the absence of their critiques these relations were invisible. The students offered new possibilities for scientific practice that honoured the sacred dimensions of being-in-relation—and solidarities. Acknowledging, through sacred acts such as ritual practice, the interconnectedness and relationality of all things has the power to shift our ontological orientations. As Bazzul (2022) has stated:

> Solidarity, as a shared sense of unity and dependence, is an aspect of collective existence that needs to be more readily extended to non-humans. Solidarity with nonhumans might seem strange, but only because a western education does not provide the creative space to think about codependence on an ontological level (p. 947)...Rethinking solidarity with nonhumans is...not just an ethical and political task, but an ontological, spiritual, and scientific one. (p. 948).

In New Zealand, the science curriculum includes the study of "the interdependence of living things (including humans) in an ecosystem" (New Zealand Ministry of Education, 2007). In the United States, the Next Generation Science Standards include a study of how "Organisms and populations are dependent on their environmental interactions both with other living things and with non-living factors" (NGSS Lead States, 2013). What is missing from these 'core ideas' is acknowledging our respons-abilities as humans who are deeply entangled with each other, within living/nonliving, human/nonhuman assemblages, as part of the study and practice of science. As Bazzul (2022) points out, "Nurturing solidarity with other beings, including water and land, is exactly what Indigenous knowledge keepers, elders, and scientists do by challenging the hubris and ethical orientations of modern western science (Kimmerer, 2013)" (p. 948–949). Nurturing solidarity means we must acknowledge harm, rather than distance ourselves from it through logical-positivist object-subject onto-epistemologies, as we stay with the muckiness of our troubled interdependent existences. Rethinking solidarity with nonhumans is an ontological and spiritual task.

There are, of course, many practical challenges and obstacles to nurturing a sacred science, especially in schools and particularly in science classrooms. We are particularly wary of how the religious beliefs and practices of a majority—particularly a hegemonic majority—could (and do) dictate what and whose spiritualities are acknowledged or 'permitted' in institutional settings. We are wary of unequivocally advocating for a sacred science, given the ways in which communities and peoples practicing minoritized spiritualities, such as land-based, womanist, and/or pagan spiritualities, have been violently oppressed for their spiritual beliefs and practices throughout history, and in settler-colonial institutions such as schools. Nor do we want to align ourselves with a hegemonic religious movement that seeks to legislate the teaching of literalist interpretations of biblical concepts such as creationism. We see such movements as a continuation of the settler-colonial project, and recognize the many ways in which religious dogma has been used to establish cultural and political hegemony, to suppress dissent and oppress the marginalized.

At the same time, we understand the students' critiques emerge from a pluriversal positioning that does not compartmentalize, but rather embraces, their spiritual-affective 'being' and 'being-with'. As we mentioned previously, they are not putting forth a refusal of science but rather a pluriversal vision for science. They draw from cultural repertoires of practice from their own communities, as well as from their embodied spiritual-affective responses to school science activities, to make visible new ontologies in/for science education (Bazzul & Kayumova, 2016). They create lines of flight from a hegemonic science toward a science that makes space for the sacred.

13.6 The Co-production of a Pluriversal Post-pandemic World

Queremos un mundo donde quepan muchos mundos. -Zapatistas

Pluriversal perspectives, like feminist perspectives, offer a transformative reimagining of science that resists hegemony and challenges the traditional dichotomies that have limited its scope:

> Pluriversality as a universal project means that the universal cannot have one single owner: the universal can only be pluriversal, which also corresponds with the Zapatistas' vision of a world in which many worlds coexist. All of us on the planet have arrived at the end of the era of abstract, disembodied universals—of universal universality. Western universalism has the right to coexist in the pluriverse of meaning. Stripped of its pretended universality, Western cosmology would be one of many cosmologies, no longer the one that subsumes and regulates all the others. (Mignolo, 2018, p. x).

If dominant science is shaped by what Mignolo describes a "universalizing tendency [that] claimed a superior position for itself" (2018, p. x) then disrupting dominance in science and science education must be multi-pronged and pluriversal. A pluriversal approach to unsettling dominant science is part of what Domínguez describes as a "*decolonial* pedagogical framework [that] actively, intentionally, seeks to *un-do* the power relations of the colonial episteme, not find parity within them. It rejects the epistemology of the White Gaze and its disembodied focus on academic standards" (Domínguez, 2021, p. 533, emphasis in original). Education that aligns with this framework, Domínguez points out, should encourage students to pursue intellectual ambition, while embracing epistemic disobedience. That is, questioning both religious and scientific singularities is integral to intellectual development.

Pluriversality transcends singularities and offers an alternative to conventional binaries (e.g., Western/non-Western; secular/religious) (Walsh & Mignolo, 2018), enabling students to engage multiple ways of disrupting, unsettling and reimagining science. A pluriversal design for science education can foreground values, such as care, interconnectedness, dignity, respect, honor, justice, and reverence, altering what it means to do and learn science (Kayumova & Dou, 2022). Nicole and Te Apanui's critiques of science offered ways to foreground care, respect, interconnectedness, dignity, and reverence. If we endorse the idea of a pluriversal design for science education, that means both secular and non-secular knowledge systems, alongside students' embodied experiences, contribute to how we come see ourselves as within, not separate from, the 'natural world.' Pluriversality is the antithesis of epistemological, cultural or religious hegemony. A pluriversal vision is one in which ethical considerations and diverse cultural values are integral to doing and thinking about science; a pluriversal vision embraces an interconnected approach that recognizes the complex web of relationships that shape our world. It is a call to cultivate an ecological consciousness that fosters a sense of response-ability towards the planet and its inhabitants.

A pluriversal science for a post-pandemic world requires rejecting "the logic of coloniality, and fundamentally changing the way we think about the world, position ourselves in schools, and in relation to marginalized youth and communities" (Domínguez, 2021, p. 553). We began this chapter with two provocations, centering students' critiques and their desires for a more sacred science. We have drawn from students' concerns to argue that a sacred science must be feminist and pluriversal. Their critiques encouraged us to contemplate, can there be a science of the sacred? We do not actually have a definitive answer. However, we would not be asking ourselves this question if it were not for the students' profound observations of the absence of the sacred in their school science experiences. Their thought-provoking observations made us think differently. They provoked us to ask questions we might not have asked ourselves otherwise—questions that surprised and challenged us. But we believe that these students (and others like them) have the power to fundamentally change the way we think about, and become with, the world, if we slow down and listen. We must have the courage to listen to and honor students' critiques and emancipatory visions, as central to the ongoing project of radically transforming science, and transforming our world.

References

Allen, C. (2001). What do you do over there, anyway?: Tales of an academic dual citizen. In M. Mayberry, B. Subramaniam, & L. Weasel (Eds.), *Feminist science studies* (pp. 22–29). Routledge.

Azarmandi, M., & Tolbert, S. (2024). A manifesto for transdisciplinary (transgressive) feminist praxis in the academy. In C. Hughes, C. A. Taylor, M. Salazar Pérez, & J. B. Ulmer (Eds.), *The Routledge international handbook of transdisciplinary feminist research and methodological praxis*. Routledge.

Barad, K. (2007). *Meeting the universe halfway: Quantum physics and the entanglement of matter and meaning*. Duke University Press.

Bazzul, J. (2022). Solidarity with nonhumans as an ontological struggle. *Educational Philosophy and Theory, 54*(7), 946–955.

Bazzul, J., & Kayumova, S. (2016). Toward a social ontology for science education: Introducing Deleuze and Guattari's assemblages. *Educational Philosophy and Theory, 48*(3), 284–299.
Berman, M. (1981). *The reenchantment of the world*. Cornell University Press.
Carson, R. (1962). *Silent spring*. Houghton Mifflin.
Domínguez, M. (2021). Cultivating epistemic disobedience: Exploring the possibilities of a decolonial practice-based teacher education. *Journal of Teacher Education, 72*(5), 551–563.
Haraway, D. J. (2016). *Staying with the trouble: Making kin in the Chthulucene*. Duke University Press.
Harding, S. (2019). *Objectivity and diversity: Another logic of scientific research*. University of Chicago Press.
Heron, J. (1998). *Sacred science: A person-centred inquiry into the spiritual and the subtle*. Endymion Press.
Higgins, M., & Tolbert, S. (2018). A syllabus for response-able inheritance in science education. *Parallax, 24*(3), 273–294.
His Holiness the Dalai Lama. (2005). *The universe in a single atom: The convergence of science and spirituality*. Harmony.
Kayumova, S., & Dou, R. (2022). Equity and justice in science education: Toward a pluriverse of multiple identities and onto-epistemologies. *Science Education, 106*(5), 1097–1117. https://doi.org/10.1002/sce.21750
Keller, E. F. (1984). *A feeling for the organism, 10th aniversary edition: the life and work of Barbara McClintock*. Macmillan.
Kimmerer, R. (2013). *Braiding sweetgrass: Indigenous wisdom, scientific knowledge and the teachings of plants*. Milkweed editions.
Krishnamoorthy, R. & Tolbert, S. (2022). On the muckiness of science ethics and preservice teacher education: Contemplating the (im)possibilities of a 'right'-eous stance. *Cultural Studies of Science Education, 17*(4), 1047–1061. https://doi.org/10.1007/s11422-022-10132-5
Liboiron, M. (2021). *Pollution is colonialism*. Duke University Press.
Lincoln, Y., & Denzin, N. (Eds.). (1994). *Handbook of qualitative research*. Sage.
Mignolo, W. D. (2018). Foreword: On pluriversality and multipolarity. In B. Reiter (Ed.), *Pluriverse* (pp. xii–xv). Duke University Press.
Nanda, M. (2003). *Prophets facing backward: Postmodern critiques of science and Hindu nationalism in India*. Rutgers University Press.
New Zealand Ministry of Education. (2007). *The New Zealand curriculum*. Available online at https://nzcurriculum.tki.org.nz/The-New-Zealand-Curriculum
NGSS Lead States. (2013). *Next generation science standards: For states, by states*. The National Academies Press.
Reason, P. (1993). Reflections on sacred experience and sacred science. *Journal of Management Inquiry, 2*(3), 273–283.
Rose, H. (1994). *Love, power, and knowledge: Towards a feminist transformation of the sciences*. Indiana University Press.
Science Learning Hub. (n.d.). Ahi pepe and tikanga. Available online at https://www.sciencelearn.org.nz/videos/1824-honouring-tikanga-and-the-sacrifice-of-moths
Stengers, I. (2001). *La guerre des sciences aura-t-elle lieu?: Scientifiction [Will the war of the sciences take place?]*. Le Seuil—Les Empêcheurs de Penser en Rond.
Sullivan, S. (2010). The secularity of philosophy: Race, religion, and the science of exclusion. In G. Yancey (Ed.), *The center must not hold: White women philosophers on the whiteness of philosophy* (pp. 153–166). Lexington Books.
Tolbert, S. (2019). Queering dissection: 'I wanted to bury its heart, at least'. In C. Taylor, C. Amade-Escot, & A. Abbas (Eds.), *Gender in learning and teaching: Feminist dialogues across international boundaries* (pp. 39–53). Routledge.
Tolbert, S., & Bazzul, J. (2017). Toward the sociopolitical in science education. *Cultural Studies of Science Education, 12*(2), 321–330.

Tolbert, S., & Bazzul, J. (2020). Aesthetics, string figures, and the politics of the visible in science and education. *Journal of Curriculum and Pedagogy, 17*(1), 82–98.
Tolbert, S., Schindel, A., & Rodriguez, A. J. (2018). Relevance and relational responsibility in justice-oriented science education research. *Science Education, 102*(4), 796–819.
Torres Olave, B., Tolbert, S., & Frausto Aceves, A. (2023). Reflecting on Freire: A praxis of radical love and critical hope for science education. *Cultural Studies of Science Education, 18*(1), 1–20.
Walsh, C. E., & Mignolo, W. D. (2018). *On decoloniality: Concepts, analytics, praxis*. Duke University Press.
West, C. (1989). *The American evasion of philosophy: A genealogy of pragmatism*. University of Wisconsin Press.

Sara Tolbert is associate professor of science and environmental education at Te Whare Wananga o Waitaha University of Canterbury in Aotearoa New Zealand. Her current work focuses on critical/feminist praxis in teacher education and education for regenerative and justice-oriented environmental present futures.

Mahdis Azarmandi is a senior lecturer at Te Whare Wananga o Waitaha University of Canterbury in Aotearoa New Zealand. Her research spans the intersections of Peace and Conflict Studies and Critical Race and Whiteness Studies, rooted in intersectional analysis, delving into how factors like race, gender, and disability intersect and impact educational systems, policies, and practices.

Part IV
Imagining Justice-Centred Futures for Post-pandemic Science Education

Chapter 14
Post-pandemic Science & Technology Education: Ongoing Challenges to Societies

J. Lawrence Bencze, Lyn Carter, and Isabel Martins

14.1 Introduction: The Pandemic and Polycrises

We live in times of profound intellectual and moral anxiety that seem to have been laid bare by the devastating COVID-19 pandemic. Accelerating climate breakdown, increasing inequality, rising fascism and retreating democracy, neoliberalism's sedimentation, post-truth populism, developing geopolitical tensions and, the Russian invasion of Ukraine, Palistinian-Israeli war and rapid advancement of generative artificial intelligence challenge us all with their overwhelming relentlessness. Adam Tooze (2022) describes our current times as a *polycrisis*, with crises and shocks so intertwined, complex, rapid, and constitutive that wholes are more than sums of the parts. Tooze's polycrisis concept has no one construct or theory to encompass its reality. Mobilised by commentators, essayists and media, as well as academic literature (e.g., Zeitlin & Nicoli, 2020), the term has entered lexicons, with Tooze (2022) arguing it has great utility in suppleness of the 'poly' prefix—not specifying a single dominant source of tension or dysfunction. Within this framing, COVID-19 added in just another shock. Nevertheless, COVID-19 has, by any account, been

J. L. Bencze (✉)
University of Toronto, Toronto, ON, Canada
e-mail: larry.bencze@utoronto.ca

L. Carter
Australian Catholic University, Ballarat, VIC, Australia

I. Martins
Universidade Federal do Rio de Janeiro, Rio de Janeiro, Brazil

C. B. Moura (ed.), *A Sociopolitical Turn in Science Education*, Contemporary Trends and Issues in Science Education 62,
https://doi.org/10.1007/978-3-031-78586-3_14

devastating.[1] The pandemic caused massive global disruption—with billions of people practising self-isolation, untold numbers of business closures and cutbacks (forcing people out of work), immense psychological distress, and many people risking their health by continuing to work in essential services (e.g., health care, food provision & transportation). More recently, infection rates have declined in some settings due to high vaccination and drug therapy regimes, but not in others.[2] The World Health Organisation continues to regard COVID-19 seriously, stating that, although "the pandemic itself is [not] over, … the global emergency it caused is—for now."[3]

Of course, all times have their crises, shocks and challenges that repudiate those populist and ideologic, nostalgic yearnings for simpler times. On the other hand, in commenting about his book with co-authors, *Unprecedented?: How COVID-19 Revealed the Politics of Our Economy* (Davies et al., 2022), Davies (2022a) argues that perhaps the difference in our contemporary COVID-related polycrisis is that:

> the pandemic wasn't just an ordinary economic crisis. In many ways, it exacerbates tendencies that were already there but by virtue of it affecting everybody, had a powerful political effect in raising consciousness and visibility of these tendencies. … With Covid, some got rich out of it, many died, and nobody could entirely ignore it. It had a clarifying and politicising effect).

One egregious example of intimate workings of power, economy and privilege of the type Davies et al. (2022) describe was the VIP 'fast track' processes developed by UK politicians to procure personal protective equipment (PPE) early in the pandemic. With many countries competitively scrambling for what was available, British Tory Peer Lady Michelle Mone set up the PPE Medpro™ company and was subsequently awarded over £203 million in contracts. Much of the resultant substandard PPE produced was never used, while £70 million was paid to Mone's husband who, in turn, paid £29 million to an offshore trust for Mone and her children. In wake of the scandal, the UK government sought the recovery of £100 million from PPE Medpro™ with a fraud inquiry then ongoing (Conn, 2022). Clearly, such public exposure of cronyism consequent to the COVID-19 pandemic provides big opportunities for Davies et al.'s (2022) 'clarifying and politicising effects' for everyone.

As science educators, we are very invested in 'clarifying and politicising effects' and other ramifications of the pandemic, in particular, and polycrises, more generally, as critical, reliable and relevant science and technology education systems go some ways towards understanding and navigating these challenges and facilitating implementation of personal and social actions to mitigate relevant problems. This includes understanding societal priorities and practices, pro-capitalist activities and power relationships that ultimately benefit the few rather than the many. Hence, in this chapter, we explore some aspects of *our* post-pandemic worlds that we have organised as discrete chapter sections. Not only does this approach reflect our areas

[1] https://covid19.who.int

[2] https://covid19.who.int

[3] https://www.who.int/europe/emergencies/situations/COVID-19

of specialisation, but each section follows a thread that, taken together, weaves a more complete picture of our vision for post-pandemic science education. Lyn begins with an overview of current global tensions exacerbated by the pandemic and productive of large ideological and geopolitical forces that are constitutive of science and technology (education). In particular, she discusses renewed state government funding for science and technology consequent to increasing China-US high-tech competition and their direct impacts on science education. Isabel goes on to argue that some consequences of the pandemic are strongly linked to structural dimensions of social life, grounded in history and in political choices made before emergence of COVID-19's SARS-CoV-2 virus. As such, they demand science educators to go beyond traditional Nature of Science pedagogical approaches towards curricula that are capable of establishing more meaningful connections between knowledge, citizenship and social justice. Larry, meanwhile, responding to ongoing concerns about adverse influences of pro-capitalist living, nonliving and symbolic networks on fields of science and technology, their educational counterparts and much else, recommends that school science focus on educating students so they develop expertise, confidence and motivation to critically analyze aspects of their world relating to science and technology and develop and implement actions to help rectify problems of their concern. The chapter concludes, after a summary of our three contributions, with suggestions offering hope for increasing global social justice and environmental vitality.

14.2 Some Realities of Our Post-pandemic Geopolitics (Lyn Carter)

Because many of our pre and post-pandemic challenges—like unbridled climate change, rising energy insecurity, unknown and unimaginable implications of generative AI and new tensions over semiconductors, amplified by competition between the US and China—are formed from global networks and assemblages of science, technology, politics, economics and education, focusing on recent moves away from globalisation and acceleration of geopolitical tensions seems a sensible entry point. Science, science education and politics have always been mutually implicative, with Mulgan (2024) arguing that science and politics have been symbiotically linked throughout history and can only be understood in tandem. Moreover, Davies (2020, p. 654) argues that science has always been constituted by and is "dependent on various political, institutional and technical preconditions." What follows is a brief review of how we have arrived at a time many commentators have signaled as concerning resurgent crises of geopolitics.

The 1989 collapse of the Berlin Wall and dissolution of the USSR in 1991 helped to sediment the next three decades of globalisation and neoliberalism. Resultant economically prosperous, albeit highly unequal, unipolar worlds led by the US prompted political scientist Francis Fukuyama (1992) to proclaim the 'end of history' with ascent of free market neoliberal capitalism and liberal democracies. At the same time, Deng Xiaoping's post Mao policies in China, with unique blends of socialism and free enterprise, continued by Jiang Zemin (1993–2003) and Hu Jintao (2003–2012), facilitated globalisation with cheaper consumer commodities, logistics and supply chains. Millions were raised out of poverty and China developed as an economic powerhouse at a rate never before seen in history. When China joined the World Trade Organisation (WTO) in 2001 to further improve its prospects, the West was hopeful China would move closer to liberal democratic governance. However, with Xi Jinping's ascendency in 2012, China has, instead, aimed at developing a more multipolar world with its own form of communist governance, authoritarian capitalist economy and spheres of influence (e.g., the Belt and Road Initiative) that lay rival to the West.

A banking crisis triggered by growth in US sub-prime mortgage defaults precipitated the Global Financial Crisis (GFC) of 2007–2008 and ultimately created severe debt crises, austerity, and generalised insecurity. Right wing populist (RWP) movements and renewed sovereign state nationalism became attractive to those in the West who felt 'left behind' by globalisation and free market orthodoxies. RWP political pressures, combined with trade and supply issues exposed by the COVID-19 pandemic, and China's competitive rise, prompted desires for 'deglobalisation' or 'decoupling'—rhetorically, if not practically. Featherstone (2020, p. 160), meanwhile, suggested that "deglobalization could well be occurring on some levels, while re-globalization intensifies on others." Similarly, Steger and James (2020, p. 194) argue that shifts have occurred from material globalisation of human movements, supply chains and commodities to intangible global processes in "communication of ideas, words, images, meanings, knowledge, sounds, electronic texts, soft-ware programs and novel cyber-assets." The COVID-19 pandemic mediated increased global digitisation and connectivity. Think Zoom™ and streaming services, even as national borders resurged and states once again became sites of primacy. In what Davies (2022b) calls a return to *statist* capitalism, the State is acknowledged as the only political and economic actor capable of printing money and issuing bonds to ease pandemic-related economic shocks, as well as fund mega projects like development of COVID-19 vaccines. This is in sharp contrast to decades of unfettered globalisation and neoliberal free market supremacy prior to the GFC when the Chairman of the US Federal Reserve from 1987 to 2006, Alan Greenspan, is reputed to have remarked 'that it didn't matter who the US President was, as the market would always work it out.'

Even as globalisation remains persistent and pliable in some forms, national security interests have now become preeminent. These sentiments were reiterated as recently as April 2023 when US Secretary of the Treasury, Janet Yellen, spoke on the US-China relationship—suggesting that, while it was in no-one's interests to deglobalize and decouple (and $11 billion investment in China by the US private sector suggests they are not), "safeguarding certain technologies from China's military and security apparatus is of vital national interest (of the US) … (and to be achieved by) forms of export controls.[4]" In a policy of containment, semiconductors and other advanced scientific and computing manufacturing items are restricted to counter the accelerating advancement of China's high-tech and science capabilities. Yellen goes on to argue that "China's economic growth, [though], need not be incompatible with US economic leadership." This point, for Tooze (2023), is a pivotal part of the polycrisis; the blind spot in the last three decades of what has turned out to be a temporary unipolar globalisation. How does the West, and particularly the US, accommodate China (and India since, together, they account for more than a third of the world's population) and make the shift to a multipolar world as these nations take their rightful places at the global table? For Tooze (2023), Yellen arrogates the right to define and limit Chinese competition and economic development so that it doesn't challenge US economic leadership. Not only is this illegitimate and unacceptable, maintains Tooze (2023), it is highly likely to be resisted. Tooze (2023) goes on to argue the provocative point that President Xi is a self-confident direct descendant of Mao, Stalin and Lenin (themselves ideological reference points for the 1920/30s as the Socialist Calculation Debate (SCD) noted above), whose resurgent communist China challenges views that the West ever won the Cold War. Four decades on from the collapse of the USSR, suggests Tooze (2023), certainties of Western liberal democracy and market capitalism's globalism, in whose embrace Western science education so confidently resides, seems to have encountered some major obstacles!

Critical to this new world of post-pandemic geopolitics is the massive surge in nation state spending on scientific and technological research and development (R&D). Mulgan (2024) describes how, in 1960, over 33% of all global R&D was funded by the US Department of Defence. This is not surprising, as it was another time of security crisis. The USSR had just won the space race with the Sputnik launch in 1957. Some of this funding overhauled science education curriculum and expanded availability of school laboratories. Mulgan (2024, p. 120) goes on:

> The equivalent proportion in 2016 was just 3.6%. The shift to business-led research is just as striking. In the US, the top five tech firms' R&D investment was now ten times bigger than the top five defence firms. … As a result, many recent technologies, including 5G mobile, artificial intelligence (AI), quantum computing, high performance batteries and biotech, have been primarily developed by business, with the military later learning how to adapt and adopt them.

This private sector science and technology R&D was driven by neoliberal globalisation's free market orthodoxies with privatisation of knowledge and commercial secrecy. The increasing geopolitical tensions and resurgence of national security concerns, however, has seen pivots back to massive state funding. For example, in August 2022, the US legislated $278 billion for the *Creating Helpful Incentives to Produce Semiconductors for America Act (CHIPS) and Science Act* for development of innovative technologies like quantum computing, AI and nanotechnologies—as well as to create new regional high-tech hubs and a bigger, more inclusive,

[4] https://home.treasury.gov/news/press-releases/jy1425

science, technology, engineering, and mathematics (STEM) workforce. The US also legislated the $370 billion *Inflation Reduction Act (IRA)* aimed at improving energy security, decarbonising their economy and accelerating clean energy transitions. STEM immigration policies have been instituted to recruit the world's best to the US, as have subsidies and other incentives for relocating innovative technology businesses. Chu and Roeder (2023) identified at least 75 major projects announced within the first 6 months of the gazetting of the IRA and CHIPS. Unsurprisingly, the US's approach is not without its controversies, with its subsidies and tax breaks attracting high tech and clean energy business away from traditional US allies like the European Union, South Korea, Australia and the UK.

Also hardly surprising is injection of funding into STEM education to both pedagogical and institutional ends. Around $13 billion of CHIPS monies have been allocated to STEM education for:

> research on effective approaches to engaging students in PreK–12, including students from groups historically underrepresented in STEM and rural students with a focus on "innovative before-school, after-school, out-of-school, and summer activities that are designed to encourage interest, engagement, and skills development in STEM" … (as well as develop) … innovations in PreK–12 STEM education by establishing Centers for Transformative Education Research and Translation that bring together partners to facilitate the widespread and sustained implementation of promising, evidence-based STEM education practices (ASTC, 2022).

The IRA will facilitate schools to become more sustainable and carbon neutral with mitigation and adaptation funding (grants, subsidies and tax credits) for electric school vehicles, installing renewable energy like solar panels, geothermal heat pumps and energy storage systems, retrofitting and building in sustainable materials and more (King & Todd Whiteman, 2022).

Of course, the US is not alone in ramping up government R&D. Huang (2023) describes how, year on year, China has massively increased its science and technology R&D spending—making it second only to the US in global terms (and in total figures for the private, non-profit and state sectors as much as can be ascertained), with the gap narrowing quickly. And it is well ahead of other countries with a Gross Domestic Product (GDP) like China's. Primarily state rather than private sector funding under its authoritarian capitalist mode, China's focus is high-tech—with foci on, for example, semiconductors, AI, clean energy and robotics to ensure its self-sufficiency and independence. As four of the world's 10 largest corporations are state-owned Chinese entities, this is not surprising. The Made in China 2025 strategy is designed to move China from the low-cost consumer orientated global manufacturer it was in the early decades of globalisation into a global high-tech manufacturer and services powerhouse for the future (Huang, 2023).

In coming to review our contemporary polycrisis, cleavage lines between superpowers like the US and China jockeying for ascendency to control and utilise extraordinarily powerful science and technologies of the twenty-first century is sobering to consider. If COVID-19 taught us anything, it was that nation states can always out spend, and ultimately control, private companies through money printing and regulation, and that we have now entered a new era of statist capitalism.

Peters (2022, p. 4) agrees when he argues that reorientation of the US's science and technology policies "reinforce a move away from the neoliberal global free-market based on contracting out toward a science and technology strategy that is masterminded, directed and supported at the federal level." These larger geopolitical realities are directly impacting science education at all levels.

14.3 Culture Wars, Disinfodemic and Management of the COVID-19 Pandemic in Brazil: The Importance of Counter-Hegemonic Perspectives in Science Education (Isabel Martins)

Apart from exposing and widening social inequality, the COVID-19 pandemic paved ways to yet other culture wars in Western societies. Politicians exploited the COVID-19 pandemic in different ways and "the political debate about COVID-19 has evolved into a bitter culture war, where arguments have become struggles between different social groups holding different beliefs about how society should be constructed and governed" (Horton, 2022, p. 346). In the young and fragile Brazilian democracy, which historically has survived attempts to be destroyed by right wing groups, these attacks included authoritarianism, disqualification of science, threats to democratic institutions and manipulation of information.

Although the Brazilian experience provides an equally emblematic and disastrous case in which denialist authoritarian rule openly disqualified and disregarded medical advice and public health recommendations about measures to counteract spread of the pandemic, it appears to resonate with and shed light on similarly unfortunate circumstances in other countries (e.g., Moura et al., 2021). The concept of culture war seems apt to characterize these conflicts because, differently to what happens in classic examples of denialism (Oreskes & Conway, 2023), arguments did not misrepresent or cast doubt on scientific evidence. Instead, they evoked debates between conservative versus progressive values, which often involved identitary issues. In fact, many statements that were widely broadcast at the height of the pandemic showed entanglements between (mis)management of the pandemic and lack of critical problematization of environmental injustice, male chauvinism and religious manipulation. Examples include claims that "Brazilian jumps into sewer and nothing happens[5]" and that "It's necessary to face the virus as a man,[6]" as well as pleas for "… a day of fasting for all the Brazilian people, in the name, obviously,

[5] https://g1.globo.com/politica/noticia/2020/03/26/brasileiro-pula-em-esgoto-e-nao-acontece-nada-diz-bolsonaro-em-alusao-a-infeccao-pelo-coronavirus.ghtml

[6] https://noticias.uol.com.br/ultimas-noticias/agencia-estado/2020/03/29/bolsonaro-diz-que-e-preciso-enfrentar-virus-como-homem-e-nao-como-moleque.htm

of Brazil being free of this disease as soon as possible.[7]" Furthermore, this culture war was fueled by fake news, incessantly and viciously spread during the pandemic by social media algorithms, online bots and websites that impersonate genuine sources. "COVID-19 vaccines can alter human DNA,[8]" "Vaccines are produced from cells of aborted fetuses[9]"; "Hydroxicloroquin is effective against Covid and free from side effects[10]" are examples of infamous claims that were incessantly and viciously spread during the pandemic and were objects of scrutiny and refutation by fact checking agencies.

Entanglement of economic and political power networks with disinformation contributed to scenarios in which narratives about nature, viruses, vaccines, and science were objects of disputes. Although struggles for hegemony are part of discursive life, they are not fair. A dramatic example was that of Manaus, a capital city in the Amazonian region, where pandemic management was the object of an official investigation by the Brazilian Senate.[11] This commission of inquiry investigated evidence that suggest that, at the height of the sanitary crisis, the Ministry of Health would have implemented an experimental strategy based upon uses of the so-called COVID Kit, composed of drugs, like hydroxychloroquine, that had no proved efficacy against the disease.[12] Delays in vaccination, neglect in the enforcement of social distance measures and shortage of basic health supplies made the city's health system collapse.

Pandemic mismanagement in Brazil, as in other parts of the globe, was also inextricably linked to regional and cultural prejudices, socio-environmental and cognitive injustices, racisms, and inequalities in access to health and education. COVID-19 helped uncover bases under which such structural elements, and their manifestations in society, had been normalized over the years, revealing numerous aspects of social injustice. For instance, death statistics from COVID-19 are lower in Africa than in Europe and North America.[13] One possible reason might be subnotification. However, this could also be related to the fact that the COVID-19 virus is more lethal in elderly people and because average life expectancy in Africa is about 20 years lower than in North American countries.[14] Also, factors like deforestation, invasions of preserved areas for economic speculation, lack of housing poli-

[7] https://valor.globo.com/politica/noticia/2020/04/02/bolsonaro-pede-jejum-para-livrar-o-brasil-do-coronavirus.ghtml

[8] https://www.bbc.com/portuguese/geral-53795050

[9] https://butantan.gov.br/covid/butantan-tira-duvida/tira-duvida-noticias/a-vacina-do-butantan-nao-e-desenvolvida-a-partir-de-celulas-tronco-de-fetos-

[10] https://amb.or.br/noticias/associacao-medica-brasileira-diz-que-uso-de-cloroquina-e-outros-remedios-sem-eficacia-contra-COVID-19-deve-ser-banido/

[11] https://www12.senado.leg.br/noticias/materias/2021/07/22/documentos-revelados-pela-cpi-sobre-kit-covid-no-amazonas-provocam-criticas-de-senadores

[12] https://jornal.usp.br/ciencias/tratamento-precoce-e-kit-covid-a-lamentavel-historia-do-combate-a-pandemia-no-brasil/

[13] https://coronavirus.jhu.edu/data/mortality

[14] https://data.who.int/countries/076

cies to support migrant populations, flexibility to get environmental licenses for industries and non-sustainable practices could be involved in both increasing uneven environmental risk and creating contexts that might favour escape and spread of the new corona virus (Tollefson, 2020). Inequality is also manifest in social stigma that viral infections are often considered diseases of poverty, and this was reinforced because people have different possibilities of protecting themselves. Actually, societal configurations made it possible for people who were not "essential workers" to protect themselves in ways that unemployed people and precarious workers could not. Another example of inequality was manifest in difficulties faced by schools in disadvantaged areas, which lacked infrastructure to carry out educational programmes that would have enabled students' participation in online teaching.

These remarks defy a supposedly democratic character of the SARS-CoV-2 virus. Often mentioned in the context of the pandemic, the image conveyed by the phrase 'we are going through the same storm, but in rather different boats' speaks eloquently about different populations being affected differently. In Brazil, as well as in other countries, social vulnerability, environmental risk and structural racism were part of conjunctures that preceded the virus. Together with questionable decisions of allocating funds to buy and distribute COVID Kits instead of purchasing vaccines and starting a mass vaccination campaign, these factors created favourable conditions for what Mbembe (2019) describes as *necropolitics*; that is, uses of social and political power to dictate how some people may live and how some must die, through creation of unique forms of social existence in which vast populations are subjected to living conditions that confer upon them statuses of the living dead.

In order to deconstruct and fight against perverse inequities, it is important to acknowledge that it has taken pretty strong ideological forces and moments to naturalise such social and environmental injustices, social inequality and power asymmetries. Centuries of slavery, colonialism and oppression lie at the background to neoliberal capitalism. They are key concepts to understand ways through which individualism, competitiveness, resilience and merit were mobilized at governance and ideology levels as necessary values to overcome undesirable effects of the pandemic (Carter, above). They were made ubiquitous and pervasive by digital technologies of information and communications that trespassed space-time boundaries and accelerated spreading of fake news and post-truths regarding COVID-19. To counteract such a 'disinfodemic,' an avalanche of "emotive narrative constructs, fabricated websites, fraudulent or decontextualized images, orchestrated campaigns,[15]" is no simple task. To fight the powerful structures and strategies for disseminating fake news in the public sphere demands the domain and appropriation of knowledge about big data, uses of digital technologies and the mobilization of both material, financial and legal resources. However, attempts that seek to combat the spread of disinformation, hatred speeches and criminal contents in the internet are often criticized on the basis that this would go against freedom of speech.

[15] https://en.unesco.org/covid19/disinfodemic/brief1

This contemporary scenario, characterized by neoliberal fallacies, renewed and disguised versions of coloniality, capitalist resilience and disinformation, poses particular challenges to science education communities. Intricate characteristics of issues like pandemics, racialisations, eco-injustices etc. focus, at least, on two points. Firstly, they point to needs for reconceptualizing curricula in ways that allow them to overcome traditional epistemological/disciplinary divides and to foster intercultural dialogues. Secondly, they point to recognition that we, science educators, are products and producers of discourses and positionings about science.

The first point is important in contexts of recent (re)conceptualizations of science education in terms of the acronym STEM. Such a change, which seems to have been uncritically accepted by many science educators, hides important connections to views of education, science and social and scientific development that were found to be bases of exploitation systems and predatory practices that survive today as legacies of Modernity (Weinstein, 2017). It seems important to question extents to which science and technology cannot be cog wheels in neoliberal capitalist models devised in the Global North (Bencze, below) but, instead, be critical and inclusive social practices committed to social justice. Valorisation of epistemological pluralism and construction of contact zones between different knowledge and value systems may provide tenets to overcome impasses like those concerning characteristics of distributive justice (Fraser & Honneth, 2003).

The second point problematises ways through which science education communities have stressed importance of exploring relationships among science, technology and society, of fostering critical thinking, and of learning about the nature of science. Likewise, although education for citizenship often is a stated goal for science education in curriculum policy documents throughout the world, that science knowledge alone will be sufficient to ensure students' informed decision making and participation in social affairs no longer stands (Martins, 2011). Not only is it necessary to expand epistemological agendas to include socio-historical dimensions of knowledge but it also seems critical to problematise political tenets in which such decisions are to be made. Moreover, we must think about roles of schools as one of many institutions that shape people's characters and experiences. As science educators, we are dialectically involved in production and consumption of discourses about science in society. This inherent reflexivity in our social practices involves "fighting from within"; that is, displacing ourselves from our comfort zones and facing up to some likely pedagogical contradictions. For example, developing students' critical thinking and activist stances towards science involves encouraging them to be questioning participants in educational practices. New pedagogical contracts are in order if we are to construct awareness of our roles as educational agents. Not only do we need to create opportunities for students' active engagement in school affairs and in curriculum enactment but also to rethink teacher education (initial and ongoing) so that construction of curriculum experiences is not detached from life experiences and cultural contexts. Our collaborative efforts involving groups of individuals who, despite intersecting identitary differences concerning age, gender, class, religion etc., share common goals, have provided us with insights as to what potentially constitutes tenets founded on mutual respect, intellectual growth and collective empowerment. This entails situating ourselves as historical beings, questioning our beliefs and privileges, deconstructing our prejudices and engaging in development of horizontal methodologies that can sustain co-teaching practices and co-authorship of semiotic resources for teaching (Martins et al., 2023).

These views were expressed as calls for acknowledging that education always is essentially a political act. Such considerations are not related to specific political parties' agendas, but to needs for analyses of characteristics, possibilities and consequences of social arrangements, of relationships between peoples and nature, and of their representation in curricula and of their role in constructing projects of society. It is about considering teachers and students as agents, education as a right, and politics as enactments of social life. In other words, establishing agendas for research and practice that allow new relationships among people, institutions and (science) knowledge to be constructed (Martins, 2011).

In summary, it is possible to claim that the COVID-19 pandemic acted as a destabilising factor that made us suspect what we used to call "normal." It exposed some bases upon which normalisation of abnormal circumstances we lived in had been constructed. This is an utterly important lesson for post-pandemic scenarios. It explains why the phrase "going back to normal" is, to say the least, a very deceptive way of describing what needs to be done because, honestly, it hasn't been "normal" for quite a while…

14.4 Critical & Activist Science & Technology Education: Opportune Contributions to Ideological Battles (Larry Bencze)

SARS-CoV-2 viruses seem to be like (many) pro-capitalist 'entities,' instigating assemblage of *dispositifs* (Foucault, 2008); that is, myriad living, non-living and symbolic entities ('actants,' in actor-network theory terms [Latour, 2005]) more-or-less in service of their own propagation at expense of wellbeing of many or most other actants. I became aware of such network power particularly after reading works like McMurtry's (1999) *The Cancer Stage of Capitalism*, Ziman's (2000) *Real Science* and Carter's (2005) *Globalisation and Science Education*. They helped explain difficulties I was having in *widely* promoting student-directed and open-ended (Lock, 1990) science inquiry and technology/engineering design projects. Perhaps pro-capitalist dispositifs were orienting school science in their favour, such as in terms of teachers' tendencies to adhere to more *Rationalist-Realist* conceptions of the nature of science in terms of Loving's (1991) *Scientific Theory Profile*.[16]

[16] The horizontal axis of the *Scientific Theory Profile* spans a continuum from *Rationalist* through *Naturalist* positions about the nature of theory negotiation in the sciences. Rationalists tend to believe in highly logical and systematic science methods in judgements about theory. Naturalists assume the conduct of science is highly situational and idiosyncratic, depending on psychological, social, cultural, political, etc. influences. The vertical continuum depicts validity of knowledge.

Teachers' such NoS positions seemed to confine student experiences to more teacher-directed and closed-ended activities (Bencze et al., 2006). I felt that such more teacher-controlled education may have compromised student intellectual independence and self-actualization. Contributions in this Chap. 14 by Lyn Carter and Isabel Martins appear to support, in different ways, such contentions. In that vein, such pro-capitalist injustices (and environmental degradation) seem likely to continue after the COVID-19 pandemic. As pointed out by Lyn Carter above, for example, it is apparent we have been and continue to be threatened by numerous and diverse intertwining crises that seem very unpredictable and, therefore, disorienting (Tooze, 2022). In Canada, where I have long worked and continue my research, on the other hand, the sociopolitical environment *appears* somewhat stable to me. Reactions to the COVID-19 pandemic did not, for instance, much resemble what was described by Isabel Martins (above). Nevertheless, there seem to be signs of increasing threats to social justice and environmental vitality. Canada appears, broadly, to function as an example of 'progressive' neoliberalism (Sharma et al., 2023); that is, a socioeconomic system that presents itself as 'progressive,' but uses that image as a kind of 'smokescreen' allowing 'behind the scenes' activities to further implement neoliberal perspectives and practices. For example, while often exhibiting somewhat progressive social policies, like supports for equity, diversity and inclusion, legislation favouring private sector interests appears increasingly prevalent (McQuaig, 2019).

In light of my desires to help address polycrises associated with pro-capitalist networks and increase social justice and environmental vitality (e.g., Martusewicz et al., 2021), I developed—with graduate students and school science teachers—the STEPWISE[17] curricular and pedagogical schema illustrated in Fig. 14.1. This schema recommends that, *before* being asked to *independently* design and conduct research (from primary & secondary sources) and research-informed and negotiated action (RiNA) projects to address harms or issues in relationships among fields of science and technology and societies and environments (STSE) of *their* interest, students should be provided with preparatory lessons and student activities. As suggested in Fig. 14.1, such an 'apprenticeship' is to consist of one or more three-phase cycles of: (i) reflecting on and expressing (e.g., in writing or orally) their current (e.g., pre-instructional) attitudes, skills and knowledge (ASK) about particular school science topics (e.g., cell biology); (ii) teacher-led lessons—in synchrony with relevant *application* activities—to help students develop ASK that the teacher and other stakeholders consider important (e.g., possibly problematic influences of corporations on science & engineering [e.g., Ziman, 2000]); and, (iii) practice activities, in which students design and conduct small-scale RiNA projects, receiving teacher supports as requested by students.

Realists tend to believe that science knowledge corresponds to reality, while (extreme) Antirealists claim that each person's constructions are valid. Moderate Antirealists believe in useful knowledge.

[17] To learn more about STEPWISE and gain access to corresponding teaching and learning resources, visit: www.stepwiser.ca

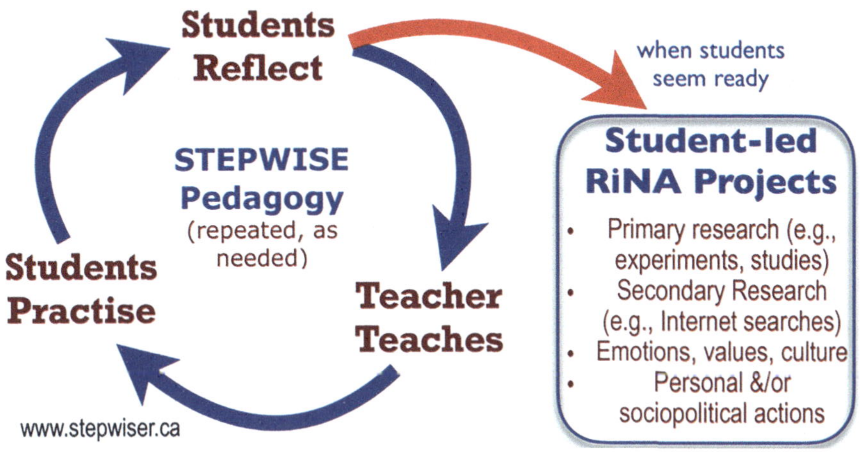

Fig. 14.1 STEPWISE pedagogical schema

In preparing students to self-develop and implement RiNA projects, STEPWISE pedagogy emphasizes *direct* teacher instruction to enable *de-punctualization* (Callon, 1991) of phenomena in science curricula. That is, in providing students with more *holistic* conceptions of fields of science and technology/engineering, they may be educated about often 'sanitized' or hidden apparent harms to wellbeing of individuals, societies and environments (WISE) that may be linked to influences of powerful people (e.g., financiers and politicians) and groups (e.g., corporations, governments, transnational groups like the World Trade Organization and free-market think tanks) on fields of science and technology and much else in societies and on environments (Krimsky, 2019). Such enlightening education also should highlight personal and networked sociopolitical actions that people might take and/or have taken. This may include references to resources they might use in designing and implementing their actions, such as uses of primary and secondary research and aspects of their prior education. It is particularly important to help students to use actor-network theory (Latour, 2005) and the dispositif concept (Foucault, 2008) for critical analyses of STSE relationships and actions. Enlightening and problematizing education of this sort seems necessary to overcome *dis-education* effects of:

(a) **popular inquiry-based learning**, which can be *discriminatory* because of many students' struggles to discover intended ASK, depending, for example, on their cultural and social capital (Bencze & Alsop, 2009),
(b) **sanitized STEM education initiatives**, which tend to omit or manipulate references to possibly problematic influences of powerful people and groups on fields of science and technology/engineering and much else (Hoeg & Bencze, 2017), which may limit teachers' education of students; and,
(c) **public propaganda**, such as through private sector suppression of science research results that may incriminate for-profit products and services and financiers and corporations responsible for them (Oreskes & Conway, 2010), which

Fig. 14.2 STEM relationships

may limit student discovery in inquiries and teachers' abilities to educate students about such research results.

Our research since about 2006 with science educators in formal primary, secondary and tertiary education contexts and in after-school clubs indicate that the schema in Fig. 14.1 has helped many students to develop significant expertise, confidence and motivation for self-directing (mostly) varied and personally meaningful RiNA projects (Bencze, 2017). Such successes suggest that the STEPWISE framework may be useful for encouraging and enabling students—as current and future community members—to analyze and critique STSE relationships and, for harms *they* determine, develop and take sociopolitical actions to overcome them. Such an education may contribute to efforts to transform societies into those giving greater priority to social justice and environmental vitality. Based on past successes with promotion of RiNA projects, it is apparent that students may contribute to social transformation in at least two broad (co-affecting) ways; that is, in both directions of translations depicted in Fig. 14.2 (Bencze & Carter, 2015); that is, in terms of:

- ***Proposed* Changes**. Perhaps as expected in school *science*, many student RiNA projects prioritize science research (World ➔ Sign translations)—while also making *proposals* for (rather than actualization of) translations from Signs ➔ World. For example, a student's research led to an actor-network map analysis of social-technical relationships involving liquid foundation make-up which, in turn, led her to develop an educational video (tinyurl.com/3j5mrwc5) that might influence numerous actants—including friends, family, members of the general public, government officials and others. Her video perhaps had much narrower 'ideological gaps' ('engineered' mis-translations) with her actual makeup than that which may be typical in cosmetics companies' advertising. Several other such examples can be found in three issues of *JASTE*: tinyurl.com/y9axcbou; bit.ly/2JGIgtf; and, tinyurl.com/3s6z59v2. With perhaps further insights from teachers (*Teacher Teaches*, Fig. 14.1) about long-standing (e.g., in the neoliberal period) adverse influences of capitalists on integrity of STEM research and publication (Krimsky, 2019) and more recent possibly dramatically increased adverse influences by right-wing populist governments (Carter et al., 2019), students may provide governments, transnational pro-capitalist organizations, think tanks, general members of the public, etc. with recommendations for increasing integrity and uses of STEM work.

- ***Actualizing* Changes.** Apparently, not so common in school science have been engineering design projects (Sign ➔ World translations, Fig. 14.2). Although recommended in many school science/STEM initiatives (Pleasants & Olson, 2018), these tend to punctualize (reduce) and de-problematize STEM processes, products and their dissemination (Pierce, 2013). Accordingly, for the last few years, graduate students, teachers and I have been conducting action research to explore possibilities for encouraging and enabling school science students to develop, implement and widely promote 'WISE' engineering designs; that is, innovations or inventions that perform desired tasks and prioritize 'wellbeing of individuals, societies and environments.' Education of students prior to their development of such 'WISE' technologies has involved instruction about problematic STSE relationships (including through uses of actor-network maps), such as those noted above about consumerism. In doing so, students also were introduced to problems linked to *compromises* that may be made (intentionally or otherwise) in engineering design—such as prioritizing profit over addressing problematic outcomes like nonbiodegradability of materials and poor labour conditions. Among student engineering designs, a student chose to design (and re-design) and build, using many recycled ('waste') materials, a candle recycler that he said could be useful to people without electricity. For other such designs, see *JASTE* at: tinyurl.com/yb45cbmv; and, tinyurl.com/3s6z59v2.

Although most students have been able to design and implement WISE technologies and have stated awareness of various compromises (mostly *during* interviews, rather than in their reports), most students limited their approaches to establishing larger, more complex, networks (dispositifs) for their innovations/inventions to 'Likes,' etc. from their 'followers' in their social media environments (e.g., Instagram™). In the context of formal schooling, it seems difficult for many (or most) students to rally—to extents similar to that of capitalists, for example—numerous actants in support of their causes (e.g., WISE technologies). Consequently, it appears that more action research is needed in this regard.

Although STEPWISE has been relatively successful in ways suggested above, it seems to be limited to contexts in which small networks of co-supportive actants ('mini-dispositifs') exist—such as: an enthusiastic and appropriately educated teacher (e.g., about capitalist influences on STEM fields); supportive colleagues and school administrators; official curricula sanctioning such perspectives and practices (e.g., MoE, 2008); appropriate instructional resources; understanding parents; and, perhaps, mentorships by one or more action research facilitators (Bencze & Krstovic, 2017). Accordingly, efforts to use pedagogical frameworks like STEPWISE to contribute to ongoing efforts at social transformation—*perhaps* more aligned with ecojustice principles—is likely to require concerted, multi-pronged, actions to encourage assemblage of *global* supportive dispositifs. Beyond educators, such assemblages likely should involve people from government, business, labour organizations, community groups and many more—as well as inclusion of new, more ecojust, technologies and systems.

14.5 Coda

The COVID-19 pandemic was, with little doubt, *devastating*—although much more so for many disadvantaged people than for advantaged ones, *ourselves* included, in different contexts around the world. Despite its shocking nature, however, such unbalanced struggles—sometimes apparently accidental and sometimes perhaps *planned*—seem *normal* (as difficult to define is that term). Indeed, as elaborated in this chapter, the 2019 viral pandemic seemed to be just one example of numerous and varied crises that, in Gestalt-like ways, are highly disorienting in their unpredictability. Having said that, some crises—like the climate emergency (Ripple et al., 2021), devasting animal farming (Monbiot, 2022) and threats of nuclear annihilation[18]—seem potentially *existential* (Ord, 2020). Meanwhile, compounding our senses of devastation, it appears we are increasingly dominated by fake news, false truths, AI-generated 'altered realities,' etc.—in a kind of *disinfodemic*.

Given range and complexity of risks (again, mostly unequally) to humans, other life forms and nonliving environments, it seems difficult to pin-point causes. Although many sources of struggle often appear to involve varying identities (e.g., regarding culture, religion, race, etc.), the pandemic seemed to de-punctualize (e.g., unmask) (Callon, 1991) dominating societal structures—perhaps especially exposing hegemonic material-semiotic systems of *political economy*. While such political and economic tensions are by no means new, there have been frequent signs of their ongoing dynamism. It is apparent, for instance, that relatively recent orientations towards global circulation of goods and materials, labour, values, etc. are being challenged by more protectionist *statist* perspectives and practices—with, moreover, apparent further subdivision of 'states' (e.g., nations) into, for example, *special economic zones* that allow 'efficient' socioeconomic production through loosened labour, environmental and materials' regulations (Slobodian, 2023). At the same time, some prominent analysts are suggesting that much power is now being concentrated into hands of ownership of major technology companies (Varoufakis, 2023).

With such complex, risky and uncertain worlds, one might wonder whether or not science and technology education should attempt to educate students about broader sociopolitical, economic, cultural, racial, gendered, religious and other factors possibly dynamically interacting with knowledge production systems. On the one hand, many STEM education initiatives have, apparently (Hoeg & Bencze, 2017), narrowed foci on student learning of knowledge production processes like science inquiry and engineering design and learning of products, such as laws,

[18] https://thebulletin.org/doomsday-clock/current-time/nuclear-risk/

theories and functioning of inventions/innovations, from them—apparently in their ongoing assistance to local and global socioeconomic competitions. We and others (e.g., Weinstein et al., 2023), however, suggest that—assuming many problems linked to science and technology result from varying power imbalances across myriad living, nonliving and symbolic entities (e.g., Foucault, 2008)—efforts must be made to rally diverse entities, such as policy makers, financial backers, education authorities, parent groups and many others, to ensure that science and technology education is structured in varying 'real-world' (as best as can be determined) contexts to prioritize social justice and environmental vitality. This stance places science and technology education in prominent societal roles. There are strong claims that societal zeitgeists are, to great extents, correlated with changes in fields of science and technology (Jasanoff, 2015)—and, as defended above, their educational counterparts. This seems like a necessary, although perhaps lofty, goal. In the near term, it seems *imperative* that students be 'vaccinated' against hegemonic power by, for example, critiquing their fields of study and larger sociopolitical contexts of which they are a part and, as well, being prepared to engage in sociopolitical actions to help overcome related harms of their concern (Hodson, 2011). At the same time, given complexities of our world and our preferences for *democratic* decision making, we must accept unpredictable outcomes that we may not support (McLaren, 2000).

References

Association of Science and Technology Centers [ASTC]. (2022). *CHIPS and Science Act authorizes significant investment in STEM.* https://www.astc.org/issues-policy-and-advocacy/chips-and-science-act-authorizes-significant-investment-in-stem

Bencze, J. L. (Ed.). (2017). *Science and technology education promoting wellbeing for individuals, societies and environments: STEPWISE.* Springer. https://doi.org/10.1007/978-3-319-55505-8

Bencze, J. L., & Alsop, S. (2009). A critical and creative inquiry into school science inquiry. In W.-M. Roth & K. Tobin (Eds.), *The world of science education: Handbook of research in North America* (pp. 27–47). Sense. https://doi.org/10.1163/9789087907471

Bencze, J. L., & Carter, L. (2015). Capitalists' profitable virtual worlds: Roles for science & technology education. In P. P. Trifonas (Ed.), *International handbook of semiotics, vol. 1 & 2* (pp. 1197–1212). Springer. https://doi.org/10.1002/sce.20124

Bencze, L., & Krstovic, M. (2017). Resisting the Borg: Science teaching for common wellbeing. In J. L. Bencze (Ed.), *Science and technology education promoting wellbeing for individuals, societies and environments: STEPWISE* (pp. 227–276). Springer. https://doi.org/10.1007/978-3-319-55505-8

Bencze, L., Bowen, M., & Alsop, S. (2006). Teachers' tendencies to promote student-led science projects: Associations with their views about science. *Science Education, 90*(3), 400–419. https://doi.org/10.1002/sce.20124

Callon, M. (1991). Techno-economic networks and irreversibility. In J. Law (Ed.), *A sociology of monsters: Essays on power, technology and domination* (pp. 132–161). Routledge.

Carter, L. (2005). Globalisation and science education: Rethinking science education reforms. *Journal of Research in Science Teaching, 42*(5), 561–580. https://doi.org/10.1002/tea.20066

Carter, J., Berman, E., Desikan, A., Johnson, C., & Goldman, G. (2019). *The state of science in the Trump era: Damage done, lessons learned, and a path to progress*. Union of Concerned Scientists. https://www.ucsusa.org/sites/default/files/attach/2019/01/ucs-trump-2yrs-report.pdf

Chu, A., & Roeder, R. (2023). Transformational change: Biden's industrial policy begins to bear fruit. *Financial Times*. https://www.ft.com/content/b6cd46de-52d6-4641-860b-5f2c1b0c5622

Conn, D. (2022, March 22). Government paid firm linked to Tory peer £122m for PPE bought for £46m. *The Guardian*. https://www.theguardian.com/politics/2022/mar/27/government-paid-firm-linked-to-tory-peer-122m-for-ppe-bought-for-46m

Davies, W. (2020). Green populism?: Action and mortality in the Anthropocene. *Environmental Values, 29*(6), 647–668. https://doi.org/10.3197/096327120X15752810323940

Davies, W. (2022a). In the British state's response to COVID-19, the unthinkable became thinkable. *New Humanist*. https://newhumanist.org.uk/articles/5922/in-the-british-states-response-to-COVID-19-the-unthinkable-became-thinkable

Davies, W. (2022b). How neo-liberal are contemporary modes of governance? *Centre for the Study of Governance and Society (CSGS)*. https://csgs.kcl.ac.uk/podcast/how-neo-liberal-are-contemporary-modes-of-governance-in-conversation-with-will-davies/.

Davies, W., Dutta, S. J., Taylor, N., & Tazzioli, M. (2022). *Unprecedented?: How COVID-19 revealed the politics of our economy*. Goldsmith Press.

Featherstone, M. (2020). Problematizing the global: An introduction to global culture revisited. *Theory, Culture & Society, 37*(7–8), 157–167. https://doi.org/10.1177/0263276420957715

Foucault, M. (2008). *The birth of biopolitics: Lectures at the Collége de France, 1978–1979* (Ed., M. Senellart). Palgrave Macmillan. https://doi.org/10.4324/9781315872193-28

Fraser, N., & Honneth, A. (2003). *Redistribution or recognition: A political-philosophical exchange*. Verso. https://doi.org/10.3366/per.2005.1.2.215

Fukuyama, F. (1992). *The end of history and the last man*. Free Press.

Hodson, D. (2011). *Looking to the future: Building a curriculum for social activism*. Sense. https://doi.org/10.1007/978-94-6091-472-0

Hoeg, D., & Bencze, L. (2017). Values underpinning STEM education in the USA: An analysis of the next generation science standards. *Science Education, 101*(2), 278–301. https://doi.org/10.1002/sce.21260

Horton, R. (2022). Offline: COVID-19 as culture war. *The Lancet, 399*(103332), 346. https://doi.org/10.1016/S0140-6736(22)00093-9

Huang, Y. (2023). *The rise and fall of the EAST*. London School of Economics and Politics. https://www.lse.ac.uk/united-states/events/2023-events/The-Rise-and-Fall-of-the-EAST

Jasanoff, S. (2015). Future imperfect: Science, technology, and the imaginations of modernity. In S. Jasanoff & S.-H. Kim (Eds.), *Dreamscapes of modernity: Sociotechnical imaginaries and the fabrication of power* (pp. 1–33). University of Chicago Press. https://doi.org/10.7208/chicago/9780226276663.001.0001

King Jr, J. B., & Todd Whiteman, C. (2022). Announcing the launch of this is planet Ed. *The Aspen Institute*. Retrieved from: aspeninstitute.org/blog-posts/announcing-the-launch-of-this-is-planet-ed/.

Krimsky, S. (2019). *Conflicts of interest in science: How corporate-funded academic research can threaten public health* (2nd ed.). Simon & Schuster.

Latour, B. (2005). *Reassembling the social: An introduction to actor-network-theory*. Oxford University Press. https://doi.org/10.1093/oso/9780199256044.001.0001

Lock, R. (1990). Open-ended, problem-solving investigations—What do we mean and how can we use them? *School Science Review, 71*(256), 63–72.

Loving, C. C. (1991). The scientific theory profile: A philosophy of science model for science teachers. *Journal of Research in Science Teaching, 28*(9), 823–838. https://doi.org/10.1002/tea.3660280908

Martins, I. (2011). Literacy as metaphor and perspective in science education. In C. J. Linder, L. Ostman, D. Roberts, P.-O. Wickmann, G. D. Erickson, & A. McKinnon (Org.). *The landscape of scientific literacy* (1st ed.) (pp. 90–105). Routledge/Taylor Francis. https://doi.org/10.4324/9780203843284

Martusewicz, R., Edmundson, J., & Lupinacci, J. (Eds.). (2021). *Ecojustice education: Towards diverse, democratic, and sustainable communities* (3rd ed.). Routledge. https://doi.org/10.4324/9780203836040

Mbembe, A. (2019). *Necropolitics*. Duke University Press. https://doi.org/10.1215/9781478007227

McLaren, P. (2000). *Che Guevara, Paulo Freire, and the pedagogy of the revolution*. Rowman & Littlefield.

McMurtry, J. (1999). *The cancer stage of capitalism*. Pluto. https://doi.org/10.7202/704266ar

McQuaig, L. (2019). *The sport and prey of capitalists: How the rich are stealing Canada's public wealth*. Dundurn Press.

Ministry of Education [MoE]. (2008). *The Ontario curriculum, grades 9 and 10: Science*. Queen's Printer for Ontario.

Monbiot, G. (2022). *Regenesis: Feeding the planet without devouring the world*. Penguin.

Moura, C. B., Nascimento, M. M., & Lima, N. W. (2021). Epistemic and political confrontations around the public policies to fight COVID-19 pandemic. *Science & Education, 30*(3), 501–525. https://doi.org/10.1007/s11191-021-00193-3

Mulgan, G. (2024). *When science meets power: The dilemmas of an accidental sovereign*. Polity Books.

Ord, T. (2020). *The precipice: Existential risk and the future of humanity*. Bloomsbury.

Oreskes, N., & Conway, E. (2010). *Merchants of doubt*. Bloomsbury Press.

Oreskes, N., & Conway, E. (2023). *The big myth: How American business taught us to loathe government and love the free market*. Bloomsbury Press.

Peters, M. A. (2022). Semiconductors, geopolitics and technological rivalry: The US CHIPS & Science Act, 2022. *Educational Philosophy and Theory, 55*, 1642–1646. https://doi.org/10.1080/00131857.2022.2124914

Pierce, C. (2013). *Education in the age of biocapitalism: Optimizing educational life for a flat world*. Palgrave Macmillan. https://doi.org/10.1057/9781137027832

Pleasants, J., & Olson, J. K. (2018). What is engineering?: Elaborating the nature of engineering for K-12 education. *Science Education, 103*(1), 145–166. https://doi.org/10.1002/sce.21483

Ripple, W. J., Wolf, C., Newsome, et al. (2021). World scientists' warning of a climate emergency 2021. *Bioscience, 71*(9), 894–898. https://doi.org/10.1093/biosci/biz088

Sharma, A., Schmeichel, M., & Wurzburg, E. (2023). *Progressive neoliberalism in education: Critical perspectives on manifestations and resistance*. Routledge. https://doi.org/10.4324/9781003224013

Slobodian, Q. (2023). *Crack-up capitalism: Market radicals and the dream of a world without democracy*. Allen Lane.

Steger, M., & James, P. (2020). Disjunctive globalization in the era of the great unsettling. *Theory, Culture & Society, 37*(7–8), 187–203. https://doi.org/10.1177/02632764209577

Tollefson, J. (2020). Why deforestation and extinctions make pandemics more likely. *Nature, 584*(7820), 175–176. https://doi.org/10.1038/d41586-020-02341-1

Tooze, A. (2022). *Chartbook #130, Defining polycrisis: From crisis pictures to the crisis matrix*. https://adamtooze.substack.com/p/chartbook-130-defining-polycrisis

Tooze, A. (2023). *Can the US solve the China dilemma?* Foreign Policy One and Tooze. https://foreignpolicy.com/podcasts/ones-and-tooze/can-the-us-solve-the-china-dilemma

Varoufakis, Y. (2023). *Technofeudalism: What killed capitalism*. Penguin.

Weinstein, M. (2017). NGSS, disposability, and the ambivalence of science in/under neoliberalism. *Cultural Studies of Science Education, 12*, 821–834. https://doi.org/10.1007/s11422-017-9844-y

Weinstein, M., Pouliot, C., Martins, I., Levinson, R., Carter, L., Bencze, L., & Sharma, A. (2023). *Science education towards social and ecological justice: Provocations and conversations*. Springer Nature. https://doi.org/10.1007/978-3-031-39330-3

Zeitlin, J., & Nicoli, F. (2020). *The European Union beyond the polycrisis? Integration and politicization in an age of shifting cleavages*. Routledge. https://doi.org/10.4324/9781003002215

Ziman, J. (2000). *Real science: What it is, and what it means*. Cambridge University Press. https://doi.org/10.1017/cbo9780511541391

Martins, et al. (2023). Building educational resources for the curricular approach to socio-scientific issues: An experience of collaboration between university and basic education school. In A. Galamba & H. Gandolfi (Eds.), *Critical pedagogies in science education: Ideas and experiences from Brazil and the UK* (pp. 150–169). Autografia.

J. Lawrence Bencze is an Associate Professor Emeritus in science education at the Ontario Institute for Studies in Education, University of Toronto. His 21-year career as a university-based science educator and researcher was preceded by a 4-year career as a school district science consultant and, prior to that, an 11-year career as secondary school teacher of science. He currently facilitates action research in different contexts of science and technology education to promote and learn about students' independent design and implementation of personal and sociopolitical actions to overcome harms of their concern in relationships among fields of science and technology and societies and environments. More about his work can be learned at www.lbencze.ca

Lyn Carter taught at the Melbourne Campus of the Australian Catholic University for many years before her recent retirement. She lectured in science and technology education to undergraduate primary and secondary teacher education students, and to postgraduate students in research methodologies and contemporary issues in curriculum. Her main research interests are in political economies shaping science education's enactment, with a particular focus on the impact of contemporary issues of globalisation, neocolonialism and ecocide. She has written for top tier journals including JRST, Science Education and RISE and published many articles, book chapters and co-edited papers. She was an Assistant Editor for RISE, and served on the Editorial Boards of JRST and CSSE.

Isabel Martins is Professor of Science and Health Education at the Federal University of Rio de Janeiro. She taught physics in Rio de Janeiro secondary schools before getting her PhD in Science Education at the University of London. She has also worked at University College London Institute of Education and at the Federal University of Minas Gerais, Brazil. Isabel has led and collaborated in several international funded projects, published in refereed journals, co-edited and co-authored academic books in Science Education. She was President of the Brazilian Science Education Research Association (ABRAPEC) and is editor-in-chief of the electronic teacher education journal Ciência em Tela. Her research interests concern the articulation of discursive, sociopolitical, cultural and epistemological dimensions of science knowledge and its implications for curriculum development, textbook analyses and scientific popularisation.

Chapter 15
Power, Knowledge, Subjectivity: Science Education and Questions of Collective Existence

Jesse Bazzul

15.1 Introduction

It's very difficult to know how much importance should be given to events that appear to change the very fabric of everyday life. Especially when such events are recent. The COVID-19 pandemic is one such event whose overall importance is still unclear. Over the last four years science education scholars have emphasized the COVID-19 pandemic's social and political contexts as being worthy of attention (Sharma, 2020; Alsop & Hoeg, 2020). The pandemic sparked a great deal of discussion around the politicization of science, collective responsibility of science in keeping communities healthy, and oppressive hierarchies that make sure some people (e.g. upper middle class people in industrialized nations) are safer than others. One of the awkward things about these discussions was the ignorance of the vast social inequalities that predated the pandemic. As if the pandemic was somehow the real problem and the oppressive hierarchies responsible for these disparities were only 'factors' of a kind. As Slavoj Žižek (2020) notes in his book *Pandemic!: Covid-19 Shakes the World* viruses are essentially 'stupid' replicating molecules. People should resist deriving too much meaning from pandemics, because even in the face of death it is ultimately the way we choose to live that matters. To focus on the meaning of the viral transmission is futile, simply because the replication of virus is a banal process that has occurred alongside human evolution for eons. Their role in evolution is much more interesting as far as events go. And while it might be

This chapter is an adaptation of an editorial in Caderno Brasileiro de Ensino de Física (Bazzul 2022a).

J. Bazzul (✉)
University of Regina, Regina, SK, Canada
e-mail: jesse.bazzul@uregina.ca

C. B. Moura (ed.), *A Sociopolitical Turn in Science Education*, Contemporary Trends and Issues in Science Education 62, https://doi.org/10.1007/978-3-031-78586-3_15

going too far to call the COVID-19 pandemic a non-event, I would argue it is only worth dwelling on insofar as it has actually changed social and ecological life for humans and nonhumans.[1] When it comes to a socially and ecologically engaged science education for all, social and ecological events that do not have implications for daily life might not be worth our collective attention as educators. Hopefully it's also clear that a case could be made for the COVID-19 pandemic recasting social and ecological life in very significant ways; and if so then it should be the focus of science educators.

Regardless of our personal connection to the COVID-19 pandemic, it certainly 'exposed' important realities of collective existence. For example, before the pandemic hierarchies between the Global North and the Global South (and let's not forget the hierarchies within—the 'North of the North vs the North' and the 'the South vs the South of the South') were already functioning to destroy some lives while allowing others to flourish (Machado de Oliveira, 2021). This is precisely why the epistemologies of the Global South should become an important focus for science education (Rezende et al., 2021). Large numbers of precarious peoples cannot subsist on the meagre provisions of capitalist social life and the long lasting legacy of colonial empire(s). Right wing populism and the systematic gutting of collective institutions were already in full swing before the People's Republic of China reported its first COVID-19 cases (indeed it's tempting to think the pandemic prevented a complete right-wing takeover of major institutions and public opinion in many countries). When it comes to engaging complex phenomena like pandemics the best thing educators can do is remain connected to the sociopolitical, historical, cultural, and ecological contexts that give them any significance and reshape daily life (for better or for worse).[2] It's through these vital contexts, and how the various aspects of daily life are enabled or disabled, that an event like a pandemic has any significance whatsoever. To put it another way, a pandemic, on its own, is a meaningless event. It gains significance only through what it changes. In this way, educators need to be 'in-tune' with a wide range of sociopolitical, ecological, and cultural contexts; otherwise they have no way of engaging how such an event matters at all.

15.2 Science, Power, Governance

One way science educators can make sense of important issues and events is to look at how scientific knowledges and practices are intricately connected to the exercising of power. More specifically how power is used to govern life in a broad sense.

[1] The anthropological work of Anna L. Tsing (2015) is invaluable for having researchers and educators think about life amongst social and ecological disturbance. See for example, *Mushroom at the end of the world: On the possibility of Life in Capitalist Ruins.*

[2] For Anna L. Tsing (2015), major disturbances, which have been an eternal part of social and ecological life on the planet, can be ethically evaluated on the basis of what they enable and/or destroy.

This involves examining the intricate relations between knowledge/science, power/governance, and subjectivity/being and the relations people have with themselves, others, and the world. The work of historian and philosopher Michel Foucault helps set the stage for this tripartite frame of understanding. Foucault was heavily influenced by the prominent historian of science Georges Canguilhem, who was fascinated with the historical connections between the biological sciences, politics, and administrative policy. While I'm not able to lay out an entire theoretical basis for understanding the relations between science, power, and subjectivity in this chapter I hope to offer some helpful starting points.

Besides exposing existing inequalities the pandemic uncovered a wide-spread thirst for collective forms of living in the form of lost salaries and wages, guaranteed hospital care, and government assistance irrespective of prior employment requirements (Žižek, 2020). Events like the COVID-19 pandemic simply reinforce the need for robust sociopolitical, cultural and historical frames for understanding in order to understand the salience of the event itself (again there's almost no salience to be found outside this framing). Just think about the 'fogginess' of this recent pandemic, whose key issues kept changing and proved resistance to simple fixes. Were the issues of global vaccine equality and science denial really pandemic problems? Or were they largely formulated outside the context of the pandemic? During the first years of the pandemic my colleague Aswathy Raveendran and I wracked our brains trying to determine the quintessential 'socioscientific' issue or key political question arising from the pandemic. And, to be honest, we were unable to pinpoint any one issue or set of questions that really got to the heart of the pandemic.

How power works in the social world is extremely complex, but especially in relation to science with its veneer of objectivity. Foucault's work draws attention to the way the exercising of power has changed in modern Western nations. From punishments that brutally target individual bodies, such as corporal punishment and public executions, to techniques of policy and police that intervene at the level of populations to 'conduct' how people live (or want to live) their everyday lives (Foucault, 1990). The development, dissemination, and deployment of 'objective' knowledges such as the biological sciences, economics, medicine, cartography, etc., allowed this shift in how power is exercised in modern Western societies. These knowledges were frequently developed alongside policy and the 'art of governing' because they gave governments the authority to justify particular practices of control. This resulted in more systematized forms of exercising power through things like: educational institutions that evaluated and sorted people according to state labour needs, census taking that monitored things like birth/death rates, and public health measures that relocated the urban poor and terminally ill. The way power is exercised over bodies today involves 'caring' for populations. Helping them grow for the purposes of economic competition, waging war, and the general caretaking and functioning of the state. Teachers, bureaucrats, and engineers are all needed for a modern nation to function. This kind of all encompassing power is sometimes referred to as *biopower*—quite literally the power over life at the level of both individuals or populations. *Biopolitics* involves all entanglements with, or resistance to, this biopower (Lemke, 2011).

The exercising of biopower can be violent and brutal, as can be seen in the treatment of colonized peoples by European powers. It can also operate through seemingly 'neutral' social arrangements such as a master-protege relationship or the provision of free healthcare for all citizen-subjects of a polity. Power simply 'does things'; it doesn't just repress (Foucault, 1990). Having said this, power has been regularly exercised over colonized and marginalized peoples for the clear benefit of dominant groups (e.g. white heterosexual European men) in ways that are openly brutal. However, educators should remember that power can be exercised bi-directionally. From above through various technologies of governance and from below through the everyday relations of collective living, which includes the relations of nonhumans (Foucault, 1982). Though nonhumans never really figured into Foucault's notion of biopower, it's easy to see, if we consider things like environmental degradation and agriculture, that power in the human social world is actually a more-than-human phenomenon.

One of the things the COVID-19 regulations allowed people to see better is how power is exercised over individual bodies and populations simultaneously. Authorities regulated the movement and proximity of individual bodies along with basic hygiene requirements such as hand-washing and the use of masks. These directives were deployed through state-corporate media, educational institutions, as well as other government institutions such as hospitals, law enforcement agencies and national border agencies. It's not as if these ways of exercising power weren't used before, it just became more obvious with increased enforcement of public health measures. Governments have always controlled the way bodies interact. One just needs to look at the social stratification of any major modern city to see that governments have exercised power over bodies long before the pandemic. However, a major point to note here is that scientific knowledges have been integral to the legitimacy of these forms of governance. But another key point about the exercising of power, even during something like a pandemic, is that the techniques of governance that seem to come 'from above', say masking rules or social distancing regulations, also come 'from below' through the legitimate demands of governments to take measures to protect the public. It must also be stated that power exercised from below in precarious times also comes as a demand for freedom from these controls—as can be seen in the way Chinese protesters likely curbed Beijing's zero-covid policy in the fall of 2022 (Davidson, 2022).

Educational institutions have been integral to the exercising of power in modern societies, and this fact behoves educators to intervene in the practices and discourses that target and shape the conduct of students, and by extension the larger community. This is especially true for science educators. Because the discourses circulating in science education have a kind of 'double-authority' in that they are most often sanctioned both by government and prevailing scientific consensus (if we're talking about state sanctioned curricula). Science education discourses carry the effects of power in ways other educational discourses do not, which means science teachers occupy a very important site at the nexus of power relations. On one hand, they are instruments of biopower because they're involved in shaping the conduct and attachments of students according to the imperatives and expectations of those wielding power (e.g. the state). However, teachers are also agents of resistance in that they, along with students and community members, can resist and make demands on institutional power. Students and teachers engaging in climate protests or refusing a prescribed curriculum are some everyday examples of this. Community engagement and activism is just one way science education is vitally important for resisting systems of oppression and environmental destruction. Science education is one key component of power relations today, as it influences how people come to understand their very being—as a member of a species, as having a sexuality, as having a responsibility to other species, and even their place in the cosmos.

The significance of the COVID-19 pandemic as a potentially major global event resides in how people might come to live everyday differently, and further to that in ways that keep them safe, sheltered, responsible, and economically viable. Again, how collective life is lived. Conduct of all kinds, whether COVID-19 related or not, is the very stakes of power relations. Foucault (1982) laid out specific criteria for such power struggles in his essay "The subject and Power". The COVID-19 pandemic provoked entanglements with power in-line with the criteria laid out by Foucault almost 40 years ago. These entanglements with power meet Foucault's criteria because they:

(i) Were international in scope in that COVID-19 affected the global community albeit differently (e.g. number of deaths and vaccine protection varied widely).
(ii) Targeted the conduct of populations in terms of their interactions and relationships.
(iii) Demanded a response from governing powers in terms of health interventions.
(iv) Invoked questions of identity for example in situations where governments needed to negotiate, prioritize, or withhold medical procedures for colonized peoples.
(v) Employed 'truth' discourses or objective knowledges, in this case epidemiology, economics, and psychology to demand, implement, or remove protections.
(vi) involved reconfiguring a subject's deep attachments, for example by raising the question of whether working in a capitalist economy is really the best thing for people.

Perhaps it's now easier to see how all the struggles with power involving precarity, capitalism, colonialism were already at play before COVID-19. Pandemic measures simply reactivated or recast various aspects of these struggles in particular ways, understanding that all these sociopolitical entanglements with power looked different in places like China, Brazil, India and Canada. During times of disaster the conduct of people, as well as the relationships and practices that comprise daily life, often become the object of intervention and scrutiny. This is because the everyday conduct, material circumstances, and identity of people are often at stake. For science educators, paying attention to the ways science is intricately connected to the governing of everyday life and the way people come to understand themselves (subjectivity) is vital when engaging important local and global issues/events.

For many, the COVID-19 pandemic was a standoff between the virus and the health care systems that struggled to contend with it. In this way, it's easy to think of our public health systems as wholly benevolent, especially by those who support collective institutions. And certainly these systems are worthy of much praise. However, their history is not just one of protection, prevention, and healing. Social medicine was also developed to control, administer, and 'harness' populations for the aims of the state (Murphy, 2017). This is especially true when it comes to the historical subjugation and erasure of black and brown bodies (TallBear, 2013). In fact, the way modern governance developed over the last few centuries was very much in relation to 'the colonies' in order to make some bodies live and others die (Stoler, 1995). Historically, interventions by a governing power aimed at containing diseases like smallpox were largely about urban sanitation and protecting the economic life of the city. Care consisted of a spatial reorganization of 'the sick', which was also a reorganization of the marginalized classes. The goal was to remove contaminated waste, loitering bodies, as well as reserve 'healthy' bodies for labour, commerce, and war (Foucault, 2003). The point is that it wasn't for reasons of benevolent care that public health was slowly introduced in the seventeenth and eighteenth centuries. Socialized medicine also did not emerge from a set of isolated medical practices joining together and somehow becoming more systematized. Rather it emerged from a concerted set of political technologies and national directives from the very beginning. If this seems hyperbolic, just consider for a moment how the mega-rich significantly increased their wealth during the pandemic while many precarious workers lost the means to support their loved ones. Practices like social distancing and business closures, while necessary, also work to designate 'who' can work and under what conditions (often those same rules don't apply to elites). Scientific knowledges developed in conjunction with these public health systems as they often drive the need for knowledges that serve the aims of government institutions and policy.

When science educators teach about public health measures in times of crisis, which they should do, they should also understand the broader political contexts that surround these measures. Pandemic measures, for example, call on subjects to embrace changes in individual conduct for the sake of the population. But educators must ask who is enabled to live and who is left to die through these measures. Why are global disparities in preventive health care allowed to exist? How complicit is the Global North in maintaining these disparities? Canada, a country whose global wealth comes directly from colonial exploitation, was one of the countries hoarding vaccines the most (Hassan, 2020)! But are Canada's actions simply in keeping with the logic of modern governmental power and Empire?

In pandemic times teachers are a little like doctors and health officials, because they're involved in disseminating official health discourses that shape the conduct, attachments, and attitudes of populations. Resistance to exploitative or abusive medicalization on the part of marginalized peoples is to be expected, but resistance also happens through demanding equal access to quality healthcare, vaccinations, and preventative health resources (Sacks, 2018). Colonized peoples have good reason to question socialized medicine, and have a long history of advocating what's best for their communities (in the Canadian context, see Daschuk, 2021).

Another interesting issue the pandemic exposed is the way modern governments rely exclusively on modern scientific knowledge as opposed to more holistic medical knowledges such as Ayurveda, Unani, Siddha and Homeopathy.[3] In countries like India these knowledges often have difficulty finding an official place in large-scale approaches to public health in times of crisis. Aswathy Raveendran and I raised this issue in an article we co-authored on public health discourses and subjectivity in India (Raveendran & Bazzul, 2021). This is a point that Flavia Rezende et al. (2021) make concerning southern epistemologies—those marginalized knowledges that are employed by Indigenous peoples that have the potential to mend the destruction caused by modern capitalism. In my view, there are two interesting things about how the COVID-19 pandemic put modern science and more traditional knowledges in relation. First, governments today *must* employ modern science to secure their legitimacy. It's not just that it's an effective thing to do; there's no choice if a government wants to keep its legitimacy. Of course, scientific knowledges are useful in times of pandemic, but the point is that politics and governance must employ them. I would go a step further and say there can be no legitimate important large-scale governmental project or directive without the deployment of modern scientific knowledges. The pandemic doesn't make this true; it simply exposes this fact.

The second point is related to the first and involves the 'otherside' of the governance coin: the individual. One's own approval or disapproval of holistic knowledge systems has largely to do with one's own relationship to knowledge and governing institutions in terms of 'who' we think we are (for example, our spiritual nature and what our purpose in the universe may be). It is this aspect of science, identity, and epistemology that Rezende et al. (2021) ask science educators to consider. That is, the way modern science has historically foreclosed Indigenous and/or local knowledges that do not fit the typical uses of modern Western science. The pandemic highlighted ongoing political, philosophical, and spiritual debates in the nature of knowledge in times of crises. Debates that underscore how all of us are subjects of the knowledge and attachments instilled in us.

15.3 Right-Wing Populism and Identity in Pandemic Times

Another issue the COVID-19 pandemic exposed is the assault on science by the far-right, with the questioning of public health recommendations and the challenging of the science of vaccination. This assault began well before the pandemic and will continue into the foreseeable future. In addition to the very important issue of global

[3] It is important not to create dualisms between modern science and different knowledge systems as they often support each other. This is not my intention here.

vaccine inequality, there was also the issue of vaccine and vaccine mandate resistance (arguably a much lesser problem but fascinating all the same). This issue was confusing for many living in North America, where there was an abundance of vaccine supplies, the infrastructure to administer them, and high levels of scientific literacy and effective science communication. In countries like the United States and Canada, vaccine (mandate) resistance was only partly to do with savvy science communication. There were many other sociopolitical factors involved. Admittedly, it was quite frustrating to see so many scientists and educators not understand that vaccine (mandate) resistance was also a matter of politics and personal identity. Vaccine mandates brought issues of governance and freedom to the surface for many communities, not least of which were white conservative communities that already had mistrust of 'government intervention' into the private (and separate) sphere of personal life. Again, these standpoints existed well before the pandemic.

In Canada, there was the infamous 'Freedom Convoy' that challenged the right of governments, acting on public health advice, to institute mandatory vaccinations and other pandemic measures such as masking, etc. Whether readers feel these protests were legitimate or not (most Canadians disagreed with the protest) it's worth remembering that the distribution of vaccines grossly favored nations in the Global North. And while there were many kinds of resistance to pandemic measures based on historical betrayals and other exploitive medical practices that targeted colonized and marginalized peoples, attitudes and positions concerning vaccine mandates were largely split along political lines in North America. The more someone identified with the political right wing the higher the chance they would oppose vaccine mandates or vaccines themselves. While people on the political left wing and centre tended to support vaccines and vaccine mandates (with plenty of exceptions of course). This made teaching in conservative communities like southern Saskatchewan, where I live and work, very interesting to say the least. It meant science educators in secondary schools and universities had to find ways to teach about science, pandemics, and politics that cut across political divides.

Teaching across political divides in North America has always been an important task. Not recognizing this is to ignore reality. During the COVID-19 pandemic, a mixture of right-wing populist and radical libertarian views gained more of a foothold in many conservative communities in Western Canada. The recent neoliberal erosion of public institutions, and with it the very notion of 'the public good', likely contributed to the apparent rise of these ideologies in cultural life today. Neoliberalism propagates this idea that *only* individuals can know their own interests, which is somewhat true. The slightly deranged neoliberal/libertarian version of this truism however rendered government scientists and medical professionals as *always* secondary in the determination of one's interests. Under neoliberal regimes the state's knowledge, or access to knowledge, is severely challenged (read disbelieved), because only individuals can truly know what is best for them in a marketplace of 'free' choices. Resistance to the mandates, in a sense, wasn't about science at all, but a fundamental challenge to the relationship between individuals, government, and the collective.

Not understanding the political stakes of COVID-19 is where the science communication folks had some of it wrong. For many, it was simply a matter of explaining things better: 'if we just communicate the science as clearly as possible people will understand what they should do'. Such an omission of politics can also be seen in the way the Freedom Convoy protests were clearly not about the best interests of the 'truckers' it claimed to represent. As a profession, Canadian truckers were highly vaccinated anyway, because they needed to be in order to gain entry into the United States (from Canada). It was clear that truckers were largely used in the service of a right-wing movement that wanted to reconfigure the very idea of social responsibility in the face of government health restrictions. Restrictions that were based on sound scientific knowledge for the public good. Here the relationship between science, power, and subjectivity emerges as a central theme as can be seen in one of the shocking slogans that emerged from this movement: "I'm not responsible for other people's health care!". Shocking, that is, for people who live in a country (Canada) where responsibility for other people's healthcare has been shared unquestionably for many decades.

Some have argued that the linkage between right-wing identity and vaccine refusal had much to do with opposition to any widespread practice that would equalize people's wellbeing and circumstances in a population (Bilewicz & Soral, 2022). For these people, the world is a place of natural winners and losers—the entitled and the unfortunate. This seems to fit with the idea that the political right-wing is interested in preserving hierarchies in societies. Not tearing them down. Therefore, one has to wonder whether vaccine resistance in North America's white conservative communities is more about a refusal of collective responsibility and an affirmation of longstanding hierarchies than anything actually related to health. If educators in North America are tasked with facilitating deep conversations about vaccines and social responsibility, how could they avoid this political dimension?

My own work regarding the pandemic took a very different, yet productive, turn. After teaching in a fairly conservative place in Canada I felt it was worth asking whether the differences between vaccine (mandate) proponents and vaccine (mandate) resistors were really so great. After all, in a politically polarizing world, it would seem like people need to find all the common ground they can. And if we take a long historical view of modern Western forms of governance it may be that the two generalized positions related to vaccine mandates, vaccine (mandate) proponents and vaccine (mandate) resistors, are not so fundamentally different. That is, it may be that these two positions share a fundamental aspect of how subjectivity has been constituted since the early days of Christianity. Although I don't have the space to lay out the entire work here, a piece has been published in the journal *Educational Philosophy and Theory* that goes over the following material in more detail (Bazzul, 2022b).

The early Christian church developed practices such as confession and baptism, which compelled individuals to tell deep moral truths about themselves to an overarching juridical apparatus (essentially elders, clergy, etc.). Christian thinkers such as Clement of Alexandria and St. Augustine of Hippo helped develop the philosophical and theological groundwork for such practices. Foucault's last volume of the History of Sexualities series, the *History of Sexuality Volume 4: Confessions of the Flesh*, attempts to locate the way a very core aspect of Christian subjects, became bound to a juridical or governing authority during the first few centuries of Christianity. It is here that Foucault establishes how one can become subject to modes of governance in ways that internalize them as modes of subjection.

The History of Sexuality Volume 4 emphasizes the fact that Christianity is not responsible for introducing restrictive ethical codes into the classical world. These restrictions already existed in the form of prohibitions around same-sex relationships and premarital sex, etc. Christianity was simply a continuation of an ethics already at play. What changed with Christianity is the way a subject was meant to relate to themselves as a 'subject of lust' and a 'subject of law'. Christianity works to further internalize discipline by entwining morality and the law in the hearts of Christian subjects (Harcourt, 2021). For Foucault the aim was to show how the Christian apparatus of obedience and confession bound the subject to telling the truth about themselves and their morality to power. Guillaume Le Blanc (2019) puts it this way:

> Confessing can be understood from the point of view of the confessor, replacing the director of conscience from antiquity, who was only there to direct action and not to pronounce his verdict on the subject. Foucault shows how this confession is an integral part of the pastoral government of individuals: this obligation to tell the truth of oneself and of one's wrongdoing becomes a mode of government. The obligation to tell the truth of one's sins is no longer a simple question of truthfulness as it was in antiquity, but rather a technology to assure a certain power over one's subjects. (para 18.)

It is interesting to note that a certain form of consent is essential when it comes to these confessional forms of discipline (and the disciplinary techniques that followed). Christian subjects came to understand that their voluntary will to avoid sin is faulty due to the sin of Adam and Eve and their exile out of the Garden of Eden. According to Christian thinkers at the time, this made it difficult for a person to achieve transparency on their own—that is to detect the desires that made them sinful. For this, Christians need to speak the truth of their sin to a moral authority, one that can invoke consequences, in order to realize the truth about themselves and receive salvation. In other words, it doesn't work to just confess to oneself (Foucault, 2021). All of their thoughts, desires, and convictions must become the subject of confessional truth telling.

While this may seem very far removed from events in the twenty-first century, processes of identification that reach to the very heart of an individual have long been at work in Western forms of governance. When certain people declare their vaccine status or stance on vaccines they are, in some sense, also declaring a deep truth about their bodies, their identity, and their 'morality'. And while it might seem like this line of thinking is outside the disciplines of science and education, it's important to remember that to engage important issues of collective existence, including those directly related to science, educators must deeply engage with the social, historical, and political contexts that frame them. This long standing need to declare a deep truth about ourselves to power, something that became systematized by Christian technologies of subjectification and control, is how I'd like to bring this chapter about power, knowledge, and subjectivity to a close.

15.4 Conclusion: Revealing Ourselves to Power

It might be said that the different ways people are brought to tell the truth about themselves today have proliferated since the early days of the Christian church. For example, people today are compelled to tell the truth about their sexualities—either proudly declaring their sexual difference or constantly reaffirming their heterosexuality. In the case of sexuality, there are both moral and legal consequences to this declaration—for example in the form of 2SLGTBQAI marriage and adoption rights or punishments for fornication that are still rendered by many religious organizations. The key point I want to make in this conclusion however is that certain aspects of how a subject is addressed by institutions of power in modern Western societies has remained surprisingly constant since the early days of Christianity. Subjects are still compelled to relate deep personal moral truths about themselves to a juridical power, whether it be through the police, through religious practices of confession, or the truth about who they really are (e.g. sexuality).

Flash forward to vaccine (mandate) debates in modern Western countries and one can see that both 'sides' are declaring a fundamental truth about themselves and their bodies in relation to a juridical power—for example in the very formative example of vaccine passports. Vaccine mandates essentially regulate bodies such that people must 'confess' the truth of their vaccine status, and, tacitly, their relative position on vaccine mandates. Their declarations are also assessed by a governing power that either approves or disapproves this declaration for the purposes of employment or access to services and social venues. While not all positions on vaccine mandates are ethically or morally 'equal', these two basic positions, support or resistance, both involve a response to the modes of subjection and control employed by governing powers. This is a good time to reiterate an important point: there's a fundamental relationship between knowledge, governance, and subjectivity playing out in the seemingly straightforward issue of mass vaccination and/or vaccine (mandate) resistance. It's not enough to just communicate the science. Educators, including public health officials must engage the detailed historical and political contexts in the interest of the public good.

Interestingly, the political right needs 'alternative' scientific facts to bolster their positions and give them legitimacy. Media literacy involves critically assessing the scientific arguments and social media posts used by the political right to achieve their political ends. These arguments from the far right value personal autonomy, individualism, and social hierarchies ahead of collective wellbeing. This literacy also involves understanding how powerful markers of identity and ideology operate and influence decisions we might think of as 'free'. Vaccine (mandate) resistance looked different in different countries, and when the next deadly pandemic strikes, educators need to be on the front lines saving lives through the communication of sound science, but also as brokers of the many relations science has to governance, power, and questions of collective existence.

I want to conclude by repeating something about the COVID-19 pandemic that seems contradictory, but is quite important. *The pandemic, in itself, has very little significance.* Rather its meaning, if it has any, resides in how it changes everyday life for better or for worse. Determining this means better understanding the entanglement of science/knowledge), power/governance), and subjectivity/relations of self. Any relevant social, ecological, or political context in relation to the pandemic existed well before the virus began spreading globally—as did all relevant questions of collective existence the pandemic provoked.

Science plays a very crucial role in engaging these questions because objective knowledge, how power works, and 'who' we think we are all intertwined in modernity. This is because knowledges with high 'truth' value have a greater capacity to wield the effects of power, and therefore have a greater capacity to mold our deepest attachments and 'who' we think we are in the world. Educators sit at the very nexus of power relations and therefore have a great ethical responsibility to consider how they both distribute the effects of power and resist them. Their proximity to scientific knowledges and practices is part of this 'power positioning'. What educators need to do when major issues or events arise is critically examine how power outlines conduct and our relation to others (both human and nonhuman) in such times. The stakes of these issues or events involve how the vast majority of people continue to conduct their everyday lives together. As educators, we can rise to this profound ethical responsibility by taking the intricate ecological, social, and political entanglements of science very seriously with our students. If we don't, all of our good teaching just might not make much of a difference to life in the long run.

References

Alsop, S., & Hoeg, D. (2020). Collegial conversations at a time of COVID-19. *Journal for Activist Science and Technology Education, 11*(2), 60–74.

Bazzul, J. (2022a). Power, knowledge, subjectivity: Science education and questions of collective existence. *Caderno Brasileiro de Ensino de Física, 39*(2), 289–297.

Bazzul, J. (2022b). Science, power, and subjectivity: Vaccine (mandate) resistance and 'truth telling' in times of right-wing populism. *Educational Philosophy and Theory*, 1–13.

Bilewicz, M., & Soral, W. (2022). The politics of vaccine hesitancy: An ideological dual-process approach. *Social Psychological and Personality Science, 13*(6), 1080–1089. https://doi.org/10.1177/19485506211055295

Daschuk, J. (2021). VACCINE HESITANCY: Indigenous Communities | KHS Research Seminar Series. URAllKinected. Retrieved from: https://www.youtube.com/watch?v=T6y5mzd6cCY

Davidson, H. (2022, November 28). *China Covid protests explained: Why are people demonstrating and what will happen next?*. The Guardian. Retrieved from: https://www.theguardian.com/world/2022/nov/28/china-protests-explained-why-are-peoPle-demonstrating-blank-piece-white-paper-a4-what-will-happen-next-zero-covid-policy-protest

Foucault, M. (1982). The subject and power. *Critical Inquiry, 8*(4), 777–795.

Foucault, M. (1990). *The history of sexuality, volume I: An introduction*. Vintage books.

Foucault, M. (2003). The birth of social medicine. In M. Foucault, P. Rabinow, & N. S. Rose (Eds.), *The essential Foucault: Selections from essential works of Foucault, 1954–1984* (pp. 319–337). NewPress.

Foucault, M. (2021). *Confessions of the flesh: The history of sexuality* (Vol. 4). Pantheon.

Harcourt, B. E. (2021). Foucault's keystone: Confessions of the flesh. *Foucault Studies, 29*, 48–70. https://doi.org/10.22439/fs.vi29.6214

Hassan, Y. (2020, December 17). *Canada: Is 'me first' COVID vaccine policy hurting other nations?* Al Jazeera. Retrieved from: https://www.aljazeera.com/news/2020/12/17/canada-is-me-first-covid-vaccine-policy-hurting-other-nations

Le Blanc, G. (2019). Why read Foucault's confessions of the flesh today? (English version). *Critique 13/13*. https://blogs.law.columbia.edu/critique1313/guillaume-le-blanc-why-read-foucaults-confessions-of-the-flesh-today-english-version/

Lemke, T. (2011). *Biopolitics. An advanced introduction*. New York University Press.

Machado de Oliveira, V. (2021). *Hospicing modernity: Facing humanity's wrongs and the implications for social activism*. North Atlantic Books.

Murphy, M. (2017). *The economization of life*. Duke University Press.

Raveendran, A., & Bazzul, J. (2021). Socialized medicine has always been political: COVID-19, science and biopower in India. *Cultural Studies of Science Education, 16*(4), 995–1013.

Rezende, F., Ostermann, F., & Guerra, A. (2021). South epistemologies to invent post-pandemic science education. *Cultural Studies of Science Education, 16*(4), 981–993.

Sacks, T. K. (2018). Performing black womanhood: A qualitative study of stereotypes and the healthcare encounter. *Critical Public Health, 28*(1), 59–69.

Sharma, A. (2020). Phronetic science for wicked times. *Journal for Activist Science and Technology Education, 11*(2), 7–15.

Stoler, A. L. (1995). Race and the education of desire. In *Race and the education of desire*. Duke University Press.

TallBear, K. (2013). *Native American DNA: Tribal belonging and the false promise of genetic science*. University of Minnesota Press.

Tsing, A. L. (2015). The mushroom at the end of the world. In *The mushroom at the end of the world*. Princeton University Press.

Žižek, S. (2020). *Pandemic!: COVID-19 shakes the world*. Wiley.

Jesse Bazzul is associate professor of science and environmental education at the University of Regina. His work currently centers on ethics and how diverse beings inhabit shared worlds. Since ethics lies at the heart of education, it is important that the discipline of education be studied as one of the arts and humanities, as well as a social science and science.

Chapter 16
Beyond the Joyful Inquiry of Feeling like a Scientist: Sitting with the Heart as Arrested, Angry, Wavering, Grieving, Hopeful, and Dreaming in Science Education

Marc Higgins and Anastasia Sanchez

16.1 Introduction

> The arrival of suffering from the edges teaches us about the difficulty of becoming conscious of suffering. It is hard labor to recognize sadness and disappointment when you are living a life that is meant to be happy but is not happy, which is meant to be full but feels empty. It is difficult to give up an idea of one's one life when one has lived one's life according to that idea. We learn that consciousness of unhappiness puts us in touch with the world allowing to pierce… the seal of happiness. So much inequality is preserved through the appeal of happiness, the appeal to happiness. It is as if the response to power and violence is or should be simply to adjust or modify how we feel; for instance, by transforming a social relation of exploitation into a personal feeling of empowerment. (Ahmed, 2017, p. 60)

In the midst of this post-peak-pandemic moment, it is becoming increasingly clear and evident that science-education-as-usual is ill-equipped to account for and be accountable to the ways in which the slipping away of futurities is accelerating, be it through climate crisis (e.g., floods, fires, droughts), ever-widening economic disparities, the rise of fascism at both the grassroots and the political level, elicitation and actualization of multiple forms of violence towards already precarious lives, amidst numerous other things. Suffering happens: it arrives into science education "from the edges [to] teach us about the difficulty of becoming conscious of suffering" (Ahmed, 2017, p. 60), if it wasn't already there because of science education (e.g., complicities to systemic colonial and racist structures; Bang & Marin, 2015; Sanchez, 2023). We begin from a simple premise: science education needs not only

M. Higgins (✉)
University of Alberta, Edmonton, AB, Canada
e-mail: marc1@ualberta.ca

A. Sanchez
University of Washington, Seattle, WA, USA

C. B. Moura (ed.), *A Sociopolitical Turn in Science Education*, Contemporary Trends and Issues in Science Education 62,
https://doi.org/10.1007/978-3-031-78586-3_16

ways to account for and be accountable to the landscape of emotion within teaching and learning within this contemporary moment (e.g., Alsop, 2016; Kayumova & Tippins, 2016; Zembylas, 2016) but also needs new ways to be felt for these to register in new ways (see Wallace et al., 2022; Tolbert et al., 2023). However, we recognize that this is not a task that is not as simple as flicking a switch, nor is it a pedagogical additive: emotion and affect have been excluded from science education by design by making them the dichotomous and co-constitutive other of scientific reason, or leveraged as resources in service of science education as usual when they are included. One significant consequence of this: too many pressing critical, social, political, and ecological conversations, if they get included at all, get reduced to epistemological objects within science-education-as-usual which do not implicate or account for their emotionality as their inclusion is reducible to the lowest hanging fruit of social change: awareness (Gilbert, 2016; Liboiron et al., 2023).

Well over a decade following the *affective turn* in social science and humanities more broadly (see Clough, 2007), which "[brings] together psychoanalytically informed theories of subjectivity and subjection, theories of the body and embodiment, and political theories and critical analysis of affect and emotion" (Zembylas, 2016, p. 541), we are noticing a subtle but significant shift in the ways in which questions of affect and emotion are being taken up. Importantly, this is not a move from absence to presence. While, as Reiss (2005) suggests, "science education has traditionally paid little attention to … emotions" (p. 17),[1] it has also been doing so for a long time (see also Fortus, 2014). However, despite questions of emotion have long been explored within spaces of science education, these investigations have focussed more on questions of dispositions and attitudes towards science, often as a problem to be resolved or a resource to be leveraged amidst and towards science-education-as-usual. Following the affective turn, we find that a quick search for science education texts that take up *affect theory* specifically, where questions of negative affect and emotion intersect aforementioned critical, social, political, psychic, and ecological analyses, is marked by paucity that tells on science education.[2]

[1] To quantify this engagement, in a meta-review of the *Journal of Research of Science Teaching*, *Science Education*, and the *International Journal of Science Education* from 2001 to 2011, Fortus (2014) suggests that affect and emotion is engaged in slightly less than 10% of manuscripts published in these first quartile (i.e., Q1) science education journals took up affect and emotion, or related themes (e.g., attitudes). Again, and further, *when* questions of affect are brought up, they are addressed on very particular terms.

[2] For example, in Bellocchi et al. (2017) edited collection *Exploring Emotions, Aesthetics and Wellbeing in Science Education Research,* the language of the affective turn and affect theory, where science education might become more accountable to the and more richly account for negative affects and their social, political, ecological, and pedagogical potentialities, show up once and twice respectively, and exclusively in Alsop's contribution to the collection. Even questions of negative emotion, which appear more frequently, are quickly juxtaposed with statements surrounding them about the importance of resolving them:

> The affective dimension may be negative to begin with, such as frustration or avoidance of a difficult task. In order to contribute to the development of interest, this negative affect must be turned into a positive which is what happens when difficulties are overcome (Petersen & Dohn, 2017, p. 194)

Nonetheless, there are some taking up the critical study of the structure of affect, and its non-linguistic and relational movements, feelings, and embodiments. Yet, despite some exceptions (e.g., Alsop, 2016; Kayumova & Tippins, 2016; Wolfe, 2021; Zembylas, 2016), we are noting that this potential is increasingly being leveraged in the service of science-education-as-usual. Specifically, questions of affect are increasingly becoming different articulations of the normative goals of science education: emotionally regulating processes of coming-to-know what scientists know as well as how they know. In other words, *feeling* like a scientist (e.g., Jaber & Hammer, 2016; Mehta et al., 2019). As more and more learners sense, feel, and grieve an accelerating extinction of possible futures and worlds, an affective landscape of science education which privileges *feeling like a scientist* seems untenable: "to not speak of affect, especially in these pedagogical contexts—possibility in all pedagogical contexts...—seems wholly lacking and unproductively limiting" (Alsop, 2016, p. 553). Further, and again, suffering happens, arriving from the edges (Ahmed, 2017). Because of this, affect must be considered beyond questions of emotional regulation and feeling the passions of a scientist in service of science-education-as-usual: as Zembylas (2016) suggests, taking affect seriously in science education is "to become not only more passionate but also more just and critical through science" (p. 549). Learners are irrevocably put into relation with relations of suffering and violence which can no longer be modulated through feeling the "right" way, especially when *feeling* like a scientist is already predicated on belonging, but a belonging whose love is conditional, unfair, and without guarantee.[3]

In this chapter, we engage with the questions such as, what if working against this willful ignorance towards feeling the futures and futurities being (fore)closed meant leaning into these negative affects? If "so much inequality is preserved through the appeal of happiness" (Ahmed, 2017, p. 60), what pedagogical possibilities are made possible by rupturing the seal of regulated emotion within the science classroom? What if we rendered difficult the habitual "transforming a social relation of exploitation into a personal feeling of empowerment" (Ahmed, 2017, p. 60) by attending not only of those who have successfully completed a science education trajectory, but also those who have not yet or might never due to the dynamics (e.g., Marker, 2019)? In order to explore these questions, inspired by other *heart*-centered

[3]This speaks to the disciplinarity of science education and the ways in which disciplinary knowledge is always already constructed within disciplinary power: shaping the norms through which difference is not only tolerated, but also made intelligible, and valuable, at the expense of difference's more difficult demands.

As a science teacher education sidebar, we collectively note the importance of allowing students to pause during learning *difficult knowledge* to allow for it to register beyond being a glass-encased shiny epistemological object detached from consequentiality in order to allow for learning with the heart: pause for reflection, for rage, for remorse. However, as Zembylas (2016) notes, such a pedagogical opening should be critically reflected upon. While allowing for the affective learner to surface is an important facet of how relational learning can be fostered, that very structure of feeling registers as, for many, conditional, unfair, and without guarantee by design. There is work to be done in attending to the *critical histories of emotion* which shape teaching and learning science (see also Alsop, 2016).

orientations to science education (e.g., Frausto Aceves et al., 2022; Moura & Guerra, 2023), we think with the critical feminist affect scholarship of Sara Ahmed (2006, 2017) to offer herein six figurations of the heart to engage this inquiry, the heart as: arrested, angry, wavering, grieving, hopeful, and dreaming. Across these figurations of the heart, we lean into the negative affects that stem from a science education whose consequentiality's time is out-of-joint as well as the positive affects of potential and preferable futures yet-to-come, futures in which there may be a world of many worlds.

16.2 Heart as Arrested

> Science educators have often had a 'difficult' relationship with emotion and its place in science teaching and learning—to say the least!—but also … it [is] interesting that the affective turn … has significantly influenced thinking and scholarship in numerous other fields of study, yet not so much science education. (Zembylas, 2016, p. 540)

In relation to science education's "difficult" relation to affect and emotion, we offer the heart as arrested as a first figuration. This isn't to say that there isn't *heart* in science-education-as-usual, but rather that the heart, the symbol of life, passion, and emotion is rendered subservient to knowledge-as-usual and purpose-as-usual. For example,[4] *knowing-by-heart* reroutes the "blood" of teaching and learning to the head to reinforce knowing what scientists know or how scientists come to know as well as why. Even the *passion* of the passionate scientist becomes another node through which science education is marked by circular relations, namely a "regime of truth" in which each truth is but a different articulation of the systems of power through which it is produced (Higgins, 2021; van Eijck & Roth, 2007). Bang (2018) suggests that science education's circularity renders it so insular, that like an Ouroboros, science education only gorges upon itself in a monstrous cycle of "greed, appetite, self-destruction, and endlessness" (p. 809). Without fresh blood, plaque builds up, the arteries harden, the heart arrests.

To animate this, in *Feeling like a scientist,* Jaber and Hammer (2016) inquire: "what forms and roles of affect are evident in accounts of professional scientists? What forms and roles of affect are evident in student inquiry? How might affect be involved in students' engagement and persistence in science?" (p. 191). However, despite the multiplicity of affects which occur within the science classroom, whether

[4]Another notable example of the ways in which the heart is weaponized in science education is through the expression *out of the goodness of our hearts* which, when put under the magnifying glass in a field largely marked by white teachers teaching unlike others, is largely about weaponizing a paternalistic imposition of settler-capitalistic futurities (e.g., uncritical arterial "pipeline" towards STEM careers, as a new form of civilizing mission and racial capitalism) in a feel-good-to-the-teacher way as a means of recuperating a teaching subjectivity.

they stem from within or beyond its porous walls, Jaber and Hammer (2016) focus on types of affect we might expect from a science with an arrested heart:

> the pleasures scientists experience in studying phenomena and the feelings involved in scholarly interactions, ... metaphoric empathy with objects of study, [and] scientists' awareness and management of these various feelings as part of their expertise and maturity in the field (Jaber & Hammer, 2016, p. 195)

Again, this is not to say that science education is without feeling, but rather that *feeling like a scientist* becomes another location in which an orientation towards science-as-usual is presented; it is an effort to contain and constrain the ways in which "emotions exert a subversive and disorganizing effect on log[o]centric ways of thinking" (Kayumova & Tippins, 2016, p. 572). On orientations, Ahmed (2017) suggests that "power works as a mode of directionality, a way of orienting bodies in particular ways, so they are facing a certain way, heading towards a future that is given face" (p. 44). Particularly, as Zembylas (2016) states, science education is caught up in orienting us away from the socio-political dimensions of affect and towards scientific rationality:

> Through carefully created omissions of the political implications of emotional responses to social injustice and inequality, one could see that a fabricated absence of the politics of emotions was crucial to the sense of a pure and unadulterated "critical" or "scientific thinking." (p. 544)

This orienting, regardless of whether affect is taken up or not, works to obscure other futures, ways-of-knowing, and -being, as well as itself: *a* path becomes *the* path. The heart-as-arrested offered here is not a stopping of the heart at all, but a relation to affect that is unsustainable: it is an abject and othering relation to affects other than those who secure science-as-usual and its futurities by grafting their raw emotionality over with its own affective directionality. On grafting, Ahenakew (2016) suggests:

> grafting is used in biology as the process of transplanting something from one organism into another (e.g., hybrid plants or cell/skin implants). Grafting, in itself, is neither good nor bad. Indeed, hybridity can be a generative process. However, in the context of ... severely uneven environments shaped by historical circumstances, ...the grafting/hybridizing does not happen as a mutual exercise, but as assimilation. Grafting, in this sense, can further contribute to ... elimination [of difference] (p. 324)

Whereas Ahenakew (2016) speaks specifically to the grafting over of Indigenous knowledges within the Western academy when it is included, albeit problematically, we believe that this metaphor holds to also describe the patchy relations of affective landscapes in science education, where we are caught up in "a kind of grafting that seems to maintain primary loyalty to accepted notions of time, progress, reality, and *being*" (Ahenakew, 2016, p. 333, emphasis ours). As the project of assimilation persists, differently echoing earlier projects of disciplinary knowledge (knowing what scientists know) and disciplinary practices (thinking like a scientist), science education finds itself in a situation where the forceful imposition of affective sameness onto difference means that there is little possibility for the requisite *fresh idea of thought* (see Bazzul et al., 2018) required for its own adaptation in an ever-changing world (see also Kayumova & Tippins, 2016). However, the violent stoppage of otherness that shapes science education is difficult to deviate from: "once a flow is directed, it acquires momentum… once a momentum is acquired it is directive" (Ahmed, 2017, p. 44).

Yet, *the heart wants what the heart wants*: affect happens beyond that which is containable and redirectable, and that which is grafted over remembers, even if that memory is but a haunting (see Higgins, 2022). In turn, in order to put science education in a different relation not only to itself, but also to our own arrested hearts, as well as the thrumming, yet threatened, livingness of human, other-than and more-than-human others, we suggest that we might need a shocking jolt as a field in order to attune to socio-ecological heartbeat of the world: break out the defibrillator paddles.

16.3 Heart as Angry

> For as we know, the experience of negation, of being stopped or feeling out of place, of feeling uncomfortable at home, does not "stop" there. It is around such experiences that bodies gather, getting together, acting, refusing this inheritance of whiteness, refusing the desire to follow that line… By accounting for the "I cannot," for the body that is stopped or held up, we also attend to the condition of possibility for the emergence of a collective form of activism… In other words, collective anger about the orientation of the world around whiteness might reorient our relation to whiteness. (Ahmed, 2006, p. 155)

As science continues to frame itself in affective neutrality, meant to "reflect" (or uphold) a political neutrality, the implicit and ongoing message is that students' affective realities and those of their communities are policed: their hearts are arrested at the door, in both senses of the word. While the project of *feeling like a scientist* is premised on a productive affective disposition that can be taught, the means by which the joy of inquiry is separated from a pluralistic affective landscape is not without cost: the conditional belonging that is proposed already presumes feeling at home within systems and structures of whiteness, coloniality, and anthropocentric domination. Further, for those who do not feel at home within science education, *happiness* becomes an imposition of feeling rather than a genuine emotion—it becomes a disciplinary and disciplining form of emotional labour. For example, as Zembylas (2016) makes explicit, "people who face systematic injustices daily generally recognize that feelings of trust and safety are not prerequisites of participation, but privileges endowed by existing hierarchies" (p. 543). As Ahmed (2017) states, in a nutshell, "happiness as a form of emotional labour can be condensed in the formula: making others happy by appearing happy" (p. 58). Further, it individualizes unhappiness by framing it as a failure to be happy.[5] However, and significantly, the affective condition which upholds science-education-as-usual operates in both senses of heart-as-arrested: because *doing* science and *being* scientist are operationalized as inseparable categories (Marker, 2016), the requirement to leave affect at the door is policed.

This stoppage, as Ahmed suggests, offers its own "condition of possibility" (Ahmed, 2006) for something else to emerge: "when separation becomes a command, willfulness is what returns; willfulness not as severance but as perseverance" (Ahmed, 2017, p. 79). In this contemporary moment, there is increasingly much to be angry about that is not easily left at the door given science education given its complicities to settler colonialism, racial capitalism, as well as ongoing practices of extractivism (Bang & Marin, 2015; McGinty & Bang, 2016; Morales-Doyle & Gutstein, 2019; Nxumalo & Gitari, 2021; Vakil & Ayers, 2019). Significantly, as Ahmed (2017) makes explicit, this willfulness takes on a particular valence for black feminists and feminists of colour whose histories at the intersections of science, education, and society is marked by "the violence of enslavement, of colonization, of empire... the requirement to give up kin, culture, memory, language, land" (p. 80). In addition to being a form of anger *on behalf of* or *with*, anger becomes the very condition of surviving systems whose pasts are passed over and still linger and lurk in the present: "some have to become willful to survive a history. We cannot 'not' start there, which is say, here. We have to become willful to say this history is still; that it has not gone; that it goes on" (p. 81). Particularly, as Marker (2019) makes explicit, the focus on success of those who are not only underrepresented but also marginalized by science education means that we not only forget to focus on the systems and structures which produce these ongoing problematics, but also contribute to the erasure of stories of strength marked by wilful refusal to become scientist on the terms that science education operationalizes. Thus, if the demand for happiness requires a halting of multiple co-constitutive forms and intensities of anger, such collective anger about the orientations of science education is galvanizing. Furthermore, part of the condition of possibility here is that folks coalesce and commiserate around such structures of feeling of dignified anger. Importantly, "willfulness too is homework" (Ahmed, 2017, p. 83), it requires

[5] Significantly, Ahmed (2017) makes explicit that feminists hold a different relation to unhappiness altogether:

> Feminist sympathy involves a different direction to unhappiness. We would understand unhappiness not as they failure to be happy and thus causing yet more unhappiness, but a refusal, a claim, a protest, or even just some ordinary thing, a texture of a life being lived. (p. 58)

In other words, unhappiness is an undesired albeit generative relation to an affective landscape that would sweep suffering under the rug in the name of and through "happiness." However, given the individualization and abject-affective framing of unhappiness, "to be sympathetic to unhappiness is often to be perceived as unsympathetic" (p. 58).

that we stay with the trouble and refuse the easy call to convert bad feelings into good feelings but rather engage through "insisting on that negativity: the charge, after all, is what keeps us proximate to scenes of violence" (Ahmed, 2017, p. 84).

16.4 Heart as Grieving

> Our happiness too might depend on what we do not notice. Perhaps we keep our happiness through a willed oblivion. We must refuse this oblivion. If something would make us unhappy, when acknowledged, we need to acknowledge it. We are willing to cause our own unhappiness, which does not make unhappiness our cause. (Ahmed, 2017, p. 259)

In being and becoming a critical science educator, it is important to not only challenge "what gives security, warmth, place, and position" (Ahmed, 2017, p. 113) premised upon ongoing structures of domination which require some bodies complacency for the comfort of others. To reiterate, a science pedagogy premised upon *feeling like a scientist* or emotional regulation in the service of cognition is to operationalize the closure of the classroom to affects which do not conform to the curricula; a closure whose repetition leads to foreclosure: "our happiness too might depend on what we do not notice" (Ahmed, 2017, p. 259). The refusal to be directed towards a numb heart to others requires "pointing out structures that many are invested in not recognizing" (Ahmed, 2017, p. 158), such as the ways that oft depoliticized and atheoretical empowering discourses in science education often reify dynamics of power they seek to destabilize (see Kayumova et al., 2019); but as mentioned before, to deviate from a firmly established path is always its own labour. Here, we suggest a double(d) grieving of "happiness" for oneself. This double(d) grieving is a grappling with both the loss of the comfort associated with the aforementioned happy but deeply arrested heart which is numb to the suffering of others, as well as the critical psychoanalytic process of grieving one's complicities.

On the former sense of grieving, for one's own comfort, regardless of the degree to which it manifested in the first place, Ahmed (2017) suggests that in refusing this willed oblivion, we become *killjoys* "whether [we] intend, however much [we] appear willing, however [we] speak or dress; to do diversity work is to receive that kind of assignment" (p. 113), albeit some receive that charge more than others. The figure of the feminist killjoy "poses a problem because [they] keep exposing a problem" (Ahmed, 2017, p. 99): the momentum acquired by dominant forms of science teaching and learning do not redirect so easily and there is always work to do in addressing and redressing the ways in which institutional efforts towards inclusion are marked by non-performativity. Ahmed describes *non-performativity*: "when naming something does not bring into effect or (more strongly) when something is named in order to not bring something into effect" (pp. 106–107). In other words, institutional accommodations are vectorially null: while their orientation deviates from dominance, they are emptied of magnitude. We see this many times over in science education with its policies that are meant to nod towards inclusion, as a means of performative protectionism: a policy facing outwards towards the public to deflect criticism and to individualize the lack of implementation when criticized, but not one that necessarily faces inwards. For example, in many Canadian provinces, we have seen the inclusion of Indigenous ways-of-living-with-Nature in provincial science curricular policy decades ago, but the ways in which they have been positioned within the documents operates as a subtle wink towards supporting educational systems' inaction: Indigenous ways-of-living-with-Nature are often presented as an orientation rather than a measurable curricular objective, alongside things such as Nature of Science, Science, Technology, Society, and Environment, amidst others (see Kim, 2022). In turn, grieving one's own comfort is in part a loss of innocence (if one was innocent towards science education): a recognition that science education never was *for all* despite slogans otherwise. Part of a being a killjoy is catching institutions in this act:

> Even when diversity work leads to the adoption of new policies, they are not necessarily enacted. Within the organization there is a gap between words and deeds, between what organizations say they will do, or what they are committed to doing, and what they are doing. Even if words become substitutes for action, these words can be useful.... If organizations are saying what they are doing, you can show they are not doing what they say they are saying. (Ahmed, 2017, p. 107)

It is uncomfortable work: it requires a certain degree of pushiness, of persistence. It is strategic work: to inventively remain in relation to modify a system to work for those who are not accommodated by that system. Significantly, even though there is joy in working towards escaping the trajectory of a science education that remains numb to the suffering of others, the attempt to subvert dominant ways-of-knowing and -being can register in numerous negative ways: frustration, intractability, untenability, exile.

Attending to the latter sense of grieving is to grieve the seductive narrative of science-education-as-usual and its framings in order to address its marked relationship to epistemic and affective ignorance. Living in the institutional disjuncture between words and deeds in order to hold to task the action behind the critical inclusion that is spelled out requires strategic thinking, but that strategy is not without its price to pay in terms of exteriority, rationality, and teleology of progress:

> When we have to think strategically, we also have to accept our complicity: we forgo any illusions of purity; we give up the safety of exteriority. If we are not exterior to the problem under investigation, we too are the problem under investigation. Diversity work is messy, even dirty work. (Ahmed, 2017, p. 94)

Just as we can no longer assume that institutions are committed to progressivism beyond performativity, we must also grieve a purity granted through exteriority: to be trained in science education is to be(come) part of the problem, albeit to differing degrees for differently positioned educators. Nonetheless, to question science education from within science education is to acknowledge and work through complicities. For example, this can entail illuminating not only the usual drive towards a particular construction of "success" within conventional science education programs, but also attending to the complicities in our own moves towards that success, how we might identify accordingly, and a recognition of how this notion of "success" is caught up in harm towards science education's Others (see Marker, 2019). In other words, these complicities are not only woven through *what* we teach, but multiple other facets; in doing the homework of willfulness required of critical and feminist science educators, a significant part of the labour is attending to the "home" part of the homework.

> The "home" part of homework can take many meanings: home as cultural, disciplinary, geographical, historical, epistemological, ontological, among others. However, *home*work is always risky as it threatens to rupture who we (think we) are, what we (think we) know, and what we (think we) do. (Higgins, 2021, p. 64)

Addressing the "home" is messy or even dirty work because epistemic and affective ignorance are not only marked by lack of knowledge or an inability to feel. Rather, psychoanalytically, it is a recognition that for new (e.g., critical, liberatory, anti-oppressive) knowledge to take hold, we must grieve the loss of knowledge already held from science-education-as-usual that acts as resistance to something otherwise: because it is often more than strictly knowledge, but also woven deeply into who we think we are and what we think we know and what we think we do. For example, "the question of how affect and emotion contribute to students' and teachers' abilities to function effectively in the science classroom is rarely considered and when it is raised, it is not a question about power" (Kayumova & Tippins, 2016, p. 572). Thus, in making the move from exclusion to inclusion, we must consider the ecology of practice which shaped exclusion in the first place. In facing complicities, there is work[6] in grieving difficult pasts, presents, and futures, and in facing these feelings that stem from the heart representing itself beyond the repressive affect of conventional science education.

16.5 Heart as Wavering

> One of the hardest things about coming up against walls: it can threaten some of our most fragile and precious, our best, our warmest connections… and this too is the risks of anger. There is so much to be against; we know this. But how easily anger can spill, can spill at those who happen to be nearby, who are closest to us. How easily in being against something we can risk those who are with us, who are for us, who are with and for; we can risk them because they are before us. Our anger, when generalized against the injustice of the world, can become directed towards those who happen to be nearest, often those who are dearest. (Ahmed, 2017, p. 172)

In coming to understand and come up against what is making science education heart-sick, we may catch ourselves, literally or metaphorically, wavering: experiencing that final hiccuping gasp that occurs after a body-shaking heart-aching cry,

[6] Significantly, the rigor in grieving is not only epistemological, but also affective: it means facing and releasing the repressed emotionality. Like a hard, chest-heaving cry, allowing the tears to create new waterways of possibility, new resiliences and refusals, all of which act as both seed and water for an otherwise to emerge.

marked by a sporadic pulling in of oxygen, an inhaling and exhaling that situates the possibility for new self-determination, which may or may not come to be. The numerous walls we might encounter come with their own consequential cost: "fragility as the wear and tear of living a feminist life" (Ahmed, 2017, p. 163). Be they in the form of systemic and structural exclusion, refusal; inaction revealing a disjuncture between policy and practice; oppressive curriculum and pedagogies being uncritically propped up by problematic concepts and orientations; individuals discursively upholding operationalized norms through turning a blind eye; or more blatant and egregious admissions and enactments of sanctioned ignorances, amidst other things—the walls we encounter reveal our own fragility, both individual and collective. Significantly, Ahmed (2017) invites consideration of affective fragility: that in encountering obstacles that feel like gaslighting, we might come to doubt ourselves, to internalize that narrative. Here, your heart may waver in the face of adversity. Granted, you might find strength and "be defiant in the face of this lack of confidence", but "confidence too can shatter" (Ahmed, 2017, p. 170) when the walls continue popping up. While "bodies need to get in the way to open up a way to others" (Ahmed, 2017, p. 184), and that this commitment to refuse to hide the effects of coming up against these walls is an important part of doing critical and feminist science education as it reveals the ways in which some bodies are not given room to navigate, there is also the invitation to not too easily dismiss and convert the negative affect of wavering, especially when it concerns those we care for the most: we can get things wrong, and at the expense of "those who happen to be nearest, often those who are dearest" (Ahmed, 2017, p. 172).

Extending the earlier discussion on grieving already held knowledge within science education, we must also grapple with questions that manifest in terms of how science educators are pedagogically positioned.[7] Particularly, cultural myths regarding teacher identity, such as *everything depends on the teacher, teacher as expert*, and *teacher as self-made* (Britzman, 1986), are exacerbated in science education where the subject matter positions pedagogues as purveyors of certainty through pedagogical content knowledge, and provides little training in matters of cultural studies (see McKinley, 2007; Wallace, 2019). As Britzman (2013) suggests:

> an interest in the emotional situation of pedagogy is difficult to sustain, particularly when we feel we have to cover material against all odds, when we believe material is our reason for teaching, and when pedagogy is solely thought of as didactic effort to the deliver explication and so rescue time from becoming wasted or meaningless. Here is the teacher's transference to the subject presumed to know, a return to the old ghost of the real teacher who does not need to learn and only knows what to do. (p. 113)

---

[7]As Bazzul (2017) states, this goes beyond simply the pedagogical: "science education seems to ignore the social, historical, and political realities that give shape to its practices, knowledges, and literatures" (p. 67). In the context of this chapter, it is worth adding to this list the *affective* realities which give it shape as well: "we need to somehow remember that science is an embodied human expression with a history within which embodiment has been embraced in different ways at different times" (Alsop, 2016, p. 543).

Such a positioning which reaffirms the notion of the educator who knows also manifests in spaces of critical science education, where science education's dominant trajectories inevitably works to push to the margins and isolate those who do not comply, and a hardening of the heart to stand up and face injustice. While this positioning productively offers protection in some instances, again, it can mean that anger spills over. In turn, as Ahmed (2017) suggests: "activism might need us to involve losing confidence in ourselves, letting ourselves recognize how we too can be the problem" (p. 175). There is a need to find a certain degree of gentleness in solidarity: critical difference can be an asset rather than a liability as we collectively hope for new trajectories in and for science education.

16.6 Heart as Hopeful

> Sadness can be an inheritance, a feminist inheritance.... It is a sadness that can be too difficult to reveal to ourselves, let alone to others, because it is sadness with the world and thus sadness in the world.... Becoming feminist puts us in touch with all that sadness, all those emotions that represent a collective failure to be accommodated to a system as the condition of possibility for living another way. (Ahmed, 2017, p. 62)

Hope in critical spaces of science education emerges in direct relation to scenes of violence and suffering, and the walls we encounter along the way in pursuing an actionable otherwise. The heart as hopeful does not emerge out of naïveté, but rather as an inheritance: the collective failure to be accommodated becomes a condition of possibility, like with the angry heart, through the coalescing and commiserating towards community, but also as a form of inheritance.[8] The sadness with the world, sadness in the world stems from ongoing histories. Echoing the previous heart figuration, it is important to remain open to how we might be (part of) the problem, particularly given how we inherit histories of violence by being and becoming science educators (e.g., Higgins & Tolbert, 2018). This is significant as:

> The word inheritance includes two meanings: to receive and to possess. In a way, we convert what we receive into possessions, a conversion that often "hides" the conditions of having received, as if the possession is "already there." (Ahmed, 2006, p. 125)

As much as there are inheritances that give rise to anger, many of which are converted to possession in the name of power, there are also many that bear remembering as received in the name of hope. Ahmed (2017) states that if "we can become attuned to the ghosts of past suffering, we can also become energized by the words that have been put down; by the collective refusal to put down our pens" (p. 77). This is certainly not an easy or idyllic hope, it is a killjoy inheritance: again, as

[8] As a rejoinder, we wish to note that the inheritance is not always one that is received. As Freire (1994) states, "in limited situations, beyond which lies "untested feasibility" alone—sometimes perceivable, sometimes not—we find the why of both positions: the hopeful and the hopeless one" (p. 9 in Frausto Aceves et al., 2022, p. 226). There is heartwork too, in receiving the conditions of possibility from those who have laboured before us.

above, it requires breaking the seal of happiness which prevents us from attuning to the ways in which suffering arrives from the edges to teach us, to impart upon us an inheritance. Once we are able to attune, "when the happiness seal is broken, when violence has intruded into scenes of bliss, we begin to hear the ghosts of feminists past. The feminist ghosts clamor around; they surround; we listen" (Ahmed, 2017, p. 63). Importantly, hauntings emerge as the cost of subjugation and violence[9]: there is the possibility in something else emerging from that relation through "attending to ghosts, and arresting widespread denial of the violence done to them" (Tuck & Ree, 2013, p. 647) and those who would be made to be future ghosts (see also Higgins, 2022). For example, this complex hope can manifest through explorations of how science and science education are always already caught up land relations which are entangled with in projects of settler colonialism, dispossession, and Indigenous erasure (see Liboiron et al., 2023). There is hope here in not letting a past already passed over pass us by again, and in representing the ways in which those before us imperfectly persisted, how they found openings, critiqued practice, deconstructed policy, created communities and spaces, got back up after setbacks, so that the next generation may too, stand ground. While negatively impacted by the scene of address they stood before and within, there was deep agency there in not allowing for things to be smoothed over in the name of happiness. Particularly when "happiness is used to justify social norms as social goods" (Ahmed, 2017, p. 254) in a disciplinary space where social norms are synonymous with axes of dominance. It is perhaps for this reason that Ahmed (2017) succinctly boils down this affective inheritance as such: "feminism: how we inherit from the refusal of others to live their lives in a happy way" (p. 63).

16.7 Heart as Dreaming

> Caring is anxious – to be full of care, to be careful, is to take care of things by becoming anxious about their future, where the future is embodied in the fragility of an object whose persistence matters…. caring about what happens, caring whatever happens. (Ahmed, 2017, p. 266)

[9] Should it need stating, the hope that is argued for here is not one in which subjugation is more inclusive such that there are more seats at the table in terms of violently securing settler futurities and white imaginaries. That hope is marked by "hopeful smiles" being cast at those exploited whose aspirational inclusion is always conditional and never fully achieved. More importantly, it does not achieve the conditioning of the heart to maintain and sustain the type of transformative and radical hope to fight not only for survivance and refusing to allow subjugation to occur *like that*; but also imagining and willfully striving towards a visionary rightful inheritance of living fully, across temporalities in which Black, Brown and Indigenous people, as well as other-than-human kin not only exist, but also thrive. It is a teeth-bearing, fierce hope, that winks and growls at oppression before it tears it apart. It does not cower, it does not fall into patterns, it is dreaming with claws made for tilling what must be tilled.

In concluding this piece, we offer the heart as dreaming as the final figuration of the heart. As dreams are an image in which near-infinite yet-to-come possibilities intermingle with the actual, we offer a critical rejoinder that some dreams sustain more than others in refusing the flat-lining that occurs when simply accepting this world is as it should be, as inevitable. Echoing Kayumova and Dou's (2022) call for a pluriversal approach so that science education may exist as a "world of many worlds"—so that pluralism is not only tolerated, or accommodated, but is also supported and in which it can flourish—but pivoting towards affect, we offer the heart as dreaming as a positive figuration, but a positivity that does not foreclose, transmute, or close its heart to the negativity experienced in doing critical work in science education. Rather, we dream of a science education where the various negative affective charges that circulate through this work (e.g., anger, sadness), as well as beyond, are not only tolerated so much as they can be converted into positive emotion, but that they are allowed to thrive and become the defibrillating jolt to unclench a heart marked by *rigor mortis* and for it to rigorously pump towards a future otherwise: what could be, what must be. For example, the body-shaking grief of processing how science-education-as-usual requires not being in kinship might allow one to perceive the heart-dampening properties of dominant capitalist or colonial narratives and the ways they persist in science education, in order to allow the blood flow of rebellion and refusal, of being a killjoy that will no longer tolerate the necromancy of status quo, towards a heart that can circulate the flow of a liberatory spirit.

Importantly, as Ahmed (2017) suggests, the negative emotions that must be allowed to be felt include the anxiety for a future otherwise who is yet-to-come: the dream. Significantly, dreaming of another future requires being careful with and enact a deep care for the fragility the present moment, that which has been achieved, and for those who persist and prevail in the moment: it is a dream of an everyday utopia, one of being able to critically and creatively face another day, individually, and as coalescing collective. It is a form of care, individual and collective, marked by picking up the pieces of breakage that occur when we run into the inevitable walls that emerge when coming up against the ways that systems of dominance mark science education. We dream of a science education that not only redirects care away from those very walls that hold up science-education-as-usual, but also cares in ways that soothes heart-ache, the disappointment when the walls feel too frequent, too difficult to overcome, so that we might critically face another day. On the former, Ahmed (2017) suggests that caring for ourselves is an apt way of refusing to care for science education's dominant enactments and understandings: "in directing care toward ourselves, we are redirecting care away from its proper objects; we are not caring for those we are supposed to care for; we are not caring for the bodies deemed worth caring about" (p. 240). On the latter, this self-care takes a collectivist nuance as dreaming to face another day is about caring for/through the fragile communities (that too, can break from the anger and negative emotion that spills over in doing critical work[10]) that emerge from/through the work. It is a care that is rooted in picking ourselves up, picking each other up through the "ordinary, everyday, and often painstaking work of looking after ourselves; looking after each other" (Ahmed, 2017, p. 240). It is about how we weave, thread, and pattern-make across the matters of care and justice that sustain us, the ways of knowing and being that enliven us, with attentiveness to difference and power within the collective (Arada et al., 2023).

If we are to continue to simultaneously face a *feeling like a scientist* which requires a voiding of grief, of heart-arresting fear, loss, pain, guilt, and remorse in seeking out a joyful inquiry; as well as refuse the more over-arching hardening our hearts through continuing to believe that standards are sufficient, best practice towards settled, reductive, inconsequential facts are enough, and that the lower scores of Black, Brown, and Indigenous learners, and their science profession under-representation are on them; if we are to speculatively imagine an actionable otherwise and justice to-come, we *must have heart*: the heartless science stays on a path of destruction because there is no affect to move us towards radical response or response-ability. So, stay mad friends: willfulness is a living and loving connection that electrifies, shocks, connects, and magnetizes.

> Loving connections are live connections, electric connections. A charge can be what you receive from proximity to others who have themselves received that charge. Proximity can be what you struggle for; separation what you fight against. In other words, the charge itself can be a connection: a way of relating to others similarly charge. The language can be our lead: if willfulness is an electric current, it can pass through each of us, switching us on. Willfulness can be a spark. We can be lit up by it. (Ahmed, 2017, p. 82)

References

Ahenakew, C. (2016). Grafting indigenous ways of knowing onto non-indigenous ways of being: The (underestimated) challenges of a decolonial imagination. *International Review of Qualitative Research, 9*(3), 323–340.

Ahmed, S. (2006). *Queer phenomenology: Orientations, objects, others*. Duke University Press.

Ahmed, S. (2017). *Living a feminist life*. Duke University Press.

Alsop, S. (2016). Encountering science education's capacity to affect and be affected. *Cultural Studies of Science Education, 11*, 551–565.

Arada, K., Sanchez, A., & Bell, P. (2023). Youth as pattern makers for racial justice: How speculative design pedagogy in science can promote restorative futures through radical care practices. *Journal of the Learning Sciences, 32*(1), 76–109.

Bang, L. (2018). In the maw of the Ouroboros: An analysis of scientific literacy and democracy. *Cultural Studies of Science Education, 13*(3), 807–822.

[10] This anger can sometimes be a misdirected consequence of having a wall come up in actionably securing an otherwise: "The future is solidified as something that can grasped or refused. When you refuse to grasp something, it is often understood as because you failed to get it." (Ahmed, 2017, p. 49)

Bang, M., & Marin, A. (2015). Nature–culture constructs in science learning: Human/non- human agency and intentionality. *Journal of Research in Science Teaching, 52*(4), 530–544.
Bazzul, J. (2017). From orthodoxy to plurality in the nature of science (NOS) and science education: A metacommentary. *Canadian Journal of Science, Mathematics and Technology Education, 17*(1), 66–71.
Bazzul, J., Wallace, M., & Higgins, M. (2018). Dreaming and immanence: Rejecting the dogmatic image of thought in science education. *Cultural Studies in Science Education, 13*(3), 823–835.
Bellocchi, A., Quigley, C., & Otrel-Cass, K. (Eds.). (2017). *Exploring emotions, aesthetics and wellbeing in science education research* (Vol. 13). Springer Nature.
Britzman, D. (1986). Cultural myths in the making of a teacher: Biography and social structure in teacher education. *Harvard Educational Review, 56*(4), 442–457.
Britzman, D. (2013). Between psychoanalysis and pedagogy: Scenes of rapprochement and alienation. *Curriculum Inquiry, 43*(1), 95–117.
Clough, P. (2007). The affective turn: Introduction. In P. Clough & J. Halley (Eds.), *The affective turn: Theorizing the social* (pp. 1–33). Duke University Press.
Fortus, D. (2014). Attending to affect. *Journal of Research in Science Teaching, 51*(7), 821–835.
Frausto Aceves, A., Torres-Olave, B., & Tolbert, S. (2022). On love, becomings, and true generosity for science education: Honoring Paulo Freire. *Cultural Studies of Science Education, 17*(2), 217–230.
Gilbert, J. (2016). Transforming science education for the Anthropocene—Is it possible? *Research in Science Education, 46*(2), 187–201.
Higgins, M. (2021). *Unsettling responsibility: Indigenous science, deconstruction, and the multicultural science education debate*. Palgrave Macmillan.
Higgins, M. (2022). Toward a hauntology of science education. In J. Beier & J. Jagodinzski (Eds.), *A human Pedagogy* (pp. 77–95). Palgrave Macmillan.
Higgins, M., & Tolbert, S. (2018). A syllabus for response-able inheritance in science education. *Parallax, 24*(3), 273–294.
Jaber, L. Z., & Hammer, D. (2016). Learning to feel like a scientist. *Science Education, 100*(2), 189–220.
Kayumova, S., & Dou, R. (2022). Equity and justice in science education: Toward a pluriverse of multiple identities and onto-epistemologies. *Science Education, 106*(5), 1097–1117.
Kayumova, S., & Tippins, D. (2016). Toward re-thinking science education in terms of affective practices: Reflections from the field. *Cultural Studies of Science Education, 11*, 567–575.
Kayumova, S., McGuire, C. J., & Cardello, S. (2019). From empowerment to response-ability: Rethinking socio-spatial, environmental justice, and nature-culture binaries in the context of STEM education. *Cultural Studies of Science Education, 14*, 205–229.
Kim, E. J. A. (2022). *Integrating indigenous and Western Education in Science Curricula: Relationships at play*. Springer Nature.
Liboiron, M., Higgins, M., & Tolbert, S. (2023). In conversation with Max Liboiron: Towards an everyday, anticolonial science (education) practice. In S. Tolbert, M. Wallace, J. Bazzul, & M. Higgins (Eds.), *Reimagining science education in the Anthropocene (Volume II)* (pp. 343–361). Palgrave Macmillan.
Marker, M. (2016). Indigenous knowledge, indigenous scholars, and narrating scientific selves: "To produce a human being". *Cultural Studies of Science Education, 11*, 477–480.
Marker, M. (2019). Indigenous STEM success stories as disquieting decolonization: thoughts on new times and, old thoughts about placeness. *Cultural Studies of Science Education, 14*, 199–204.
McGinty, M., & Bang, M. (2016). Narratives of dynamic lands: Science education, indigenous knowledge and possible futures. *Cultural Studies of Science Education, 11*, 471–475.
McKinley, E. (2007). Postcolonialism, indigenous students, and science education. In S. K. Abell & N. G. Lederman (Eds.), *Handbook of research on science education* (pp. 199–226). Lawrence Erlbaum.

Mehta, R., Keenan, S., Henriksen, D., & Mishra, P. (2019). Developing a rhetoric of aesthetics: The (often) forgotten link between art and STEM. In M. S. Khine & S. Areepattamannil (Eds.), *STEAM education: Theory and practice* (pp. 117–141). Springer Nature.
Morales-Doyle, D., & Gutstein, E. R. (2019). Racial capitalism and STEM education in Chicago Public Schools. *Race Ethnicity and Education, 22*(4), 525–544.
Moura, C. B., & Guerra, A. (2023). Dwelling in borderlands: A conversation between two science teachers-researchers. *Cultural Studies of Science Education, 18*(1), 257–275.
Nxumalo, F., & Gitari, W. (2021). Introduction to the special theme on responding to anti-blackness in science, mathematics, technology and STEM education. *Canadian Journal of Science, Mathematics and Technology Education, 21*, 226–231.
Petersen, M. R., & Dohn, N. B. (2017). Interest and emotions in science education. In A. Bellocchi, C. Quigley, & K. Otrel-Cass (Eds.), *Exploring emotions, aesthetics and wellbeing in science education research* (pp. 187–202). Springer Nature.
Reiss, M. J. (2005). The importance of affect in science education. In *Beyond Cartesian dualism: Encountering affect in the teaching and learning of science* (pp. 17–25). Springer Nature.
Sanchez, A. (2023). The social focus framework: Anti-racist and anti-colonial conscientization. In S. Tolbert, M. Wallace, J. Bazzul, & M. Higgins (Eds.), *Reimagining science education in the Anthropocene (Volume II)* (pp. 121–142). Palgrave Macmillan.
Tolbert, S., Wallace, M. F., Higgins, M., & Bazzul, J. (2023). *Reimagining science education in the Anthropocene (Volume II)*. Palgrave Macmillan.
Tuck, E., & Ree, C. (2013). A glossary of haunting. In S. Holman Jones, T. E. Adams, & C. Ellis (Eds.), *Handbook of autoethnography* (pp. 639–658). Left Coast Press.
Vakil, S., & Ayers, R. (2019). The racial politics of STEM education in the USA: Interrogations and explorations. *Race Ethnicity and Education, 22*(4), 449–458.
van Eijck, M., & Roth, W. M. (2007). Keeping the local local: Recalibrating the status of science and traditional ecological knowledge (TEK) in education. *Science Education, 91*(6), 926–947.
Wallace, M. F. (2019). Showtime: The biopolitical performance of 'effective beginning science teacher'. *Cultural Studies of Science Education, 14*, 963–980.
Wallace, M. F., Bazzul, J., Higgins, M., & Tolbert, S. (2022). *Reimagining science education in the Anthropocene*. Palgrave Macmillan.
Wolfe, M. (2021). *Affect and the making of the schoolgirl: A new materialist perspective on gender inequity in schools*. Routledge.
Zembylas, M. (2016). Making sense of the complex entanglement between emotion and pedagogy: Contributions of the affective turn. *Cultural Studies of Science Education, 11*, 539–550.

Marc Higgins is an associate professor in the Faculty of Education at the University of Alberta and is affiliated with the Aboriginal Teacher Education Program (ATEP). At the intersections of Indigenous education, science education, and creative and critical educational research methodologies, Marc's research investigates the complexities and complications that occur through interfacing Indigenous and Western modern ways-of-knowing (i.e., epistemology) and ways-of-being (i.e., ontology) within educational spaces. Marc works to unsettle education and educational research differently around contested curricular concepts (e.g., what "counts" as science) toward an ethical otherwise.

Anastasia Sanchez is a PhD candidate in the Learning Sciences and Human Development program in the University of Washington, College of Education. She is currently working with K-8 teachers to promote Antiracist and Anticolonial science education through a critical Social Focus framework. Anastasia's research focuses on the reimagining and reconstruction of science education to better grapple with consequential "wicked problems" in support of collective thriving. Anastasia's research is inspired by her Chicana roots and 12 years of experience as a middle school science teacher, language acquisition specialist, and Title 1 curriculum specialist.

Chapter 17
From a Multicultural to Multinatural Science Education: Perspectives from an Amerindian Perspectivism for Post-pandemic Scenarios

Matheus Monteiro Nascimento, Bruno Ferreira dos Santos, and Cristiano B. Moura

17.1 Introduction

According to Roberts (2007), the field of science education underwent (and was undergoing, in 2007 when the article was published) a significant tension that pitted two perspectives against each other. Roberts (2007) systematizes these perspectives as visions of scientific literacy that represent two opposite ends of a spectrum. He calls them Vision I and Vision II. For the author, Vision I gives meaning to scientific literacy by looking "inward at the canon of orthodox natural science" (Roberts, 2007, p. 730), that is the products and processes of science itself. This approach envisions literacy within science, valuing scientific knowledge 'in itself'. Against that, Vision II derives its meaning from the character of "situations with a scientific component, situations that students are likely to encounter as citizens" (Roberts, 2007, p. 230). This perspective envisions literacy about science-related situations in which aspects other than science have a significant role, connecting such knowledge with situations where it could play a role in society and human life in a broader sense. By examining recent issues of the leading research journals in science education (excluding, perhaps, more disciplinary journals focused on the teaching of physics, chemistry, biology, and earth sciences specifically), it is evident, in a early

M. Monteiro Nascimento (✉)
Universidade Federal do Rio Grande do Sul, Porto Alegre, Brazil
e-mail: matheus.monteiro@ufrgs.br

B. Ferreira dos Santos
Universidade Estadual do Sudoeste da Bahia, Vitória da Conquista, Brazil
e-mail: bf-santos@uesb.edu.br

C. B. Moura
Simon Fraser University, Burnaby, BC, Canada

C. B. Moura (ed.), *A Sociopolitical Turn in Science Education*, Contemporary Trends and Issues in Science Education 62,
https://doi.org/10.1007/978-3-031-78586-3_17

examination, that this duality has undergone substantial changes in recent years (Odden et al., 2021). Today, the growth of fields such as gender perspectives, race, and identities, as well as the consolidation of fields such as Science, Technology, Society and Environment (STSE) education, and cultural perspectives in science education (cf. Odden et al., 2021), are examples that suggest that we might be in a different moment that recognizes the contextualized and humanistic view of science education as a fundamental pillar. Or, if we go even further, with activist pedagogies (Alsop & Bencze, 2014; Bencze, 2017; Reis, 2014) and the rise of perspectives aligned with the third view of scientific literacy (Sjöström & Eilks, 2018; Aikenhead, 2007), conceptions of science education strictly focused on unproblematic instruction without considering contexts or a broader perception of the goals of science education (Colucci-Gray, 2021) seem disconnected from the demands of the present time.

Aligned with contextualized perspectives of the science curriculum, the debate on multiculturalism has already made significant progress in the field (Hodson, 1993), and its deepening has largely led to the integration of indigenous knowledge and cosmogonies into the science curriculum. Aikenhead and Elliott (2010) describe a cross-cultural integration experience in the Canadian curriculum in the province of Saskatchewan, which included the revision of curriculum documents, and changes in textbooks, among other actions aimed at incorporating the perspective known as "two-eyed seeing". Aikenhead and Elliott (2010) argue that this process brings benefits to both Indigenous students, who gain essential cultural capital to access positions of power in a world dominated by Eurocentric perspectives, and non-Indigenous students, who would gain insight into their own culturally constructed Eurocentric world. A similar example is given by Stewart (2005) regarding the curriculum in Aotearoa New Zealand, where the author documents and analyzes the engagement of school science with "pütaiao," the Maori term for science. Furthermore, the integration of indigenous knowledge into science education has been the subject of long-standing debate in the field of science education (see Kim et al., 2017 for a critical synthesis). Within the debate on the post-pandemic directions of science, science education, and society as a whole, it is important to note that the signs of Earth's depletion, of which pandemics like the one we are experiencing are symptoms (Moura, 2022), have fueled the debate on decolonization and integrating Indigenous knowledge as a program for our society. Krenak (2020a, b), speaking from the wisdom of the Krenak people, criticizes the course of consumer society, exploring how this society leads the Earth to collapse and how a post-pandemic horizon points to reconnection with the Earth, based on the concept of "florestania" as an alternative to life in cities proposed by Western thought.[1]

Although the debate on multiculturalism in science education, primarily led by countries in the Global North, has international reach with repercussions in various countries, including Brazil, where two of the authors write, and where the third

[1] The experience of forestania in Krenak (2022) stresses the notion of citizenship (*cidadania* in portuguese) and life in the city as a possibility of another urbanity that nourishes the power of life against a model of civilization that distributes hunger and poverty.

author has roots (e.g., Pinheiro, 2017; Nascimento & Gouvêa, 2020), it is important to note that certain perspectives (notably some perspectives from the Global South) on this debate have not been internationally discussed within the science education community. Here, in particular, we focus on an Amerindian viewpoint (cf. Viveiros de Castro, 2017), which has gained momentum in recent decades, especially in anthropology, and which, as we intend to argue, can offer a different lens for the multiculturalism debate in science education. More than just a new perspective, we understand that the ideas of multinaturalism and Amerindian perspectivism, as we will present and discuss below, challenges the very dichotomy of nature-culture that haunts multicultural perspectives in science education.

In summary, the idea we aim to put forth here is this: if multicultural science education represents a shift towards the development of cross-cultural curricula (Aikenhead, 2001), the integration of knowledge and curriculum indigenization (Aikenhead & Elliott, 2010; Machado & Coppe, 2022), what would be the consequence for science education of adopting a perspective that values not multiculturalism but multinaturalism? Our objective in this chapter is modest: we seek to initiate a conversation on this theme and introduce some concepts (multinaturalism and Amerindian perspectivism) that can be productive for post-pandemic science education, which aims to deepen sociopolitical perspectives and appreciate different ontological views on the nature-culture relationship.

The text is organized into three parts: in the first part, we provide a brief history of the discussion on multiculturalism in science education and extract the fundamental ideas from this debate. Following that, we present the concepts of multinaturalism and Amerindian perspectivism, as outlined by Viveiros de Castro (2015, 2017), delving into the essence of these concepts and subsequently connecting them to science education. In this connection, we aim to highlight avenues for advancement in the dialogue on multiculturalism, originally mobilized and aiming at a post-pandemic, justice-oriented science education that can contribute to the flourishing of ontological alternatives for living in/with/on the world.

17.2 Universalism and Multiculturalism

The notion of culture as a diverse, complex element specific to different human groups has not always been recognized in social sciences (Geertz, 1973; Rapport & Overing, 2002). During the eighteenth and nineteenth centuries, various cultures were categorized based on their level of development and progress, according to evolutionary and hierarchical perspectives that aimed to explain cultural diversity. These evolutionary theories of culture suggested that more "advanced" and "civilized" cultures were superior to "primitive" or "savage" cultures. Western European societies were believed to be at the top of the hierarchy of cultures, while other cultures were seen as lower and backward. These hierarchical and evolutionary perspectives were largely influenced by Charles Darwin's theories on biological evolution and, later, the social theories of Herbert Spencer and Lewis Henry Morgan

(Rapport & Overing, 2002). Such theories were used to justify European colonialism and imperialism, which were seen as a means of "civilizing" and "improving" other "inferior" cultures. This stance can be observed since the invasion of the Americas by Europeans (Dussel, 2005), and continues as a project of the positivist philosophy inaugurated by Auguste Comte, in France at the beginning of the nineteenth century.

In this perspective, human nature is controlled by natural laws that can be discovered through observation and systematic analysis of human behaviors and interactions (Aron, 1969). These natural laws govern human behavior and determine the way people relate and interact with each other. For positivism, human nature is seen as something that can be improved through education, science, and technology. Positivism holds that education, science, and technology can improve human nature. As such, understanding human nature is crucial in making the world a better place. It was thought that by relying on reason and science, we could gain a deeper understanding of the world and human nature, ultimately leading to progress for both society and humanity.

For Auguste Comte, humanity would be an organic whole, governed by natural laws, and should be studied scientifically, seeking to understand its nature and evolution. Comte believed that humanity developed through three evolutionary stages: theological, metaphysical, and the positive (Comte, 1929 [1851]). In the theological stage, the explanation of natural and social phenomena takes place through supernatural entities. During the metaphysical stage, explanations were given using abstract concepts like natural forces or abstract ideas. However, in the positive stage, explanations are based on observable facts and scientific laws. According to Comte, humanity has reached the most advanced stage of evolution, the positive stage, due to the progress of science. And for humanity to continue to evolve, it needed to be governed by a positive social science, capable of studying social phenomena objectively and scientifically. This science could bring people together and overcome the divisions and conflicts that exist between people (Aron, 1969). Thus, Comte's thesis on human unity proposes that humanity should be seen as a unified whole, governed by natural laws that can be understood scientifically. Through science and reason, it would be possible to overcome differences and conflicts between nations and achieve a more just and harmonious society.

However, this dominant point of view typically reduces human nature to simple natural laws, ignoring the complexity and subjectivity of human beings (Myhre, 2006; Rodseth, 2018). Still, in the nineteenth century, new philosophical propositions argued that human nature would be influenced by factors such as culture, history, and personal experiences, which cannot be completely explained through objective scientific analysis (Aron, 1969). Positivist ideas widely influenced the social sciences, as we can see in the emergence of evolutionary anthropology, which argued that cultures evolve in a hierarchical, linear manner, from primitive to civilized. The first opposition to the thesis of human unity systematically appeared at the end of the nineteenth century and the beginning of the twentieth century. For instance, the German anthropologist Franz Boas (1928), when studying the Inuit people, argued that each culture is unique and should be studied on its terms, highlighting the importance of cultural diversity. Boas also pointed out that the practices and beliefs of a culture must be understood in their cultural context, without judgment of values or prejudices, giving rise to what came to be known as cultural relativism. Another contribution by Boas was his strong opposition to the concept of cultural assimilation, defending the preservation and respect for minority cultures and highlighting the importance of cultural diversity for a more just and egalitarian society, that is, the foundations of multiculturalism.

Multiculturalism in the area of research in Science Education gained prominence, especially following a special call for publication on this topic in the journal *Science Education* (Aikenhead, 1993). Since then, many works have been published to define multiculturalism and its relationship with Science Education (e.g. Atwater & Riley, 1993; Hodson, 1993); discussing its potential and also its limitations (Ogawa, 1995). One of the first definitions places multiculturalism as "a construction, a process, and an educational reform movement to provide equitable opportunities for culturally diverse student populations to learn quality science in schools, colleges, and universities" (Atwater & Riley, 1993, p. 664). On the other hand, Hodson (1993) states that "'multicultural science education' can mean many things to many people. Relative emphasis will need to vary from country to country, region to region, and even from school to school and class to class" (p. 688). Regardless of the definition, however, there is in common the recognition of cultural diversity. This awareness of cultural diversity in science education appears as a central issue in the multicultural perspective. Thus, the notion of multicultural science education that researchers assume in their investigations is closely related to the way they conceptualize the relationship between science and cultural diversity.[2] Some may focus on the contribution of many cultures to science in their historical context, or on the contribution of non-Western scientists in the context of current scientific activities. Others may focus on the needs of students from minority ethnic backgrounds.

This first phase of multiculturalism in Science Education, therefore, refers to the recognition and respect for the diversity of cultures, worldviews, and forms of knowledge that are brought to the classroom by students and teachers. This perspective involves recognizing how culture shapes science teaching and learning, and ensuring that students of all cultural backgrounds have access to equitable and effective science education (Luft, 1998). Students from different cultural backgrounds often face barriers to learning science, including difficulties with language, cultural mismatches between the science curriculum and their own cultures, and the lack of representation of scientists from diverse backgrounds in the curriculum (Nieto & Bode, 2007). Multiculturalism thus aims to address these problems by promoting inclusive and culturally responsive teaching practices that incorporate students' cultural backgrounds and worldviews into the science curriculum (Gay, 2002). Another important aspect of multiculturalism highlighted by the area of

[2] In Brazil, the idea of critical multiculturalism (Moreira, 2002) has spread widely and influenced science education.

science education is the greater diversity in the representation of scientists and scientific perspectives in the curriculum; this includes highlighting the contributions of scientists from different cultural backgrounds and belonging to minority groups (Carlone & Johnson, 2007; Barton, 2003). It also involves recognizing how culture shapes scientific knowledge and practices and participating in critical discussions about the social and cultural implications of science.

Nevertheless, although multicultural perspectives play an important role in including discussions about cultural diversity in science education, its limitations have been pointed out by researchers in the area over the last thirty years (Rodriguez, 2015; Boda, 2019). Highlighting, the main question raised is that while "cultural diversity claims that culture should be viewed in a relativistic perspective, why is science itself never viewed in such a relativistic perspective? Science is always treated as absolute" (Ogawa, 1995, p. 584). This question brings back the debate on whether science is universal, an object of study in the philosophy of science since the beginning of the twentieth century. As we have seen, being the result of a historical process, modern Western science was perceived as "correct" and "universal", being imposed on other countries and cultures during European exploration and colonization (Le & Matias, 2019). The common belief in Western societies is that science is universal and objective, which hides the reality that it was a very particular social group that shaped this scientific paradigm. Although some researchers may argue that theories and knowledge of the natural world are independent of any specific culture (Le & Matias, 2019), Stanley and Brickhouse (1994, p. 389) remind us that "our knowledge of reality will always be mediated, incomplete, and distorted".

For us, multiculturalism is premised on the recognition of different cultures built by different groups throughout history, with consequences for the perception of the history of science to be taught, as well as for the ways of teaching science to different groups with different cultural backgrounds. Even perspectives that consider different sciences, such as that of Ogawa (1995)—with consequences for the construction of cross-cultural curricula (Aikenhead, 2001)—do not consider an ontological debate regarding the existence of different realities from different points of view (whether human or not). In this sense, we understand that there is a need to expand this debate, raising the notion beyond multicultures and thinking about multinatures. In the next section, we present the main theoretical assumptions of multinaturalism as proposed by Viveiros de Castro and, subsequently, we draw some parallels with science teaching.

17.3 Perspectivism and Multinaturalism

According to Viveiros de Castro (2017), the last few decades have witnessed the emergence of a new image of the indigenous Amazon, an image, in his words, "characterized by an emphasis on the complexity of social forms and the diversity of the region's natural physiognomy" (p. 278). At the same time, this occurs within a new theoretical context that favors a more nuanced understanding of the relationships between society and nature in the study of societies that do not differentiate between an exclusive sphere of human beings and another to which human beings would be relegated other beings (dos Santos & Tola, 2016). In their theoretical reflections, anthropologists such as Viveiros de Castro (2017) and Philipe Descola (2021) share the attempt to escape the dichotomy or opposition between Nature and Culture, understood as two separate categories. In this dichotomy, the first assumes a fixed and universal character, and the second, a variant and particular character (Picelli, 2016, p. 56). According to new understandings, nature (or, to be more precise, the more-than-human worlds—the worlds of the different beings co-dwelling on Earth) starts to be seen as a subject participating in social relations, and there is no ontological discrimination between humans and non-humans. This movement, which has been characterized in Anthropology as the "ontological turn", connects metaphysical questions with French post-structuralist ideas (Reynoso, 2015). Based on the idea of an ontological turn, the objective of this discipline is no longer the study of cultural representations. Instead, it is concerned with the manner in which each society defines existing beings in the world and the relationships between them (dos Santos & Tola, 2016, p. 72). Simultaneously, the ontological turn in Anthropology will also mean a reaction to the ecological crisis and the Anthropocene (dos Santos & Tola, 2016).

As Candau (2008) notes in the context of multiculturalism/interculturality, the shifting focus away from humans (and, therefore, the emphasis on culture as opposed or simply detached from the more-than-human world) is not limited to the academic realm. It also manifests in various domains of human activity, such as the arts, laws and philosophical perspectives. As an example, in 2008 the new constitution of Ecuador (Tanasescu, 2013) recognized rights to nature, or Pachamama, in its constitution. Concepts born at the Global South, as the idea of 'living well' (Buen Vivir) were also adopted as main rationale in constitutions of Ecuador and Bolivia (Gudynas, 2011). Dozens (perhaps hundreds) of art exhibitions around the world, driven by the climate emergency, have also brought the issue of the Anthropocene and the boundaries between human and non-human into debate in a variety of ways. Highlighting a specific aspect of this debate, we will explore, based on the thoughts of Viveiros de Castro, the idea of Amerindian perspectivism.

17.3.1 What Is Perspectivism?

According to Viveiros de Castro, perspectivism is a quality of Amerindian thought according to which "the world is inhabited by different species of subjects or people, human and non-human, who apprehend it from different points of view" (2017, p. 301). The world would then be the set of "a multiplicity of points of view" (2015, p. 42). Amerindian perspectivism derives from the cosmopolitics of Amazonian societies that "affirm a multiplicity of perspectives intrinsic to reality" (2015, p. 35).

According to this thought, "the manifest form of each species is a covering ('clothes'/"roupa" in the original) hiding an internal human form" (2017, p. 304). All existing bodies—human and animals—are considered clothing or envelopes of an internal form or soul that becomes "the subjectivity formally identical to human consciousness, materializable, so to speak, in a human bodily scheme hidden under the animal mask" (2017, p. 305). According to Amerindian perspectivism, all beings represent "the world in the *same* way – what changes is the *world* they see" (2017, p. 328—italics are the author's). The point of view, therefore, is in the body and the difference in points of view occurs in the specificity of bodies and not in the soul: "Animals see different things in the same way as we do because their bodies are different from ours" (2017, p. 329). The body, as Viveiros de Castro understands it, goes beyond a particular physiology or anatomy, but is considered a set of ways of being that constitute a habitus; the body comprises a set of affections and capacities, and the origin of different perspectives is found there. Therefore, perspectivism according to Viveiros de Castro is a corporal mannerism: "the indigenous theory of perspectivism emerges from an implicit comparison between how different bodily modes ("species") 'naturally' experience the world as affective multiplicity" (2015, p. 87).

For Viveiros de Castro (2017), the origins of Amerindian perspectivism are located in the Amazonian ontology of predation, which assumes a symbolic valorization of hunting, and the importance of shamanism. The first derives from a vector of spiritualization of animals, in the form of a "hunter ideology". Shamanism in the Amazon is defined by him as "the ability manifested by certain individuals to deliberately cross bodily barriers and adopt the perspective of allo-specific subjectivities, to manage relationships between them and humans" (2017, p. 310). According to Viveiros de Castro, in shamanic practice "the meeting or exchange of perspectives is a dangerous process, and a political art – diplomacy. Therefore, if Western 'multiculturalism' is relativism as public policy, shamanic perspectivism is multinaturalism as cosmic policy" (2017, p. 310). He adds that shamanism implies a particular way of producing knowledge, which is at the antipode to the objectivist epistemology of Western modernity. In Amerindian shamanism, "to know is to personify, to take the point of view of that which must be known – of that, or, rather, that; because shamanic knowledge aims at 'something' that is 'someone', another subject or agent" (2017, p. 311). For the Amerindian interpretation, the object to be known as "an incompletely interpreted subject", because, to be known, the objective needs to be personified, which means, in turn, having a point of view, a perspective. All existing people, in this way, share self-awareness or reflexivity which, therefore, ceases to be an exclusively human attribute.

17.3.2 *What Is Multinaturalism?*

The term multinaturalism highlights, in Viveiros de Castro, "one of the contrasting features of Amerindian thought about modern 'multiculturalist' cosmologies". (2017, p. 302). For him, the Amerindian conception, contrary to multiculturalism, has as its starting point "a unity of the spirit and a diversity of bodies", which entails, as a consequence, that culture (or the subject) is the form of the universal and nature (or the object) is the form of the particular (2017, p. 303). The virtually universal notion in this thought "is that of an original state of undifferentiation between humans and animals, described by mythology" (2017, p. 307). Indigenous myths narrate how "animals lost the attributes inherited or maintained by humans" (2017, p. 308). Indigenous thought assumes that animals and other beings in the cosmos that were once humans continue to be so, albeit in a non-evident way (2017, p. 309).

In the naturalist ontology proposed by Viveiros de Castro, the relationships between human society and nature are themselves natural, as opposed to the social separation between nature and society in the animist ontology (2017, p. 316). Therefore, human society itself is considered one natural phenomenon among others in a naturalistic ontology. In it, the society/nature interface is natural: "Humans are organisms in 'ecological' interaction with other bodies and forces, all regulated by the necessary laws of biology and physics" (2017, p. 316). According to Viveiros de Castro, for multinaturalism the notion of nature does not need the notion of culture to be defined (p. 316), since for Amerindian thought, nature and culture "are part of the same sociocosmic field" (2017, p. 320). Culture would then be the nature of the subject, that is, "it is how every agent experiences its own nature" (2017, p. 324). Thus, multinaturalism is the result of the presumption of the simultaneous existence of multiple natures.

17.3.2.1 Perspectivism and Relativism

As Viveiros de Castro argues, perspectivism is not relativism but multinaturalism because cultural relativism or multiculturalism

> presupposes a diversity of subjective and partial representations, incident on external, single and total nature, indifferent to representation; the Amerindians propose the opposite: a purely pronominal representative or phenomenological unity, applied indifferently to real diversity. One 'culture', multiple 'natures'; constant epistemology, variable ontology – perspectivism is multinaturalism, as a perspective is not a representation (2017, p. 329).

As a corollary, it can be said that a perspective "is not a representation because representations are properties of the spirit, but the point of view is in the body" (2015, p. 65). Perspectivism would, therefore, be a kind of "bodily mannerism" (2015, p. 66).

Following this line of argument, if the world is composed of multiple points of view, perspectivism, as a consequence, is arranged on "an axis orthogonal to the opposition between relativism and universalism" (Viveiros de Castro, 2015, p. 42). Perspectivism, in this sense, "is not relativism, but relationalism" (2017, p. 331).

For Freitas (2018), Amerindian perspectivism not only implies the singularity of the points of view associated with each existing species but that any ontological regime will always be positional and provisional. With these multiple regimes, the idea of an absolute point of view, "independent of entities", or of an external and unified nature represented by different subjects loses strength (Freitas, 2018, p. 396). A perspective, therefore, is the result of relationships between entities, hence the idea of relationalism. For ontological perspectivism, a type or category of object depends on the relationship of that object with something else (Paleček & Risjord, 2012).

The ideas of multinaturalism and the Amerindian perspective derive, as a whole, from new understandings about the cosmopolitics of Amazonian peoples, which question the separation between Nature and Culture, a separation already postulated in other spheres and social practices beyond the academic (Latour, 1993). Amerindian perspectivism assumes the multiplicity of reality based on the differentiated apprehension of bodies and their associated points of view, which produces multiple natures. By tensioning with multiculturalism, and by centralizing relationships between humans and non-humans, Amerindian thought and multinaturalism impose on us the exercise of thinking about its consequences for Science Education. This is what we will cover in the next section of this chapter.

17.4 Implications, Ideas, and Convergences for Science Education

We have seen that the area of Science Education in recent years has been concerned about questioning the hegemonic perspective (and discourses) of science, bringing multicultural perspectives to the educational debate. However, such approaches, in general, as much as they contribute to recognizing the multiplicity of aspects of "culture", still support a singular view of "nature". In this view, the category "nature" encompasses everything that exists independently of human beings, and whose entities constitute resources that can be exploited by the political and socio-economic criteria of societies that have the power and the means to establish such criteria. As elaborated by Krenak (2022), the view that nature would be a large "warehouse" of so-called "natural resources" available to humans is something deeply rooted in the practices and worldviews of Eurocentric capitalist society and that needs to be problematized by education, and particularly science education as we maintain.

That is why we seek inspiration in the works of Viveiros de Castro and indigenous authors to establish connections between Amerindian thought, multinaturalism, and the movements taking place in our area, seeking to reshape science and scientific education for a post-pandemic world. We understand that the current moment is one of crisis, similar to what we experienced with the economic, scientific, and cultural revolutions of the eighteenth and nineteenth centuries. At the time, new methods of interpreting social reality emerged, with broad changes in the field of knowledge, that is, an epistemological turn (Schmitz, 2019). In current times, however, we defend a turn, but now ontological, that would effectively take into consideration our relationships with the more-than-human world. In these terms, the entities of nature abandon the passive and agency-less character that allows them to be exploited by humans for the extraction of resources and begin to be taken as subjects participating in social relations, and there is no ontological discrimination between humans and non-humans. But the question remains: how can this new multinatural understandings might impact research and education in science? In the scientific field, Bruno Latour (2007) adopts a similar perspective when dealing with scientific practices and delving into these issues is beyond the scope of this text. In science education, however, as it is still an incipient topic, we are unable, at this moment, to present answers based on consolidated theoretical or empirical studies. This does not prevent us, however, from discussing the possible implications for our area. Thus, we seek to point out consequences in three axes.

Firstly, assuming the notion of ontological turn, the objective ceases to be the study of cultural representations and begins to focus on how each society defines existing beings in the world and the relationships between them. Therefore, scientific education cannot be limited to teaching static definitions and taxonomies created by specific groups. On the contrary, it is essential to expand the discussion to include elements of nature within the scope of social relations, including science, without ontological discrimination between human and non-human actors. In Amerindian thought, nature should not be substantialized, but understood in a relational configuration, a mobile perspective, a point of view. In this perspective, there is no separation between humans and non-humans, body and spirit. Humans see themselves as humans and see animals as animals; animals see us as non-humans, and themselves as humans (Viveiros de Castro, 2015). The same understanding is valid for non-inanimate beings. A river can be understood as an ancestor, like a grandfather (Krenak, 2022), for example. This relationship provides the greatest contribution to science teaching, which is committed to the diversity of cultures and the multiplicity of worlds that the radical alterity of Amerindian thought provides us.

It is worth noting that within the scope of Science Education, recent advances have analyzed and questioned the human-nature relationship in research and practices in the area, in different ways. Sharma and Buxton (2015), for example, evaluated the discourse of a science textbook from Georgia State in the US using critical discourse analysis and found some problematic traces in this textbook, such as the attribution of responsibility and agency to "generic humans" or "people", avoiding the specifics of how certain groups of humans have been involved in creating, modifying, or sustaining certain human-nature relationships. Also, they found that the text was emphasizing the material distancing of human beings from the natural conditions of their existence, and over-simplifying this theme, in general. Still, Sharma (2017) noted that the idea of nature thought of as a separate entity from humans is a construction of modernity, further emphasized by a neoliberal rationale, that commodifies the so-called "natural resources". Decentering humans from science education discourse and drawing from posthumanist scholarship, Jeong et al. (2021) challenge the notion of scientific literacy and the mainstream project of (science) education of "making" humans, towards a new perspective in which students can understand our entangled relationship with the more-than-human world. Guerrero and Torres-Olave (2022) applied a critical discourse analysis to curricular documents for secondary education in Chile to investigate the concepts of scientific literacy and agency. They found tensions between different visions of both concepts and student agency and a deliberate banking model of teaching for teachers in the documents analyzed. Also positioning their work in the broader dialogue on the Anthropocene, Moura and Guerra (2022) propose a revised historical approach to science in science education to rediscuss the human-nature relationship. All of those examples (and many others that are not explored here) have challenged the common and sometimes underlying assumptions about human-nature relationships. The Amerindian perspective we bring to this chapter also adds to this ongoing conversation about how we could redraw (or erase, more precisely) the boundaries between the human and more-than-human worlds.

This task is not only conceptual but a very practical problem of how to teach science. In this sense, teaching in the context of multiple ontologies, especially highlighting the ontologies of Amerindian peoples, allows students, who would be taught based on a Western Eurocentric pedagogy, designed based on capitalist and extractive ideas, to develop another relationship with the world that live. For example, it is common when teaching science from the hegemonic perspective of nature to say that water "is an inorganic compound with the chemical formula H_2O. It is a transparent, tasteless, odorless, and nearly colorless chemical substance, and it is the main constituent of Earth's hydrosphere and the fluids of all known living organisms" (Franks, 2012). This definition leads people to see rivers and lakes only as energy potential for building dams or as a volume of water to be used in agriculture (Krenak, 2022). The indigenous worldview, which understands the river as an ancestor, as a being with an anthropomorphic essence, has the potential to create another relationship with these natural beings, different from that provided by capitalist extractivism.

In other words, not just other ways of thinking about the world, but different pedagogies need to be mobilized for this task. Such pedagogies are less focused on knowing and knowing for (which does not mean a break from content and concepts), and more focused on ways of learning to live in/with the world. If, according to Viveiros de Castro (2017), we are all part of the same spirit and diversity of bodies, this idea therefore invokes another set of pedagogies that are less obsessed with the demarcation, classification, and hierarchization of bodies and more focused on a holism that allows us to live in greater harmony with the more-than-human world. Lessons can be taken here from, for example, place-based pedagogies (Gruenewald, 2003), wild pedagogies (Jickling et al., 2018), outdoor education, among others, which are often informed by an experiential education perspective, privilege the construction of ties and relationships with space and the more-than-human world around.

A third axis of discussion concerns projects to decolonize science education. In recent years, several works (e.g. Monteiro et al., 2019, Gandolfi, 2021, Moura et al., 2023) have advocated for a perspective that highlights, from historical and contemporary approaches to science, the marks of coloniality that persist in the production of science and scientific education. Such pedagogies open space for epistemic humility, as it allows recognizing that there is knowledge and the construction of knowledge beyond modern Western science. In the same sense, recognizing another ontological perspective opens space for the recognition of these other ontologies and the possibility of learning from the more-than-human world, based on the recognition of another positioning of that world. As an example in environmental education, Ford and Blenkinsop (2018) report an encounter with a river and the various more-than-human agents that appear in this encounter. In what ways can we encourage such encounters in science education? How can we recognize the wisdom not only of those who had been crushed by the colonial project but also, based on the knowledge of these humans, embark on a journey of deeper learning in the more-than-human world?

We could present many other examples taken from science classes to enrich our argument. The central idea, however, is to support a multinaturalist pedagogy so that students develop this ancestral or minimally closer and more familiar relationship with the elements of nature. The ontological opening provided by these new pedagogies should lead us to understand—in a broad way—the protection of the environment cultivated by the Amerindian peoples and a way of thinking that is distant from the mercantile logic of our society.

17.5 Final Remarks

This chapter presents concepts from currents in contemporary Anthropology and the discussion of the impacts and consequences of their adoption by Science Education. In doing so, we define the notions of Amerindian perspectivism and multinaturalism based on the work of Viveiros de Castro, a Brazilian anthropologist who elaborates on these notions based on his and other researchers' research with people who inhabit the Amazon forest. The implications of these ideas for science education are just beginning to be glimpsed, and they permeate our system of moral and social values and our knowledge-production practices. In the field of science education, the demand for a more humanistic and contextualized vision in curricula has already managed to open space for the introduction of indigenous knowledge and cosmogonies in different countries, based mainly on the idea of multiculturalism. Initiatives like these confirm the need for Science Education to expand its articulations towards non-Eurocentric knowledge and theories, linked to the Global South, becoming more plural.

Envisioning possibilities in a post-pandemic world, we present a theoretical essay that aims to bring some provocations, discussing how non-hegemonic knowledge systems can inspire the future of science education, helping to create new and less destructive modes of existence. The different multicultural curricular experiences already carried out have shown the challenges and limitations, but now we ask ourselves the consequences of multinatures for Science Education. For Viveiros de

Castro (2017), the idea of Amerindian perspectivism does not carry with it the relativism that multiculturalism implies, but brings relationalism, since perspective results from diversity and the relationship between different points of view on reality. The presence of multinaturalism blurs the boundaries between Nature and Culture, between human and non-human beings, which implies facing important conceptual and pedagogical issues in Science Education. With this, the ontological turn in science education has the potential to stimulate political imagination towards new and radical ways of inhabiting the world.

References

Aikenhead, G. (1993). Foreword: Multicultural issues and perspectives on science education. *Science Education, 77*, 659–660.

Aikenhead, G. S. (2001). Integrating Western and aboriginal sciences: Cross-cultural science teaching. *Research in Science Education, 31*, 337–355.

Aikenhead, G. S. (2007). Expanding the research agenda for scientific literacy. In C. Linder et al. (Eds.), *Promoting scientific literacy: Science education research in transaction* (pp. 19–34). Geotryckeriet.

Aikenhead, G. S., & Elliott, D. (2010). An emerging decolonizing science education in Canada. *Canadian Journal of Science, Mathematics and Technology Education, 10*, 321–338.

Alsop, S., & Bencze, L. (2014). Activism! Toward a more radical science and technology education. In L. Bencze & S. Alsop (Eds.), *Activist science and technology education* (pp. 1–19). Springer.

Aron, R. (1969). Les étapes de la pensée sociologique. *Revue Philosophique de la France Et de l'Etranger, 159*, 397–404.

Atwater, M. M., & Riley, J. P. (1993). Multicultural science education: Perspectives, definitions, and research agenda. *Science Education, 77*(6), 661–668.

Barton, A. C. (2003). *Teaching science for social justice*. Teachers College Press.

Bencze, L. (Ed.). (2017). *Science and technology education promoting wellbeing for individuals, societies and environments*. Springer.

Boas, F. (1928). *Anthropology and modern life*. Norton.

Boda, P. A. (2019). Conceptualizing the margins in science education: The limits of multicultural analyses. *Cultural Studies of Science Education, 14*, 493–514.

Candau, V. M. (2008). Human rights, education and interculturality: Tensions between equality and difference. *Revista Brasileira de Educação, 13*(37), 45–56. https://doi.org/10.1590/S1413-24782008000100005

Carlone, H. B., & Johnson, A. (2007). Understanding the science experiences of successful women of color: Science identity as an analytic lens. *Journal of Research in Science Teaching, 44*(8), 1187–1218.

Colucci-Gray, L. (2021). For a science curriculum that matters. *Science Teacher Education, 90*, 22–27.

Comte, A. (1929). *Système de politique positive; ou, Traité de sociologie, instituant la religion de l'humanité* (5th ed.). Société positiviste.

Descola, P. (2021). *Outras naturezas, outras culturas [Other natures, other cultures]*. Editora 34.

dos Santos, A., & Tola, F. (2016). ≃Ontologías como modelo, método o política? Debates contemporáneos en Antropología [Ontologies as model, method, or policy? Contemporary debates in Anthropology]. *Avá Revista de Antropología, 29*, 71–98.

Dussel, E. (2005). Europa, Modernidade e Eurocentrismo [Europe, Modernity and Eurocentrism]. In: E. Lander (Org.), *A colonialidade do saber: Eurocentrismo e Ciências Sociais. Perspectivas*

Latinoamericanas. [The coloniality of knowledge: eurocentrism and social sciences. Latin-american perspectives] (pp. 25–34). CLACSO.
Ford, D., & Blenkinsop, S. (2018). Learning to speak Franklin: Nature as co-teacher. *Journal of Outdoor and Environmental Education, 21*, 307–318.
Franks, F. (Ed.). (2012). *The physics and physical chemistry of water* (Vol. 1). Plenum Press.
Freitas, A. S. (2018). Contribuições do perspectivismo ameríndio Para as pesquisas em Filosofia da Educação [Contributions of Amerindian perspectivism to research in Philosophy of Education]. *Revista Brasileira de Estudos Pedagógicos, 99*(252), 387–403. https://doi.org/10.24109/2176-6681.rbep.99i252.3731
Gandolfi, H. E. (2021). Decolonising the science curriculum in England: Bringing decolonial science and technology studies to secondary education. *The Curriculum Journal, 32*(3), 510–532.
Gay, G. (2002). Preparing for culturally responsive teaching. *Journal of Teacher Education, 53*(2), 106–116.
Geertz, C. (1973). *The interpretation of cultures: Selected essays*. Basic Books.
Gruenewald, D. A. (2003). The best of both worlds: A critical pedagogy of place. *Educational Researcher, 32*(4), 3–12.
Gudynas, E. (2011). Buen Vivir: Today's tomorrow. *Development, 54*, 441–447.
Guerrero, G. R., & Torres-Olave, B. (2022). Scientific literacy and agency within the Chilean science curriculum: A critical discourse analysis. *The Curriculum Journal, 33*(3), 410–426.
Hodson, D. (1993). In search of a rationale for multicultural science education. *Science Education, 77*(6), 685–711.
Jeong, S., Sherman, B., & Tippins, D. J. (2021). The Anthropocene as we know it: Posthumanism, science education and scientific literacy as a path to sustainability. *Cultural Studies of Science Education, 16*, 805–820.
Jickling, B., Blenkinsop, S., Timmerman, N., & Sitka-Sage, M. D. D. (2018). *Wild Pedagogies*. Palgrave Macmillan.
Kim, E. J. A., Asghar, A., & Jordan, S. (2017). A critical review of traditional ecological knowledge (TEK) in science education. *Canadian Journal of Science, Mathematics and Technology Education, 17*(4), 258–270.
Kopenawa, D. (2013). *The falling sky: Words of a Yanomami Shaman*. Harvard University Press. https://doi.org/10.4159/harvard.9780674726116
Krenak, A. (2019). *Ideias para adiar o fim do mundo [Ideas to postpone the end of the world] (Kindle Edition)*. Companhia das Letras.
Krenak, A. (2020a). *O amanhã não está à venda [The tomorrow is not for sale]* (Kindle Edition). Companhia das Letras.
Krenak, A. (2020b). *A vida não é útil [Life is not useful]* (Edição do Kindle). Companhia das Letras.
Krenak, A. (2022). *Futuro ancestral [Ancestral future]* (Edição do Kindle). Companhia das Letras.
Latour B. (1993). *We have never been modern* (trans: Porter, C.). Harvard University Press.
Latour B. (2007). *Reassembling the social: An introduction to actor-network-theory*. Oxford University Press.
Le, P. T., & Matias, C. E. (2019). Towards a truer multicultural science education: How whiteness impacts science education. *Cultural Studies of Science Education, 14*, 15–31.
Luft, J. (1998). Multicultural science education: An overview. *Journal of Science Teacher Education, 9*(2), 103–122.
Machado, V. F., & Coppe, C. (2022). Educação Científica contra o Preconceito: da Natureza às Multinaturezas [Scientific Education Against Prejudice: From Nature to Multinature]. *Revista Brasileira de Pesquisa em Educação em Ciências, e32489*, 1–23. https://doi.org/10.28976/1984-2686rbpec2022u119141
Monteiro, B. A. P., Dutra, D. S. A., Cassiani, S., Sanchez, C., & Oliveira, R. D. V. L. (2019). *Decolonialidades na Educação em Ciências [Decolonialities in Science Education]*. Editora Livraria da Física.
Moreira, A. C. (2002). Currículo, diferença e diálogo [Curriculum, difference, and dialogue]. *Educação & Sociedade, 18*(79), 15–38.

Moura, C. B. (2022). Educação em ciências no "portal covid-19": Problematizando a relação humano-natureza por meio da história e da filosofia da ciência [Science education in the "COVID-19 portal": Problematizing the human-nature relationship through the history and philosophy of science]. In F. Garay et al. (Eds.), *Encuentro de ciencias básicas 5: el COVID-19 y sus efectos: las clases remotas y la deserción: investigaciones enmarcadas en resultados de aprendizaje*. Universidad Católica de Colombia.

Moura, C. B., & Guerra, A. (2022). Rethinking historical approaches for science education in the anthropocene. In M. F. G. Wallace, J. Bazzul, M. Higgins, & S. Tolbert (Eds.), *Reimagining science education in the Anthropocene* (pp. 115–129). Palgrave Macmillan.

Moura, C. B., Alsop, S., Camel, T., & Guerra, A. (2023). Science education in a world in crisis: Contributions from the South to a defense of a cultural–historical approach in science teaching. *Cultural Studies of Science Education, 18*(3), 669–693. https://doi.org/10.1007/s11422-022-10129-0

Myhre, K. C. (2006). The truth of anthropology: Epistemology, meaning and residual positivism. *Anthropology Today, 22*(6), 16–19.

Nascimento, H. A., & Gouvêa, G. (2020). Diversidade, Multiculturalismo e Educação em Ciências no Enpec [Diversity, Multiculturalism and Science Education at Enpec]. *Revista Brasileira de Pesquisa em Educação em Ciências, 20*(u), 497–521. https://doi.org/10.28976/1984-2686rbpec2020u497521

Nieto, S., & Bode, P. (2007). School reform and student learning: A multicultural perspective. In J. A. Banks & C. A. M. Banks (Eds.), *Multicultural education: Issues and perspectives* (9th ed., pp. 258–274). Wiley.

Odden, T. O. B., Marin, A., & Rudolph, J. L. (2021). How has science education changed over the last 100 years? An analysis using natural language processing. *Science Education, 105*(4), 653–680.

Ogawa, M. (1995). Science education in a multiscience perspective. *Science Education, 79*(5), 583–593.

Paleček, M., & Risjord, M. (2012). Relativism and the ontological turn within anthropology. *Philosophy of the Social Sciences, 43*(1), 1–23. https://doi.org/10.1177/0048393112463335

Picelli, P. (2016). O perspectivismo de Viveiros de Castro: proposta de uma nova antropologia [The perspectivism of Viveiros de Castro: Proposal for a new anthropology]. *Revista Alteridade, 2*(1), 53–63.

Pinheiro, P. C. (2017). The construction of the science in the community website: Antecedents, foundations, hybrid narratives and epistemological content. *Revista Brasileira de Pesquisa em Educação em Ciências, 17*(1), 271–297. https://doi.org/10.28976/1984-2686rbpec2017171271

Rapport, N., & Overing, J. (2002). *Social and cultural anthropology: The key concepts*. Routledge.

Reis, P. (2014). Promoting students' collective socio-scientific activism: Teacher's perspectives. In S. Alsop & L. Bencze (Eds.), *Activism in science and technology education* (pp. 547–574). Springer.

Reynoso, C. (2015). *Crítica de la antropología perspectivista: Viveiros de Castro – Philipe Descola – Bruno Latour [Critique of perspectivist anthropology: Viveiros de Castro – Philippe Descola – Bruno Latour]*. Editorial Sb.

Roberts, D. A. (2007). Scientific literacy/science literacy. In S. Abell & N. G. Lederman (Eds.), *Handbook of research on science education* (pp. 729–780). Lawrence Erlbaum.

Rodriguez, A. J. (2015). What about a dimension of engagement, equity, and diversity practices? A critique of the next generation science standards. *Journal of Research in Science Teaching, 52*, 1031–1051. https://doi.org/10.1002/tea.21232

Rodseth, L. (2018). Hegemonic concepts of culture: The checkered history of dark anthropology. *American Anthropologist, 120*(3), 398–411.

Schmitz, F. (2019). *O círculo de Viena [The Vienna Circle]*. Contraponto.

Sharma, A. (2017). Countering the neoliberal ontology of nature: The STEPWISE option. In L. Bencze (Ed.), *Science and technology education promoting wellbeing for individuals, societies and environments* (pp. 625–638). Springer.

Sharma, A., & Buxton, C. A. (2015). Human–nature relationships in school science: A critical discourse analysis of a middle-grade science textbook. *Science Education, 99*(2), 260–281.

Sjöström, J., & Eilks, I. (2018). Reconsidering different visions of scientific literacy and science education based on the concept of Bildung. In Y. Dori, Z. Mevarech, & D. Baker (Eds.), *Cognition, metacognition, and culture in STEM education* (pp. 65–88). Springer.

Stanley, W. B., & Brickhouse, N. W. (1994). Multiculturalism, universalism, and science education. *Science Education, 78*(4), 387–398.

Stewart, G. (2005). Mäori in the science curriculum: Developments and possibilities. *Educational Philosophy and Theory, 37*(6), 851–870.

Tanasescu, M. (2013). The rights of nature in Ecuador: the making of an idea. *International Journal of Environmental Studies, 70*(6), 846–861. https://doi.org/10.1080/00207233.2013.845715

Viveiros de Castro, E. (2015). *Metafísicas canibais [Cannibal metaphysics]*. Cosac Naify.

Viveiros de Castro, E. (2017). *A inconstância da alma selvagem e outros ensaios de Antropologia [The inconstancy of the wild soul and other essays in anthropology]*. Ubu.

Matheus Monteiro Nascimento Professor in the Physics Department at the Federal University of Rio Grande do Sul since 2019, with a PhD in Physics Education (2019). Worked as a Visiting Professor at Ludwig Maximilian University of Munich (08–12/2024) and Hacettepe University, Turkey (01–03/2025). Research interests include the intersection of social sciences and science education, focusing on the sociocultural dimensions of teaching and learning

Bruno Ferreira dos Santos Full Professor in the Sciences and Technologies Department at Southwest Bahia State University, with a PhD in Human and Social Sciences (2010). Research interests include sociological approaches to science education and the commitment to equity and social justice in schooling.

Cristiano B. Moura is an assistant professor at the Faculty of Education at Simon Fraser University, Canada. He has a PhD in Science, Technology and Education from CEFET-RJ, Brazil. His main research interest is fostering justice-centred science education through historical approaches, drawing mainly from Cultural History of Science, Science Studies and Decolonial theories. He was recently appointed editor-in-chief for the Science & Education journal.

Chapter 18
Painting Vital Entanglements: Science Education as Science Poethics

Thiago Ranniery

> Science accurately describes from outside; poetry accurately from inside. Science explicates; poetry implicates. Both celebrate what they describe. (Le Guin, 2017: M16)

> In Manet, there is a decisive force, a kind of strategic instinct of pictorial action. In his best paintings, he achieves poetry, that is, the summit of art, through what I will allow myself to call... the resonance of the execution.
>
> – But how one speak of painting? (Válery, 2012: 43)

18.1 Field Diary Fragment I

*Brazil. May 2022. It was Sunday, and I felt somewhat disoriented. The forced distance due to the Covid-19 crisis had turned my home into a self-imposed exile. It felt I was living the experience of a lonely city, as if the city of Rio de Janeiro were a necropolitical companion to Olivia Lang's (*2016*) cartography of New York. As I delved into Martin Cerda's (*2017*), an Argentine essayist who distinguishes between loneliness and isolation, I began to appreciate the complexity of the former and its inherent dangers. It reminded me of my teenage years when I used to listen to a song called Porto Solidão [Loneliness Pier]. I was born eight years after the song's released at an extinct former national festival of Brazilian popular music organized by a major television network in the country in 1980. Personally, I discovered this music thanks to my mother, who used to play her vinyl record repeatedly. The chorus goes something like this: "the loneliness that lingers and throws me against the pier". The lyrics suggest that loneliness can be a special place, a destination to*

T. Ranniery (✉)
Universidade Federal do Rio de Janeiro, Rio de Janeiro, Brazil

C. B. Moura (ed.), *A Sociopolitical Turn in Science Education*, Contemporary Trends and Issues in Science Education 62,
https://doi.org/10.1007/978-3-031-78586-3_18

which one is thrown in the midst of a turbulent life, a port to which one is taken, to which one arrives and to which one goes.

Thrown out of my house, I began a research project on the pedagogical life of virology laboratories. Finally, I received permission to follow an interdisciplinary group of scientists in the face of Covid-19. Although I claim to be a science education researcher, I was originally trained as a marine biologist. My specialty was the ecology of estuaries, those heavily anthropized areas of the Brazilian coast. At some point in those early scientific years, my advisor coordinated a research project on microbiological infections in estuarine fishes, with the goal of investigating the extent of anthropogenic impacts. There, where she saw the human precipitation as the final explanatory cause, I was seduced by the submersion of the "human in global networks of consumption, waste and pollution, capturing the strange agencies of the ordinary things of our lives" (Alaimo, 2016*: 117).*

The invitation to join the group was made on the basis of this unique training of mine. The expectation was that I would be able to support the relationship between researchers acting as consultants to education departments and public-school teachers in the intense political struggle to return to face-to-face teaching. I didn't want to explore what such an experience would simply mean for virologists, molecular biologists, and physicians within an epistemic culture that relegates viruses and molecules to objects of study from other fields of knowledge. I was interested in signs and affections, ontological processes, axiological creationism, speculative pragmatics that accumulate around scientific practices, and in short, how they function as pedagogical forms that deflect an ethics and a poetics. However, when I was challenged to carry out this translation exercise between parcels of worlds in friction, a question was rekindled: how could I turn around in this pressurized encounter between a cartography of scientific virological practices and the politics of bad life that would well name the processes that answer for the existence of monsters that pressurize Brazilian public schools[1]*? I assumed that the only place to live was on the border between these worlds.*

*Accepting the invitation, I crossed the city to a laboratory located on the campus of a large public university built on an estuarine island. As I walked through the corridors of a labyrinthine building built during the Brazilian military dictatorship, the urgency of responding to the disaster imposed on the sciences made me feel that all the descriptions of laboratories I knew were too monotonous. The complexity of laboratory life that Latour and Woolgar (*1986*) brought to light is even more surprising today, and the perspective offered to the foreign navigator does not allow*

[1]In Brazil, several state and municipal governments formed working and advisory groups with health scientists and researchers. There have been many conflicts between these groups and teachers. Authorizations for the return of face-to-face activities in public schools were not always accompanied by policies on school infrastructure or working conditions for teachers. The catastrophic management of the federal government also classified educational activities as an essential service. With the demands of families to open schools, this movement created accelerated social and political pressure on the lives of teachers, administrators and students, reshaping education in economic terms. On this point, see Macedo (2001).

him to be located on a perpendicular axis. To say that this ethnographic experience was intense is an understatement. There is no way to approach science education in a post-pandemic world without shaking our cosmological coordinates in a variety of ways, whether intellectual, affective, or existential. Although it was by no means a comfortable experience, I began to wonder if Virginia Woolf (1990) was right, if there is nothing more to experience than what is in front of one's eyes, if loneliness does not in fact lead one to consider some larger questions about what it means to be alive.

18.2 The Heart of Things

This text is a thought experiment. Around an ethnographic situation experienced in a virology laboratory, I rearrange fragments of a field diary written in the wake of the Covid-19 crisis. My construction is closer "a puzzle after losing the box top" to take advantage of Labatut's (2020: 69) description of the young physicist Heisenberg. In "When we cease to understand the world", Labatut (2020: 69) reproduces Heisenberg's conviction, after a summer conversation with Niels Bohr, according to which to talk about atoms "language could serve as nothing more than a kind of poetry". Just as for Heisenberg the perception of the radical otherness of subatomic particles requires a new language that creates metaphors and connections, the problem I pursue here is the place of poetry in the elaboration of educational responses in a post-pandemic world. In a world exposed to the radical otherness of viral worlds, we are challenged to rethink the relationship between poetry and science and to stage a performative lexicon that can "return to be enchanted by the cosmos and spread this enchantment to society", to follow Novello's (2022: 13) suggestion. In this text, I dream of entering the "dark nucleus at the heart of things", as Labatut (2020: 64) describes Heisenberg's perception.

This essay is also indebted to Ingold (2007), not only for an anthropological archaeology of lines, but also for the permission he offers us to overlap drawing, writing, and sound, that is, to refuse the boundaries that isolate visual and aural perception, and to dare to tell a story that overlaps very different materials. In this sense, my text creates a kind of contact zone between scientific practices of molecular biology and indigenous artistic practices in order to trace a play of resonances. My hope is to describe ways of education that will allow for a more speculative, inventive, and creative re-engagement with the sciences. For somewhat obscure reasons, the text became a disjointed web, following a narrative voice that embodies the process of disorientation on many levels. Not only because I was bombarded with diverse, contradictory, and confusing statements and recommendations from numerous sources, but also because of the uncertainty of what the act of writing would become as it crossed such seemingly disparate worlds. However, if disorientation could be described here, in the words of Ahmed (2006: 163), as the "becoming oblique" of the world, then I am grateful for this disruption, for it has helped me to make the familiar strange and perhaps "even to find joy and excitement in the horror of disorientation" (Ahmed, 2006: 4).

As I realized, intimate writing and experimenting with records that are not necessarily those of the critical essay or the dense description imposed themselves as a daily routine of resistance to the stupor of the events of a time that must not be forgotten. Perhaps this is why I could not separate the writing of the field diary from the analysis of the research material. My attempt to immerse myself in what was existentially affected me and to think from there was indistinguishable from my theoretical experience because it is much less like a box of resources than the skin I inhabit. It would be more accurate to say that the text began to configure a fabric of restlessness rather than an evaluative critique. Its argument was transformed into the evocation of an atmosphere. In this climate, the effect of the resonance between two images is an opening towards a *science poethics*, an explicit exposition of the performative dimension of scientific practices, and a reclaiming of the enchantment of the cosmos, which, in affinity with Amerindian cosmological agencements, points to an extended mesh of relationships. Science education is the name we could give to the work of figuring it out, that is, the exercise of diving into this mesh, touching it, listening to it, writing it. It is a painting of vital entanglements.

18.3 Field Diary Fragment II

Two images side by side. The first (Fig. 18.1) is the result of a 3D rendering created by a computer algorithm from a set of X-ray, nuclear magnetic resonance, and cryo-electron microscopy data. According to the website of the Cellular Landscape Project, led by scientific visualization researchers and visual artists Evan Ingersoll and Gael McGill, the image is a model of a eukaryotic cell, the rendering of which was programmed from the paintings of cell biologist and plastic artist David Goodsell. It is nothing more than "an attempt to visualize the great complexity and beauty of the molecular choreography of the cell".[2] The second (Fig. 18.2) is a painting by indigenous activist and artist Jaider Esbell entitled "A Conversa das Entidades Intergaláticas para decidir o futuro da humanidade [The conversation of intergalactic Entities to decide the future of humanity]". These two were introduced to me by Olga, a graduate student in immunology, with the astute question: don't they have resonance?

When we met, Olga was working on her dissertation on immune responses of the cell membrane. She invited me to participate in the activities of a chair called "Images of Real and Imaginary Worlds". The goal of the chair was to bring together research on image and sound in the biomedical sciences. The inaugural lecture of the 2022 academic year was given by Kurt Wüthrich, who received the 2002 Nobel

[2]The information is available on the project's website: https://gaelmcgill.artstation.com/projects/Pm0JL1

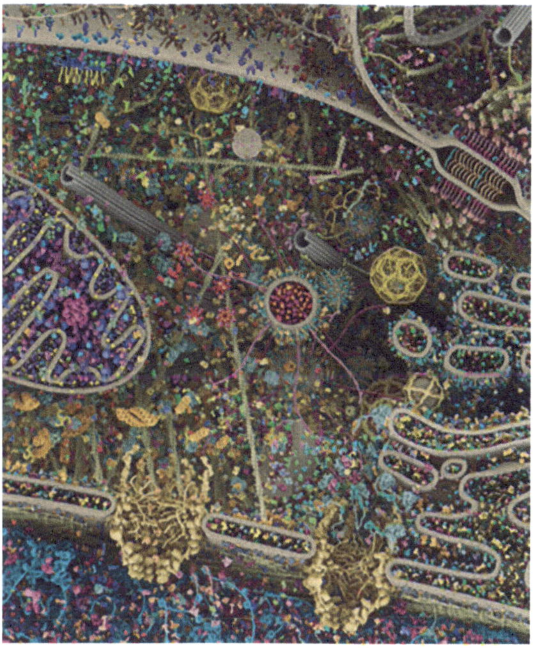

Fig. 18.1 Cellular landscape cross-section through a eukaryotic cell. (By Evan Ingersoll & Gael McGill)

Prize in Chemistry for his contributions to the imaging of macromolecules using nuclear magnetic resonance (NMR). In a review of this literature, Billeter et al. (2008)*, for example, describe how the field of NMR has produced significant findings that "extend the characterization of proteins from atomic-resolution three-dimensional structural models obtained by NMR or X-ray diffraction to dynamic and binding properties". The image is not simply a more realistic and detailed representation of a cell, as described in the captions of the Portuguese papers.*

"It is more like a painting, a computer score, than a photographic record", Olga told me. I couldn't hide my surprise at the use of the words "painting", "score", and "choreography", which made me think about the ontological status of the aesthetic experience of computational molecular biology and its search for a visual language. In the words of Goodsell (2021*: 402), "the power of SciArt*[3] *has perhaps its strongest manifestation [...] where the things we study are particularly amenable to visual representation". By creating watercolors for scientific dissemination from the Protein Data Bank (PDB), a global resource that archives 3D structural data of biological molecules, Goodsell* (2016a*: 22) describes the paintings of vibrant cellular landscapes as "simulating a view of the cellular mesoscale, taking a slice of the cell and depicting its molecular composition". In a world that nurtures scientific practices based on alliances with the arts, a visual level is opened to an invisible life that is as intensely small as it is expansively large. The power of the image is spread by a mesoscalar network (and even by the cosmos), and the poetic does not separate*

[3] This is abbreviated to mean Science and Art.

Fig. 18.2 The Intergalactic Entities' Conversation to Decide the Universal Future of Humanity. (By Jadier Esbell)

into a pole of maximum and another of minimum effectiveness. Don't they have a resonance?—I asked myself.

As I pondered the answer, I saw friends sharing the news of the death of Jaider Esbell, whom I could describe as an articulator of worlds. A member of the Macuxi people, Esbell was born in the indigenous land of Raposa Serra do Sol, on Mount Roraima in the far north of Brazil. This people are descended from the Pemon people of Caribbean linguistic affiliation (Santilli, 2001*). Together with the Taurepang and Wapichana peoples, the Macuxi and Pemon are the descendants of Makunaimã, the indigenous deity of immemorial time, who is part of the sacred of some indigenous peoples who live in their care (Goldemberg,* 2019*). Associated with an emblematic generation of young indigenous artists, Esbell went to the stars when he went to the Guarani village of the Silveira River in Bertioga, a city in the state of São Paulo, on November 2, 2022. More than the cause of death—the discomfort with civilization, we could say with Freud (*1962*)—I am interested in its meaning: the cosmogonic principle of communicating spheres between nature, culture and spirits, carried out by Esbell. In the myth, Makunaimã leaves the earth and passes through other celestial landscapes, tired and disappointed by man.*

*I recognize my limitations in consolidating my ethnographic account as carried out by Narby (*1999*) and his exploration of the correspondences between the hallucinatory and mythological spheres of shamanic peoples in the lowlands of South America and the DNA images emerging from scientific practices. However, one of Narby's observations provides a clue to the problem of resonance. Commenting on the figure of the cosmic serpent in indigenous graphics, Narby (*1999*: 74) states that "the distance between molecular biology and shamanism/mythology was an optical illusion produced by the rational gauze that separates things". During the 34th Biennial of São Paulo, Brazil, Esbell installed two large 24-meter floating snakes in the lagoon of the city's Ibirapuera Park. Entities—the title of the installation—move* *between the visible and the invisible, between the material and the immaterial. The scientific paintings of the strange, iridescent cellular landscape also resonate.*

18.4 In the Darkness

Two images side by side. In his *Cosmic Manifesto*, Novello (2022: 31) warns of a necessary dialogue between the different ways of knowing nature—the ways of scientists and the companions of other knowledge that have been disqualified throughout the history of the sciences—in order to "carry out together this great task of renewing the scientific order of the cosmos". The gravity of the present situation is such that the sciences can no longer consider themselves to be the only access to truth. This does not mean that the sciences have lost access to truths of fundamental importance, but only that scientific practices are becoming open to connection with other practices that are themselves grounded in other modes of existence. As Kopenawa (2013) points out, white environmentalists (*napë*) must yield and begin to listen to the teachings of the Yanomami shamans, who are able to perceive the invisible world of the spirits (*xapiripë*), the guardians of the land-forest. Perhaps this is the point of Stengers' (2005) cosmopolitical proposal: to open up scientific practices to a complicated conversation with other practices, which they have always judged to be tied to speculation and imagination.

However, I cannot proceed without emphasizing that I do not intend to affirm an abstract identification between scientific practices and contemporary indigenous art, nor will I project some kind of indigenous ethnoscience embedded in the paintings of Brazilian indigenous artists. I do not want to exaggerate any similarities. In fact, I use these different negatives because they allow me to focus both on the correspondence between indigenous and scientific notions and on the expectation that indigenous artists will achieve what some people are content to imagine. My desire is fraught with the risk of violent reduction inherent in any comparison. I seek to rethink science education by contrasting it with contemporary indigenous art in an exercise that Viveiros-de-Castro (2004) calls controlled equivocation rather than equivalence. That is, it is possible to explore how science education can be read from the contact between practices of different orders and natures and the shared experience of impossibilities. This entails a prior difficulty about the status of what we call aesthetics when it does not correspond to the place we designate for art in the Western white world. As Cesarino (2013) says, indigenous arts are multimodal arts. They do not constitute an autonomous discursive modality and are not dissociated from other practices such as shamanism, ritual, cosmological explanation, and even entertainment. *In indigenous society, everyone is an artist* is the title of an interview by Esbell (2019) on the nature of his work.

So how do we commit ourselves to an art of encounter? In *The Story of the Lynx*, Levi-Strauss (1995) suggests that myth and science, separated for so long, came together in the second half of the twentieth century. The sciences have been transformed: quantum theory and relativity, the consideration of the infinitely small and the infinitely large, move away from ideas of immutable and unchanging laws and begin to speak of incompleteness, undecidability, strange worlds that coexist. The mythical imagination was no longer alien to the sciences, which began to rethink their certainties and search for new narratives. I would like to add the imaginative agenda of molecular biology to this list, using the available means to emphasize that "forms of animism are immanent in mechanistic logics", as Myers (2015: 19) suggests in her ethnography of protein christolography laboratories. While the term animism owes a debt to "a (post)colonial geography in which some humans were represented as unable to [...] control language and experience through self-reflexive reason" (Povinelli, 2016: 49), the important conclusion to draw from this encounter is that a constitutive tension between science, imagination, myth, and magic is inherent in scientific practices and performs a mode of knowing that embraces and exposes a strange unknown in a multivocal landscape.

Indeed, I began my research project with virology laboratories, following Blackman (2019) and his focus on practices discarded by the history of science, hoping to illustrate the value of working in science education with what is considered strange, bizarre, anomalous, and even with "threshold experiences" that are described as having an "alien phenomenology". Since then, I have been fascinated by pushing the boundaries of queer theory, in which I was trained in the tropics,[4] into the realms of the sciences, undertaking different cartographies of "a desire for another way of being in the world and in time" (Munõz, 2009: 142). Therefore, animism would not describe "another" ontology, which would make it completely external to scientific experience, nor is it a single new truth or ontology that would overthrow any other. More importantly, it would experiment with it and make it work in scientific practice. But I barely got my hands dirty, and I realized that I would never read any of Goodsell's works on molecular life as I had before. After reading Esbell (2020a), who writes about what he calls contemporary indigenous art, a perspective imposed itself on my reading. A diplomatic provocation by Esbell is worth noting[5]:

> […] the question of cultural appropriation has been much discussed, but what about the essential expropriation that was done to create worlds, and in this case to create the colonial world? The vast design of the colonial world, it is made of shells collected on our beaches to decorate this wonderful panel. [...]. So, we are claiming, at least, the statement that most of what they call Art in the West is the result of the expropriation of other cultures, of other peoples, and that the very construction of the building that is called culture is a break with

---

[4] Far from translating a global queer into local specificities, Pereira (2019) and Pelúcio (2014) propose to articulate a queer theory in the tropics. I am indebted to this decolonial turn of queer in Brazil. However, I feel that there is an attraction to how "non-Western collectivities" think about questions of gender, body, sexuality, and life that do not always turn to what is considered "Western", such as the sciences. The background to my work has been an interest in how overcoming modern dichotomies (subject/object, organism/environment, culture/nature) implies a reconceptualization of scientific practices.

[5] Transcription of Esbell's speech at the launch of the exhibition *Moquiém_Surarí*, 2022 Available: https://www.youtube.com/watch?v=zT5q2zID2Ac&t=2537s&ab_channel=MAM-MuseudeArteModernadeS%C3%A3oPaulo

> the balance of the live of peoples, [...] peoples who have lived their poetic journeys, their subjectivities, creating worlds for thousands of years [...]. So, who knows, we are also animating this chorus of people who are able to blow up the frame.

Therefore, I do not re-move this path by the many theories critical of modern Western scientific epistemologies or by the findings of the suicidal character of the capitalist sociotechnical matrix in a post-pandemic world. I am not convinced that the strength of biotechnologically produced images will be nullified by the new anthropological contract that Stevens' (2013) ethnographic account offers, for example. By going through the technological transformation of science laboratories, Stevens (2013) detects the progressive centrality that images have gained to present complex biological models based on an increase in the computer tools, but he draws attention more to a radical change in political economy and in the epistemic practices that biostatistics and cybernetics have produced. In a certain sense, I do not write about the impacts of this transformation, but rather under their effects.

In a seminal essay, Haraway's cyborg (1985) had already risked figuring out how technology should be seen neither as salvation nor as damnation. This figure offered "a myth system waiting to become a political language to ground one way of looking at science and technology and challenging the informatics of domination – in order to act potently" (Haraway, 1985: 98). Mobilizing Chinese cosmological thinking, a recent version of this theme is inflected by Hui (2020) and his requirement to create capacities to reposition technology in broader realities. Criticizing the current association of technology as synonymous with improving the human or worse from its defeat, Hui (2020: 156) argues that "the development of new senses capable of sustaining other forms of knowledge" is still a work to be done. It is in the encounter with art that, the author suggests, scientific thought—focused on the improvement of techniques and technologies—and philosophical thought—focused on the creation of meaning—can explore a shared sensitive agenda.

In this way, I refuse to accept that the reference to Capital and its counterparts, such as development and modernization, or the most popular reference to money in its licit or illicit forms, are inescapable resources for talking about science, technologies and scientific education in a post-pandemic world. Far from ignoring the fact that the sciences have been and continue to be globally embedded in the colonial capitalist machinery, what I have intuited is that this deity called Capital is a difficult image to dissociate from an operation of legitimizing the deadly devastations attributed to it. It is a poisonous image that paralyzes sensitivity and thought. If the system centered on the *Anthropos* and its state of ontological exception has brought us to this point, if the avatars of Man have their genesis, their deadly power, their justification, their propaganda and their narcotic poison in this system, why on earth would it be in our interest and our duty to remain prisoners of this matrix? Why and by whom should we be obliged to serve as its propagators in the registers of thought, sensibility and meaning? In my opinion, it would be necessary to conjure this blockade by working to pay attention to the lines that unfold before us and the worlds that flourish painted by the left hand of darkness—Le Guin's work title

(2019)—that give visual concreteness to a cosmopolitan making of scientific practices that touches and sees us before we can touch and see it.

For Le Guin (2016), if science and poetry are different—one sees from the outside, the other from the inside—both celebrate what they describe. "We need the languages of both science and poetry to save us from merely stockpiling endless 'information' that fails to inform our ignorance or our irresponsibility" (Le Guin, 2016: M16). For his part, the night gardener Labatut (2020: 34) also said: "even scientists no longer comprehend the world". In his meditation on quantum physics, the character states: "It's as if the theory had fallen to earth from another planet, and we simply scamper around it like apes, toying and playing with it, but with no true understanding" (Labatut, 2020: 134). To perform a language that makes us intimate with darkness is to develop reminders and invitations so that the dances of the universe with chance, unpredictability and metamorphosis are not solitary. Accompanied by the epigraph of Válery (2012), I suggest that the pictorial action of the scientific practices of molecular biology, by changing the medium of expression to painting and changing the scale, by figuring what is so small that it is large, plays with the lines of strangeness and unfamiliarity of the obscure landscapes of the invisible worlds. But this strangeness is not external, it is an unsettling, alien fickleness that shines within ourselves—a subtle phenomenon of resonance that is so important in shaping the method of imaging in molecular biology.

Science education could be this *science poethics*. In this dis/junction, I propose an amalgamation of poetic action, this doing with language, and an ethics of friendship with the unknown that exfoliates the social skin of the *Anthropos*. In another moment, Le Guin (2017: 87) suggests that to turn the key to human domination, what matters is an "acceptance of impermanence and imperfection, a patience with uncertainty and the makeshift, a friendship with water, darkness, and the earth". To transform the world through language, it is necessary to transform language itself, something close to what Aristóteles (2015) advised poets to acquire: a language elevated in relation to ordinary language. The difference is that it is not a matter of elevating language to poetry, of distancing it from life, but of making poetry and the magical life of the sciences coincide. As is well known, the association between poetry and magic is a common one. Perhaps one of the most sophisticated conceptual formulations comes from Cortázar (2008). For the author, the poet is defined as a "maker of ontological exchanges" by "fulfilling the magical form of the principle of identity and being something else" (95). It is contrary to the logical principle of identity because it is an "exploration of that which is not man" (96), of "ontological qualities that are not those of man", "[because] 'the other' is in fact that which can grant you degrees of being beyond the specific human condition" (98).

This process transforms the status of scientific education because of the different plans or natures involved in the changes. To reimagine scientific education is to describe it as the work of reclaiming and exchanging the animated worlds of science, rather than accepting once and for all the tacit engulfment of a single world. As Krenak (2020: 29) writes, "when we depersonalize the river, the mountain, when we strip them of their meaning — an attribute we hold to be the preserve of the human being — we relegate these places to the level of mere resources for industry and extractivism". In other words, when we discourage the world, we leave it at the mercy of a deadly power. Reclaiming animism emerges as a demand for engagement, a chance to rebuild life and meaning in the post-pandemic era.

18.5 The Question, Once Again

Do pictures not have resonance?—the question returned. Although Gómez-Barris (2018) had already offered me different clues about queer arts and political ecologies in colonial extraction zones, it took the reorientation of my encounter with these images to make me realize how much they alter and shatter the existence of discrete invisible entities, embodying "to perceive those things that are not usually available to the naked eye" (Gomez-Barris, 2018: xiv). I am referring to the ways in which Amerindian collectives have experienced dynamics of composition in which graphics are mobilized in artistic performance to expand access to the diversity and complexity of Amerindian thought (Lagrou & Van Veltheim, 2018). Regarding indigenous graphics, Lagrou (2011: 762) points out that they "produce a body in a constructive relationship with the flows that pass through it"; their compositional techniques are "mediating techniques to concretize the passage between different points of view" (Lagrou, 2013: 75–75). Therefore, the "lines draw attention to what connects and not to what separates bodies and different beings, it is an art of between-two" (Severi & Lagrou, 2013: 14), an ontological choreography.

This was the moment when the oblique becoming opened the antrum excavated by the image. By merging scientific dissemination with protein painting, Goodsell (2016b: 12) insists that "computer graphics images are our primary way of exploring and understanding the structure of biological molecules, and the pictures we create are the evidence that we use to document our discoveries". He asserts that "the molecules in our cells perform their jobs in a strange, unfamiliar world [...]. The principles that guide objects in our everyday world – gravity, friction, temperature – are different at the molecular scale and often have surprisingly different effects" (Goodsell, 2009: 5–6). The image that went viral on social networks is therefore an experiment in the double sense of the word, both scientific and artistic, combining the scientific fact with the speculative. It is well known the approximation that Latour (2010) makes between the idea of fetish in African religions and the idea of fact in modern sciences. A fetish is an object that people worship and at the same time acts upon them. In a laboratory, molecules and cells are not simply objects, but unpredictable agents. If, as Latour (2010) suggested, scientific facts are not just "data", the images that echo each other go beyond showing how scientific facts depend on networked interactions and articulations. They suggest a kind of aesthetic view of these tangled networks within and beyond us.

Lagrou (2011), in a study of Amerindian aesthetics, emphasizes that its fluidity of form "allows for a bricolage work in which otherness is altered, but the exogenous origin of substances, images, and flows is never silenced" (762). In this mode of composition, "agency, action, and contemplation become inseparable" (739), and "aesthetic efficacy" results in "the capacity of an image or a way of acting on the world, thus creating and transforming it" (749). There is an agency of signs: "they are not symbols that speak of and represent a reality external to themselves" (762). Such procedures of Amerindian aesthetics can lead us to consider how the image of the cellular landscape is fabricated through an interweaving of different image data that collide, making it the documentary expression of a composite formation. It is an image that does not represent the general cell or the actual existing cell, but functions as such. The idea of an image that represents only the eukaryotic cell is ruined by the work of rendering. Analogously, the similarity is eroded by the work of condensing multiple data that the rendering performs. The strange cellular landscape lies in the very set of relationships and connections. Not only does the computerized scientific practice render an image of the choreographic metamorphosis that makes the cell an entity, but the metamorphosis also gains a visual body in the image.

There is an explicit invitation: to see the work of an interstitial difference, a kind of transversal ray between different levels, embodying an enchanting crossing between the absolutely invisible and the visible, between the inside and the outside. It is an image that evokes its "biopoetic" (Coccia, 2018: 75) molecular bonds, drawing a tangle of lines in each landscape. Departing from the Western canonical model of representation, the image of the cellular landscape produced by computer algorithms does not simply make present what cannot be seen, nor is it characterized as a mere credible vicarious presence of the represented object. There is an inversion: when it comes to illuminating a zone of impossibility for vision, an agent atmosphere is perceived and captured, with the image remaking the possible relationships with it. In the vocabulary of Dolezel's (2000: xi) literary theory of possible worlds, these scientific practices engage in "transworld travel". In another paper, Esbell (2019) argued that indigenous art and shamanism produced transworldly journeys.

The notion of representation is perhaps even inadequate, because it is not simply a matter of representing another level (the invisible, the molecular), but of presenting it in the same performative act in which it reciprocally presents itself or presents this level to the other who sees it. In perspective, by allowing exogenous sensible qualities to merge in a process of visualization, the image materializes a body with dynamic binding properties, in which signs refer to a psychic potentiality distributed horizontally in the cosmos. It's a painting of a vital entanglement. Regarding the entanglements, Barad (2012: 41) refers to them as "enfoldings of spacetimematterings". Painting vital entanglements is a way of saying that scientific practices make transworldly travels, that is, without separating space, time, and matter, and far from the fantasies produced by culture that make us return to the figure of the anthropos. Perhaps it would be more appropriate to affirm that what is at stake is a *transpresentation* of a living matter, a mutually present rendering of its various levels of composition, and at the same time making present the crossings of these levels, the folds that Deleuze (1986) commented on when talking about how Francis Bacon's painting freed itself from representation.

What I call *transpresentation* comes from Margulis and Sagan's (1995: 86) answer to the question of what life is: "Life is the representation, the 'presencing' of past chemistries, a past environment of the early Earth that, because of life, remains on the modern Earth". There are two dimensions I want to emphasize with this name: difference and time. If these experiments are spaces of epistemic uncertainty, blurring distinctions between real and imaginary, visible and invisible (Blackman, 2019), then the contact between subject and object, inside and outside, is reconfigured. Alloa (2021), for example, notes how images of the body in the history of the West dismantle the epidemiological anxieties that made otherness an invader. Similarly, if there is an obscure, unknown universe beyond and below, before and after, outside and within us, the image makes the boundary between inside and outside—the barely visible cell membrane—not only uncertain. "The opening made by the intruder is not the opening through which he enters, but the opening through which I cannot return to myself" (Alloa, 2021: 16). When Ahmed (2006: 163) insists that it is a matter of irresolvable tension to know whether "the strangeness is in the object or in the body that is near the object", the image of the cellular landscape makes queerness a crossing without fusion, an effect of the different levels that overlap, that is, a transworld travel in which the oblique becoming leads to an indeterminacy of matter and being and to an impossibility of controlling and directing nature.

To use a term popularized by Latour (1998), the black box of the sciences is broken by the staging itself, since the image is also an agent of a transtemporal dimension, a folding through different temporalities. The image of the cell—that "watery, membrane bounded encapsulation of space-time" as described by Margulis and Sagan (1995: 86)—encapsulates different time scales: cosmic, historical, organic and quantum. Therefore, the cellular landscape appears as a materialization of space-time relations in knots, passages of strange shapes that crystallize. In other words, the image makes the mythical imagination present, not in the simple sense of a return to mythical time itself, but in the sense of an updating of the virtuality of mythical space-temporality that pervades the queer materialization of the world and the cosmos, which makes it possible to fulfill plans without the present (historical) time ceasing to exist. When they mobilize the remarkable pictorial affinity, the poetics of the sciences become a way of speaking and transferring vision from the molecular level to the pictorial level (and vice versa), while still referring to this crossing of times, spaces, and materials. It is an image that implies not only a transit between different dimensions, but also the overlapping of them, producing a transformation in this world through its performance.

Although Olga, the immunology student, and my interlocutors are reticent about the cosmic possibilities mediated by technological equipment, in the resonance between images we can affirm that a truly animistic concept of the cell breaks out within scientific practices. Indeed, Cusicanqui (2021) suggests that in colonial contexts that privilege the textual, images would allow for the exploration of senses not censored by official language and what is only described when viewed. What is at stake is what Taddei (2017) has described as a metaphysical phagocytosis—rather than an ontological conflict (Blaser, 2013)—between world-making practices, which emerges through a public performance of the aesthetic consideration of images in scientific practices. It is a chance to see the naturalistic ontology of the sciences as nothing more than a subset of an animistic ontology that transforms science into a strange poetic cartography of the unknown. This procedure serves to say that the strangeness lies in the formless work of subjectivation of the universe.

In her essay for a collection on the art of living on a damaged planet, Le Guin (2016: M16) argues that "we need to relearn our being in [the world] (…) our kinship as animals with animals". She criticizes the modern narratives of animals as machines (Descartes), of plants as non-sentient beings (Aristotle), and proposes to subjectify the universe: it is possible to hear trees and rocks. Le Guin (2016: M16) warns that subjectifying does not mean colonizing, but "involves a great reach out of the mind". When it comes to subjectifying rather than objectifying, "knowing is personifying" (Viveiros-de-Castro, 2002: 358). Knowing means transforming what one wants to know into a fabulative narrative, a mythical story. In order to act on something through language, it must be inscribed in the language of subject-action. As Stengers (2011) writes, when we speak of practices disqualified by modern sciences, we should not simply commit an act of tolerance. To go beyond the curse of tolerance is to take seriously indigenous claims that there is an intergalactic conversation—that is, multiple worlds, planes, and natures—between entities—cells, technologies, images, and us—to decide the future of humanity. I argue that the vocation of science education is to continue this exercise in the subjectivation of worlds, to suspend it as one of the tributary branches of a philosophy of understanding, and to bring it back into the sphere of the sensible.

18.6 Field Diary Fragment III

*The leaves are coming to an end. Maybe I should get a new notebook. Maybe I tend to exaggerate the use of such diverse materials and references. Maybe I use "maybe" too much. The fact is, I don't have anything like a conclusion. I only have loose threads about science education in the face of the situation we have been drawn into, which is the time of catastrophes generated by the Anthropocene in its various aspects. While studying nuclear magnetic resonance, I found my old copy of an essay by Nancy (*2014*) on a philosophy of listening. In it, Nancy (*2014*) considers resonance as a singular particularity of listening that can serve as a basis for all the senses. I know I am talking about images, but resonance is also the name of this method of biotechnological production of images of invisible worlds, a visual plane that transforms and transposes the movements of electromagnetic waves into images, transferring them from one plane to another. When waves become images, the image is an echo. Looking at the images rendered by computer programs that propose to trace dynamic links and unstable connections, the obsession to show them in their multiple vicissitudes succeeds in realizing a verb-visual theater of the sciences.*

*Thus, the painting of vital entanglements does not emulate an unknown ritual, it performs it. Esbell (*2020b*) says that it "goes beyond the point of crossing borders and boundaries, which are fragile in many respects, such as legitimizing a* *self-identifying claim or miscegenation". The way is to mark the unpayable debt, in the words of Ferreira da Silva* (2022), *and accentuate the indigenous gift. At this point, I am overcome by a restlessness: is there anything new in postulating a re-enchantment of the world? In this respect, it is worth recalling Adorno's* (1991) *skepticism about Rilke's poetry: does not the attempt to give things a voice sing the victory of objectification? To return to the images, would this re-engagement take place through a consciousness-raising? Is art a vehicle for the consciousness of scientific education? Or is it rather a means of idiomatic creation? Can science education offer itself as a revelation of life under the body of the world, that is, does it become poetry obliquely? In the midst of the elaboration of modern legacies and the critical formalization of the present, can science education twist the language and the conceptual apparatus that underpin its existence and open up to the difficult coexistence of thoughts and perspectives?*

The questions are many. Therefore, I write with two cautionary principles. The first concerns the celebratory tone of a more or less generic indigenous thought in the service of suggesting other (postcolonial) points of departure for an alternative (decolonial) genealogy of science education. To refer practices to a supposed culture—even if they are conceived as lines of escape from official ideology, seeking to emphasize cultural difference or the authenticity of identity—is to steal their political purse. When Esbell insists on "contemporary indigenous art", I reject the grammar that encapsulates it in an archaic, primitive, ideal, essential symbology that needs to be recovered. The second principle stands out in the objectification of the "indigenous". The theme appears in the academic pages, already burdened with centuries of guilt and bad conscience, dealing with a just reparative historicism. This historical vision of the indigenous collectives reproduces above all its own milestones: the scene of the colonial encounter. If, on the other hand, we think with interest about the creative styles of these collectives, we begin by displacing the meaning of such a history, subscribing it to myth. And myths are actualized in rituals, including scientific practices.

By expressing caution and doubt, we could cultivate a sensitivity to rethink the synaesthetic link between scientific practices and artistic practices of indigenous collectives. In this resonant execution, painting vital entanglements perform an "integrally relational ontology" (*Viveiros-de-Castro,* 2002*: 385*)*. There is no universal referent or perspective of totality, a finished world, as the queer events of the world are filled with materializing space-time dynamics that are "immediately relational multiplicities"* (*Viveiros-de-Castro,* 2015*: 67*)*, apprehended differently by each perspective. The composition of the images does not define insurmountable contours: it goes beyond us, beyond the line, beyond the end (of the world):*

> *When we call something indigenous, I suspect that this something still falls into a common field of reference [...]. It should no longer be associated with pejoratives, but with the more singular aspects of desired beyond* (*Esbell,* 2020a*: 24*)*.*

Just as when I write alone, but sometimes comment aloud on an impression, the resonance testifies: we are not alone in the world. Both images are thus less ambassadors of an invisible world than vital paintings of an intrusive, inexplicable, transcendent otherness. In fact, this beyond can only be painted.

References

Adorno, T. (1991). *Notes to literature – vol I*. Columbia University Press.

Ahmed, S. (2006). *Queer phenomenology*. Duke University Press.

Alaimo, S. (2016). *Exposed*. Universtiy of Minnesota Press.

Alloa, E. (2021). *O corpo da imagem, A imagem do corpo* [The body of the image, the image of the body]. Zouk.

Aristóteles. (2015). *Poética* [Poetics]. Editora 34.

Barad, K. (2012). Nature's queer performativity. *Kvinder, Køn & Forskning, 1–2*, 25–53.

Billeter, M., Wagner, G., & Wüthrich, K. (2008). Solution NMR structure determination of proteins revisited. *Journal of Biomolecular NMR, 42*(3), 155–158.

Blackman, L. (2019). *Haunted data*. Bloosmary Academic.

Blaser, M. (2013). Ontological conflicts and the stories of peoples in spite of Europe: Toward a conversation on political ontology. *Current Anthropology, 54*(5), 547–568.

Cerda, M. (2017). *Palabras sobre palabras* [Words on words]. Ediciones de la Biblioteca Nacional de Chile.

Cesarino, P. (2013). Cartografias do cosmos: Conhecimento, iconografia e artes verbais entre os Marubo [Cartographies of the cosmos: Knowledge, iconography and verbal arts among the Marubo]. *Mana, 19*(3), 437–471.

Coccia, E. (2018). *A vida das plantas* [The life of plants]. Cultura e Barbárie.

Cortázar, J. (2008). *Valise de cronópio*. Perspectiva.

Cusicanqui, S. (2021). *Ch'ixinakax utxiwa*. n-1 edições.

Deleuze, G. (1986). *Francis Bacon*. Continum.

Doležel, L. (2000). *Heterocosmica*. The John Hopkins Press.

Esbell, J. (2019). Na sociedade indígena, todos são artistas [In indigenous society, everybody is an artist]. *Arte & Ensaios, 27*(41), 4–48.

Esbell, J. (2020a). Literatura indígena brasileira contemporânea: autoria, autonomia e ativismo – o que dizer e para quem? [Contemporary Brazilian indigenous literature: Authorship, autonomy and activism – What to say and to whom?]. In: J. Dorrico, F. Danner, & L. F. Danner (Eds.), *Literatura indígena brasileira contemporânea* [Contemporary Brazilian indigenous literature] (pp. 20–25). Editora Fi.

Esbell, J. (2020b). *A arte indígena contemporânea como armadilha para armadilhas* [Contemporary indigenous art as a trap for traps]. Avaible: http://www.jaideresbell.com.br/site/2020/07/09/a-arte-indigena-contemporanea-como-armadilha-para-armadilhas/

Ferreira da Silva, D. (2022). *Unpayable debt*. Stenberg/MIT Press.

Freud, S. (1962). *Civilization and its discontents*. W.W Norton & Company Inc.

Goldemberg, D. (2019). *Makunaimã*. Elefante.

Gómez-Barris, M. (2018). *The extractive zone*. Duke University Press.

Goodsell, D. S. (2009). *The machinery of life*. Springer.

Goodsell, D. S. (2016a). Cellular landscapes in watercolor. *Journal of Biocommunication, 40*(1), 22–26.

Goodsell, D. S. (2016b). *Atomic evidence*. Springer.

Goodsell, D. S. (2021). Art as a tool for science. *Nature Structural & Molecular Biology, 28*, 402–403.

Haraway, D. (1985). A manifesto for cyborgs: Science, technology, and socialist feminism in the 1980's. *Socialist Review, 80*, 65–108.

Hui, Y. (2020). *Tecnodiversidade* [Technodiversity]. Ubu.

Ingold, T. (2007). *Lines: A brief history*. Routledge.
Kopenawa, D. (2013). *The falling sky*. Harvard University Press.
Krenak, A. (2020). *Ideas to Postpone the End of the World*. Anansi Press.
Labatut, B. (2020). *When we cease to understand the world*. Pushkin Press.
Lagrou, E. (2011). Existiria uma arte das sociedades contra o Estado? [There would be arts of societies against the state?]. *Revista de Antropologia, 54*(2), 747–780.
Lagrou, E. (2013). Podem os grafismos ameríndios ser considerados quimeras abstratas? Uma reflexão sobre uma arte perspectivista [Can Amerindian graphics be considered abstract chimeras? A reflection on perspectivist art]. In C. Severi & E. Lagrou (Eds)., *Quimeras em diálogo* [Chimeras in conversation] (pp. 67–110). 7 Letras.
Lagrou, E., & Van Veltheim, L. (2018). As artes indígenas: olhares cruzados [Indigenous arts: A cross-section of perspectives]. *BIB, 87*, 133–156.
Lang, O. (2016). *The Lonely City*. Cannongate Books.
Latour, B. (1998). *Pandora's Hope: Essays on The Reality of Science Studies*. Havard University Press.
Latour, B. (2010). *On the modern cult of the Factish gods*. Duke University Press.
Latour, B., & Woolgar, S. (1986). *Laboratory life*. Princeton University Press.
Le Guin, U. (2016). Deep in admiration. In A. Tsing, H. Swanson, E. Gan, & N. Bubant (Eds.), *Arts of living on a damaged Planet* (pp. M15–M23). University of Minnesota Press.
Le Guin, U. (2017). *No time to spare*. Houghton Mifflin Harcourt.
Le Guin, U. (2019). *The left hand of darkness*. Ace Books.
Levi-Strauss, C. (1995). *The story of lynx*. The Chicago University Press.
Macedo, E. (2001). #Stayathome #Fiqueemcasa: Opportunities for new governances of public education in Brazil. *Prospects, 51*(1–3), 393–405.
Margulis, L., & Sagan, C. (1995). *What is life?* University California Press.
Munõz, J. E. (2009). *Cruising Utopia*. Duke University Press.
Myers, N. (2015). *Rendering life molecular*. Duke University Press.
Nancy, J. L. (2014). *À escuta* [Listening]. Edições Chão de Feira.
Narby, J. (1999). *The cosmic serpent*. Penguim.
Novello, M. (2022). *Manifesto cosmológico I e II* [The Cosmic Manifesto – I and II]. n-1 edições.
Pelúcio, L. (2014). Possible Appropriations and Necessary Provocations for a Teoria Cu. In. E. Lewis, R. Borba, B. Fabrício & D. Souza. (Eds). Queering Paradigms IV: South-North Dialogues on Queer Epistemologies, Embodiments and Activisms (pp. 31–52). Peter Lang.
Pereira, P. P. (2019). Reflecting on Decolonial Queer. *GLQ: A Journal of Lesbian and Gay Studies, 25*(3), 403–429.
Povinelli, E. (2016). *Geontologies*. Duke University Press.
Santilli, P. (2001). *Pemongon Patá*. Editora UNESP.
Severi, C., & Lagrou, E. (2013). Introdução [Introduction]. In: C. Severi & E. Lagrou (Eds)., *Quimeras em diálogo* [Chimeras in conversation] (pp. 11–24). 7 Letras.
Stengers, I. (2005). The cosmopolitical proposal. In B. Latour & P. Weibel (Eds.), *Making things public* (pp. 994–1003). MIT Press.
Stengers, I. (2011). *Cosmopolitics II*. University of Minnesota Press.
Stevens, H. (2013). *Life out of sequence*. The University of Chicago Press.
Taddei, R. (2017). *Meteorologistas e os profetas da chuva* [Meteorologists and the prophets of the rain]. Terceiro Nome.
Válery, P. (2012). *Degas dança desenho* [Degas dance drawing]. Autêntica.
Viveiros-de-Castro, E. (2002). *A inconstância da alma selvagem* [The inconstancy of the Indian soul]. Cosac Naify.
Viveiros-de-Castro, E. (2004). Perspectival anthropology and the method of controlled equivocation. *Tipití, 2*(1), 3–22.
Viveiros-de-Castro, E. (2015). *Metafísicas canibais* [Cannibal metaphysics]. n-1 edições.
Woolf, V. (1990). *The waves*. The Hogarth Press.

Thiago Ranniery is a Full Professor of Science Education at Federal University of Rio de Janeiro, Brazil, where he is a Vice-Director of the Faculty of Education. As Fellow of the Foundation for Research Support of the State of Rio de Janeiro and National Research Council of Brazil, he coordinates a research team on Curriculum, Ethics, and Difference. Currently, he serves as Treasurer of the Brazilian Association of Homoculture Studies and is a member of the Technology Committee of the International Association for the Advancement of Curriculum Studies. His research areas are curriculum, difference, science, and technology.

Chapter 19
Exu and Einstein Dancing Around a Bonfire: Towards a Cosmopolitical Science Education

Nathan Willig Lima and Isabel Martins

19.1 Introduction

The COVID-19 pandemic, beyond its dimension as a public health catastrophe, has exposed, even more distinctly, profound contradictions, ruptures, and vulnerabilities in our political organization, at both local and global levels, which are still present in the post pandemic world. These rifts in our social fabric can be summarized in three interrelated dimensions: (i) increasing political tension with the rise of the far-right, (ii) escalating social inequality intersected by ethnic-racial and gender markers, and (iii) scientific denialism and the propagation of fake news.

The first crucial point for understanding the contemporary political landscape is the tension between a conservative far-right and other political factions present in different countries. Bruno Latour even suggests this new phase of the far-right as a new pole that would reshape the current political landscape (Latour, 2020). It is widely known that this new far-right has been using *fake news* and *post-truth* strategies to create confusion in the public opinion in relation to important socioscientific issues such as global warming and the pandemic. In the specific case of the pandemic, there is evidence that countries with the worst performance in combating the disease, such as Brazil and the United States, were those with politically divided populations (Jasanoff et al., 2021). In these cases, different positions on how to deal with the pandemic were deeply interconnected with polarized political perspectives, making the dialogue impossible.

N. W. Lima (✉)
Universidade Federal do Rio Grande do Sul, Porto Alegre, Brazil
e-mail: nathan.lima@ufrgs.br

I. Martins
Universidade Federal do Rio de Janeiro, Rio de Janeiro, Brazil

C. B. Moura (ed.), *A Sociopolitical Turn in Science Education*, Contemporary Trends and Issues in Science Education 62,
https://doi.org/10.1007/978-3-031-78586-3_19

Secondly, the pandemic proved to be particularly harsh on socially vulnerable populations who could not adhere to social distancing policies when vaccines were not yet available (Santos, 2020). The number of job terminations due to deaths in Brazil, for example, surged by 70% in 2021, with truck drivers, cleaners, and salespersons leading the ranks.[1] Social inequalities are on the rise worldwide (UNO, 2006; OXFAM, 2023). In Brazil, they are historically associated with specific markers like gender and ethnic-racial identity. For instance, after the pandemics, black women have had even fewer jobs and lower salaries when compared to white women and men.[2] In the peripheries, the deaths of young black individuals are a brutal reality that plagues the country.[3] As a result of the complex ways poverty intersects with ethnic-racial issues some public security policies have become instruments of state-sponsored death, solidifying what can be understood as necropolitics (Mbembe, 2018). The discourses that underpin these practices, as well as the popular notion that Brazilians are a peaceful and racially mixed people (Gonzalez, 2010), provide support for the perpetuation of violence and the maintenance of social and cognitive injustice, which places knowledge, wisdom, and culture of Afro-diasporic populations in an inferior position. In the most extreme cases, this cognitive injustice manifests as concrete violence against African-derived religious centers.[4]

The third key point associated with the development of the COVID-19 pandemic is the increase in scientific denialism, the spread of fake news, and the use of falsehood as a political strategy, which has given rise to the notion of a post-truth era (Mcintyre, 2018). In the case of the pandemic, specifically, the former President of Brazil downplayed its effects, calling it a “mere flu”[5] and suggesting a connection between the vaccine and the development of AIDS.[6] The substantial portion of the population who adhered to the President’s suggestions demonstrated how fragile trust in science and institutions is in contemporary times. This mistrust in science has prompted numerous publications worldwide, such as the special issue of the Science & Education journal titled “Why trust Science?” (Erduran, 2022).

The perception that we are living in times where it is necessary to restore trust in science casted doubt on views that considered the conception of science as an organized body of knowledge that grows over time, moves away from falsehood, and is

[1] For instance, https://g1.globo.com/trabalho-e-carreira/noticia/2021/10/02/em-ano-de-pandemia-desligamentos-por-morte-saltam-70percent-motoristas-de-caminhao-faxineiros-e-vendedores-lideram-registros.ghtml

[2] See, for instance, https://g1.globo.com/jornal-nacional/noticia/2022/07/25/cai-participacao-de-mulheres-negras-no-mercado-de-trabalho-em-relacao-ao-periodo-pre-pandemia.ghtml

[3] See, for instance, https://www.cnnbrasil.com.br/nacional/negros-somam-80-das-mortes-violentas-de-jovens-no-pais-aponta-estudo/

[4] See, for instance, https://g1.globo.com/jornal-nacional/noticia/2022/11/15/quase-metade-dos-terreiros-do-pais-registrou-ate-cinco-ataques-nos-ultimos-dois-anos-mostra-pesquisa.ghtml

[5] See, for instance, https://www.bbc.com/portuguese/brasil-55107536

[6] See, for instance, https://www.cnnbrasil.com.br/politica/bolsonaro-vira-alvo-de-inquerito-no-stf-por-ligar-vacina-contra-covid-a-aids/

capable of providing satisfactory answers to various human problems, to be naïve. Moreover, it challenged notions that scientific literacy should promote students' autonomous independent critical thinking so that they would not depend on third-party advice, including experts' (Auler & Delizoicov, 2001). Actually, during the pandemic, social media platforms were flooded with science education researchers, repeating slogans like "trust in science," "science saves lives," and "listen to the experts." Not only did the pandemic reveal a complex relationship between society and experts but it also showed that the challenge of building a society that takes into account the knowledge and wisdom from various collectives while listening to science remains a distant reality. Nowadays, it seems that neither society respects other forms of knowledge nor values science.

We view education as an ontological-political practice (Silva, 2010), that is, a practice that aims to shape an "ideal being". It is ontological because it deals with "being" and it is political because the definition of "ideal" is collectively determined, reflecting the organization of a particular collective and responding to the demands and interests of that collective, typically those who hold power within it. The challenge is, therefore, to promote educational principles and practices that allow students to face up to the dilemmas presented by contemporary scenarios.

What kind of education would that be? What would be its organization? What would be its goals? How would science have to be presented and integrated into this educational proposal given, on the one hand, the tension between valuing other forms of knowledge in the pursuit of social justice and, on the other hand, combating post-truth? In other words, what could be the sources to help us think about post-pandemic science education? Would they be extensions and/or adaptations of current models or do we need to seek other models, structures, and grammars to think about education in science?

In this text we entertain the possibility of a radical departure from Western Modern models of both Science and Pedagogy by inviting readers to displace themselves from comfort zones to explore an alternative point of departure. The proposal of this text is to start from the ontological turn in Anthropology (de Castro, 2018), which is in line with both discussions in the field of Science Studies (Latour, 2004a) and the cosmopolitical proposition (Stengers, 2007; Blaser, 2016; Latour, 2004b), to think about educational challenges posed by contemporary scenarios. Adopting a cosmopolitical perspective for science education implies explicitly acknowledging the political dimension of scientific practice and education itself. However, this proposition is even more radical in that it assumes that the set of elements considered real (the cosmos) is the result of a political process, just as the elements of the cosmos also influence politics (Latour, 1999). In other words, cosmopolitics views the collective of humans and non-humans as a complex network that constructs nature and society in a continually partial and provisional way. Based upon this reflection, we propose the concept of Cosmopolitical Pedagogy (Lima et al., 2023), aiming to construct a material and cognitive world (Santos, 2019) that is both fair and inhabitable for all (Latour, 2020).

Cosmopolitical pedagogy assumes that nature and society are not fixed but are what is to be built throughout the pedagogical process. A comfortable path to pursue a supposedly cosmopolitical education would be to sustain the Western cosmovision and to introduce elements of other collectives. However, a truly cosmopolitical perspective demands from us a more symmetrical approach. Instead of simply looking at other collectives from our cosmovision, we should look at our collective assuming other cosmovisions. With that in mind, we explore the possibility of drawing from worldviews and concepts derived from Afro-diasporic propositions to think about (science) education, an idea conveyed by the symbolic image of a dance between *Exu* and Einstein around a bonfire. This attempt is motivated by analysing the potential of decolonial related perspectives (Monteiro et al., 2019) in providing alternative ways to think about and to position ourselves with respect to ruptures, dilemmas and contradictions we have been facing as science educators. As such, our proposition is inspired by Luiz Rufino's Pedagogy of *Encruzilhadas* (Crossroads) (Rufino, 2019), which brings an Afro-diasporic cosmoview to think education as an ethic and aesthetic process.

We discuss the possibility of reconciling knowledge of different collectives and emphasize the importance of science in the process of building a fair world. This is done by proposing an encounter between *Exu*, an *orixá*[7] from Afro-diasporic mythologies, and Einstein, as the representation of a science expert, as a possible foundational and structural agent for our cosmopolitical science education. We also introduce the image of a dance around a bonfire as a symbolic figure of the cosmopolitical pedagogical process. The bonfire is associated with the notions of Justice and Balance, linked to another *orixá*, *Xangô*, and the dance around represents the celebration of life, emphasizing that cosmopolitical pedagogy should embrace joy and eudaemonia[8] as vital elements of its process.

At the end, we discuss the potentialities and challenges of such a proposition to face the post pandemic challenges. Also, we sketch some important questions and themes to be addressed in future research.

19.2 Cosmopolitics and Cosmopolitical Pedagogy

In a general sense, Western Modern thinking is underpinned by an absolute ontological separation between Nature and Society (Haraway, 1991; Latour, 2013). In particular, the period of the development of modern science is marked not only by the defense of a given specific set of methodologies considered valid in the process of knowledge production but also by the consolidation of the concept of "fact" (Whitehead, 1925), which assigns a group of propositions an absolute ontological status independent of any specific local context.

---

[7] Although many Afro diasporic mythologies refer to one creator (*Olorum*), it is manifested in different divinities, the *Orixás*.

[8] This concept is explained further in the text.

Different thinkers in the twentieth century challenged this notion by presenting relational metaphysical perspectives. Whitehead's Philosophy of Propositions (Whitehead, 1978) and Gabriel Tarde's Monadal Sociology (2007), for instance, were instrumental in informing research in the field of Science Studies that portrayed "facts" in science as a heterogeneous network of humans and non-humans with varying degrees of stability (Latour, 2005). Similar metaphysical perspectives have been discussed by different authors (Lima, 2023) such as Holbraad et al. (2014) and Kohn (2015). For instance, according to de Castro's Amerindian perspectivism (2018), entities do not possess an absolute essence: what they are depends on their relationship with the other. Thus, reality becomes a relational property. Similarly, for Latour, the reality of a proposition is a continuous property determined by the extension and stability of the network supporting it (Latour, 1999). This concept implies that the cosmos is not a predetermined set of objects and facts ready to be discovered but is something dynamic, creative, and limitless in its composition.

We suggest that similar insights can also be found in some Afro-diasporic knowledge and worldviews.[9] In these cosmoviews, for instance, *orixás* are worshipped. *Orixás* can be associated with specific beings, narrated in specific stories, or more generally interpreted as divinities, externalizations of the creator Olorum. Moreover, the *orixás* are inseparable from Nature itself, with their "points of power" existing in Nature. It's not that the *orixá* rules over a particular element of Nature; the element is inseparable from the orixá *itself.* This concept promotes a reenchantment of Nature, which, from this perspective, takes on a sacred and divine dimension. Such a relationship with nature is also typical of Quilombola culture (dos Santos, 2015) and indigenous Brazilian peoples (Krenak, 2019), despite having their own specificities. This worldview motivates a different relationship with and care for nature, which can be essential in addressing the ecological crises of the Anthropocene (Latour, 2004a).

Thus, unlike in Modern thinking, where Nature is viewed as an objective and inert entity, separate from the social milieu, and ready to be discovered and exploited, in Afro-diasporic conceptions, Nature is seen as alive, divine, and capable of influencing the social milieu (through the action of the orixás) and responding to human demands and offerings. In this view, Nature is not there to be exploited but should be revered, honored, and respectfully engaged with.

Furthermore, the worship and activation of the *orixás* involve material offerings (such as fruits, flowers, food, and drinks) and the magical activation of the individual performing the ritual. In other words, the power and strength of natural elements are activated by the semiotic and discursive dimension. A fruit is just a fruit, but when offered to an *orixá*, it becomes a source of strength and power that activates

[9] It is important to note that Afro-diasporic knowledge and worldviews form a complex and heterogeneous set of propositions. In Brazil, for example, they have evolved in various ways through encounters with Western Christian and indigenous Brazilian propositions. However, according to Lopes and Simas (2021), they present unity across diversity. This unity is manifest in different common notions, which express convergences in thinking and practices.

nature and changes the course of individuals in the world. Therefore, there is a deep interconnection between Nature and Society, and discourse (speech) is recognized as an essential element in the creation of reality.

Rufino (2019) proposes a Pedagogy of *Encruzilhadas,* in which he discusses the orixá Exu as an important principle for Education. In this sense, Rufino (2019) proposes a deep decolonial education, assuming an Afro-diasporic cosmoview and grammar to speak about Pedagogy.

In this work, we adopt a cosmopolitical perspective to think about alternative possibilities of developing science education approaches for the post-pandemic world. In particular, inspired by Rufino (2019), we draw from Afro-diasporic worldviews and language to contemplate educational phenomena and propose extending them to consider particular aspects of science education contexts.

It is important to emphasize that this proposal does not advocate religious education. Considering education from the Afro-diasporic worldviews does not involve teaching religious concepts. Our goal is to understand and to explore some of the underlying metaphysical and epistemological objectives, methods, structures, and stylistics of this this afro-diasporic cosmovision and language, especially those concerning the intricate forms of relationship between nature and society, as starting point to discuss science and science education in/as a cosmopolitical process.

19.3 A Cosmopolitical Science Education and the Afro-Diasporic Cosmoviews

The challenge we undertake in this section is to propose a cosmopolitical science education structured based on metaphysical conceptions presented in the previous section. We believe that this approach holds the potential to envision an education powerful enough to address the dilemmas discussed in the introduction of this work.

First and foremost, the cosmopolitical proposition does not position itself as a directive political proposal but rather as enabling broader dialogues, which are capable of (re)building the world from diverse perspectives (Stengers, 2018). This, in itself, serves as a counterpoint to looming fascist tendencies to shut and outlaw dissonant voices. Secondly, the cosmopolitical proposition, as conceived in our works (Lima & Nascimento, 2021, 2022), involves bringing knowledge, issues, and needs of various collectives to the forefront of political and scientific discourse, in line with with the construction of "matters of care" (de la Bellacasa, 2011), allowing us to move towards a more just society. Latour discusses "matters of fact" and "matters of interest" by arguing that all facts are constructed by collectives and, thus, must be able of interesting and affecting actor in the first place. On the other hand, de la Bellacasa (2011) claims that we must move to "matters of care", so we can discuss who is allowed and who is forgotten in these "matters of interest". In this sense, to build up a fair society we must allow different interests, including needs of those that do not have access to power, to be considered in the public arena.

Finally, the cosmopolitical proposition, despite acknowledging the coexistence of multiple realities, facilitates the comparison of the extent and stability of different networks, enabling us, in specific cases, to choose one network of propositions over others, thus helping us escape pitfalls of the post-truth era (Latour, 1999; Lima et al., 2019).

In a previous work where we discussed the role of the history of science in post-pandemic education, we outlined several characteristics of education (in a broader sense) that we refer to as non-narcissistic.

1. It does not advocate for a predetermined sociopolitical organization.
2. It enables the recognition and conception of various forms of epistemic organization and production, drawing on the histories of the sciences.
3. It emphasizes the opening of black boxes and values sociotechnical networks as anchors of reality.
4. It rediscovers episodes and actors erased from hegemonic history.
5. It empowers participants in the pedagogical process by enabling mediation of actors across various spheres of power.
6. It paves the way for eudaemonia as an alternative or complement to hedonist consumption (Lima & Guerra, 2022, 396).

We would like to explore the extent to which some aspects of Afro-diasporic cosmovisions would not only resonate with these points but allow to deepen and expand them. By doing so, it is possible to acknowledge science's immense potential to address contemporary challenges, but also to emphasize that other, often marginalized, collectives also offer crucial knowledge from which we have much to learn. These premises agree with both Freire, who sees knowledge and sensitivity as essential to act upon the world (Freire, 2021) and Latour, for whom affecting the body is necessary for mobilization (Latour, 2020). In this sense, we argue that there is much to learn from the knowledge and sensitivity of Afro-diasporic collectives and indigenous peoples, as their relationship with nature is fundamental to construct a new order for the Anthropocene. In the following sections, we delve deeper into these points through the lens of an Afro-diasporic cosmology, using a metaphorical dance between Exu and Einstein around a bonfire to illustrate elements of a possible cosmopolitical (science) education.

19.3.1 Exu as a Metaphor of a Pedagogic Principle

Exu is considered, in different narratives, the first *orixá*. *Exu* was born even before his mother. He is the lord of power, strength, and vitality, the owner of *axé* (*axé* is the term used to refer to this "life force" that can be transmitted). He is often associated with the trident, which symbolizes his tripartite (positive, negative, and neutral) or tri-temporal (past, present, and future) nature. *Exu* transcends space and time. He walks crooked paths along straight lines, and he killed a bird yesterday with a stone that he will only throw tomorrow.

Thus, *Exu* is a fundamental symbol that opposes cherished premises of the modern colonial project (Rufino, 2019). He represents a "dribble" taken at Cartesianism, Western-centric and positivist thinking. An education centered on *Exu* is an education that places the pursuit of diversity as its driving force, the possibility to construct and act within the incompleteness, doubt, and the ability to coexist with others without annihilating all that is different. *Exu* is the mouth that devours everything and the body that gives everything. As Rufino (2019) states, through *Exu*, we do not seek to destroy the other but to ingest, incorporate, and spit them out in a different form (Rufino, 2019). *Exu* is the principle of collective, open, creative, plural, and relational construction, and the possibility of constructing the world, fully embodying the cosmopolitical proposition that has been presented.

As the dynamic principle of communication, *Exu* lives in "*encruzilhadas*" (crossroads) (Rufino, 2019). It represents the passage, the crossing between the material and the spiritual, life and death, the concrete and the symbolic. *Encruzilhada* is where paths intersect and from where paths emerge. *Encruzilhada* is the meeting, communication, creativity, and the creation of something new. Thus, Pedagogy of Encruzilhadas is an education that seeps through the cracks, moves in curves, and allows the convergence of different forms of knowledge, which can be likened to the construction of an ecology of knowledge (Santos, 2019).

In contrast to colonial project, which either annihilates[10] or subjects other forms of knowledge and ways of being in the world to an inferior status, in a Pedagogy of *Encruzilhadas* has the potential to explore the diversity and plurality of knowledge, methods and ways of being in the world. Associated with this concept, Rufino (2019) introduces the notion of an epistemological "rolé" (spin), that is, the reconciliation and dialogue of different perspectives that allows for novel ideas. Thus, education at the crossroads does not advance towards propaedeutic teaching but invests in the possibility of creating worlds, creations that are always incomplete.

Drawing on Exu as a pedagogical principle implies a shift in discourse, which involves moving away from categorical assertions to questionings, that is, to embrace questions and doubts. Such a proposal is, therefore, in line with Freire's pedagogy of question (Freire, 2021), which declares questions are the primary path to knowledge production, much more than the transmission of statements. In other words, the pedagogical process should focus on the production of good questions as well as on the rigorous and collective search for good answers. Thus, Freire (2021) explains that his pedagogy does not advocate for a specific social model. The idea that social organization must be collectively built is also in line with the cosmopolitical proposition we have been advocating.

As *Exu* is the *orixá* of life, it is clear that the perspectives and epistemologies that are part of this epistemological "*rolé*" are those that affirm the existence of the other and that are not complicit with the colonial necropolitical project. Therefore, Exu and epistemological "rolé" are metaphors that resonate with points (1), (2) and (5) described in our proposal for a cosmopolitical pedagogy. Moreover, the fundamental goal of a cosmopolitical scientific education is to promote life, justice, and plurality. In other words, it should allow reconciliation with the other and present itself

---

[10] A process often referred to as "epistemicide" (Santos & Meneses, 2014).

as an unfinished proposal. Finally, cosmopolitical education must empower individuals, enhancing their strength and ability to act in the world.

It is, therefore, in this sense that we envisage a cosmopolitical (science) education based upon both a pedagogy of questions and a pedagogy of encruzilhadas. As such, it enables participants in the pedagogical process to truly incorporate the knowledge produced, to change their agency in the world, to transform their ability to construct reality, and to act towards creating a better world. In other words, such a cosmopolitical (science) education would impart *axé* to those who participate in it.

19.3.2 Einstein (the Specialist) and Science in a Cosmopolitical Education

In our attempts to envisage ways toward a cosmopolitical education we have brought forth *Exu* as a metaphor that encapsulates principles advocated by both the Pedagogy of Crossroads (Rufino, 2019) and the Pedagogy of Questions (Freire, 2021), such as plurality, creation, and incompleteness. However, neither Freire or Rufino explicitly address issues concerning science or science education. What would be involved in establishing such a dialogue or, to use a concept of the Pedagogy of Encruzilhadas, in this epistemological rolé?

This is where the challenge seems to become more complex. While Pedagogy of Crossroads admits plurality and incompleteness, modern science strives for objectivity and reliability. Besides, Modernity established itself as a hegemonic project by subordinating other forms of knowledge and ways of existing in the world. As Latour (2004a) points out, the idea that scientists have privileged access to reality has been used as a political instrument, where the "scientifically proven" is used to silence a multitude of understandings about reality—an attitude that undermines the very democratic project.

We have chosen Einstein as our representation of scientists for a number of reasons. Not only is Einstein the most emblematic contemporary example of a science specialist, but he is also a scientist engaged with social concerns, who took a political pacifist stance. Moreover, his Theory of Relativity allowed us to conceptualize fundamental principles of space and time in terms of its inter-relationships, the consequences of which are in many ways counterintuitive to everyday experience. He also saw the importance of addressing the general public in order to communicate scientific ideas.

Is it possible to bring scientific knowledge into an epistemological *rolé* with other forms of knowledge? Would it be possible, then, to bring a scientist, here symbolized by Einstein, to an *encruzilhada*? What would an encounter between Einstein and *Exu* be like?

Our understanding is that this *rolé* is indeed possible. To do so, we need to distinguish between what can be understood as official science and unofficial science (Latour, 2013). Official science corresponds to Western Modern Science discourse and presents itself as absolute, universal, neutral, and a collection of facts.

However, when we investigate the history of science or look into science practice as it happens in laboratories, we encounter another science (unofficial science). This unofficial science of practices is profoundly human, full of controversies, disputes, interests, emotions, mistakes, doubts, and misconceptions.

That is, when we look at what science does in its practices (i.e., adopt an empirical perspective to build our knowledge about what science is), we realize that it produces propositions by articulating complex networks of humans and non-humans (Latour, 2013). There is no moment in scientific practice where a direct confrontation between discourse and reality allows us to have a proposition deemed absolutely true or false. Nonetheless, even though provisional and mutable, can be characterized by their extension, stability and aptness to address reality. For example, in the case of the pandemic, there was a debate in Brazil about whether people should get vaccinated or not. There was a network in favor of vaccination and one against it. Both brought political interests and both had representatives from the scientific community. However, there was a more stable network, the one in favor of vaccination, and this network was the one that allowed the preservation of life, gave the *axé* of life, and therefore was also the path that the crossroads allowed us to choose.

Thus, there is no contradiction in proposing an epistemological *rolé and, at the same time,* to value and credit science as a powerful worldview, which capable of providing important solutions to contemporary problems. This is completely different from saying that science is an absolute, factual, and superior form of knowledge a priori.

Also, in a cosmopolitical pedagogy, science is welcome when it gives the *axé* of life, when it produces enchantment, opposes mortality and necropolitics. To mobilise yet another metaphor from Afro-diasporic cosmovisions, science has its own "*mandingas*" (spells). For instance, wasn't the vaccine a way to dribble a project which would have the death of the most vulnerable ones as a consequence?

Finally, we must stress that science is neither inherently good nor inherently bad; it carries the values of the collectives and contexts in which it is formed. This is jot to say that such values and meanings cannot be reconfigured (Auler, 2018). Thus, involvement of different parties in cosmopolitics and cosmopolitical education is justified as long as they are committed to Life, Justice and to the construction of a fairer world. In this sense, we need an education that explicates processes and values involved in the participation of different actors in the articulated networks that are entailed in science (education) practices (elements 3 and 4 mentioned in our proposal).

19.3.3 The Dance Around the Bonfire

Part of our challenge has already been achieved: bringing Einstein and Exu into the same scenario, a meeting typically considered improbable though not impossible at a crossroads. Now, one might wonder: what will these two accomplish?

Would Einstein encourage Exu to study the Theory of Relativity? Or would Exu, the "lord of space and time", invite Einstein to dance?

Einstein represents a scientist who has given singular contributions of utmost importance to Science. Nonetheless, he is also a scientist who positioned himself against racial prejudice and in favor of civil liberties. The Russell-Einstein manifesto was the foundation stone of Pugwash Society, an organization dedicated to freeing the world from weapons of mass destruction. Einstein is therefore a symbol that science must be committed to life as well as to "dialogue across divides".[11]

Besides, if we take *Exu* as a pedagogical principle, it must have precedence over any other principle. In other words, it is the defense of life, plurality, difference, alterity, construction, and communication that guides the process, not the other way around. This is why they must dance around a symbolic fire. The fire is associated with the *orixá Xangô*, who governs Justice and Balance, so that the dance around it represents Justice and Balance between different kinds of knowledge held by all members of this gathering. In other words, any process, method, structure, or knowledge that is brought into the pedagogical process should be driven and warmed by the fire of Justice and Balance.

The dance is therefore, yet another metaphor that conveys both the possibility of expression and joyful interaction amongst participants. At the same time, it involves individual expression and group coordination, attention to the other, sensitivity to interpret and respond to signals. It is a fully relational and creative type of semiosis that involves encounters. Therefore, in our explorations, dance carries a particular meaning in helping us think about pedagogical processes which would allow for blurring ontological between so-called social and natural beings. Furthermore, they should enable mutual forms of both interactions and recognition between participants, kinds of knowledge, and power systems, as well as an increased ability to move in the world. That could also promote a greater awareness of expanded sources of existential fulfillment as opposed to impoverished ways of existing in the world promoted by consumer societies where nature is reduced to a set of resources to be exploited. Finally, the dance metaphor recuperates the embodied nature of knowing in a way that can be related to axé as a force of life that provides joy, personal fulfillment, and collective belonging that is manifest in and inseparable from a bodily experience (earlier referred to as Eudaimonia in item 6 of our proposal).

19.4 Final Remarks: Implications for (Science) Education

We have argued for a view of (science) education that assumes a cosmopolitical perspective, recognizing that both Nature and society (the cosmos as a whole) are in the process of construction and emerge from the reconciliation of different actors from different collectives. This proposition allows us to conjoin different cosmos,

[11] https://pugwash.org/1955/07/09/london-launch-of-the-russell-einstein-manifesto/

making plurality and difference a path to collective construction, without falling into a radical relativism that could prevent us from taking concrete actions in specific cases, such as the pandemic.

We started by identifying three significant challenges: (i) the rise of the extreme right, (ii) increasing social inequality, and (iii) scientific denialism. Education and science education ought to be able to address such problems and provide possible paths to contribute to the construction of an inhabitable world.

Although there may be a recognition that if hegemonic education and science education do not address the political aspects of science nor recognize and value the knowledge of different cultures, they will be incapable of dealing with challenges (i) and (ii), this is not a straightforward task. For instance, in multiculturalist perspectives, where in many cultures, only one nature is assumed, which allows for the notion of a transcendental entity that mediates all cultures (Latour, 1999). In this sense, either there is no privileged access to such Nature, and all cultures are equivalent, or there is one privileged culture (science). Whereas the first proposition leads us to radical relativism, the second recovers an absolutist perspective. In both cases, the possibility of action in severe situations, like climate change mitigation, are impaired.

Cosmopolitical Pedagogy, however, recognizes the value of Science without having to deny other cultures and other natures (multinaturalism) (de Castro, 2018). This allows us to choose a specific action in drastic situations, which would correspond to what Latour would call the widest and most stable network, and, at the same time, to assume the coexistence of different cosmos as an important element for a truly fair world. The union of these aspects allows us to build a participative and collective world, which could protect us from the authoritarian rule of the extreme right project, and to fight against social injustices while opposing post-truths and science denialism.

In this text, we suggest the possibility that the Pedagogy of Crossroads (Rufino, 2019) can provide metaphysical and epistemological assumptions to think about what would be involved in the cosmopolitical pedagogical process. The argument was developed around the metaphor of Exu and Einstein enacting a heartfelt, constructive, and meaningful interaction around a bonfire that represents justice and balance. In this sense, a cosmopolitical education should be an education capable of producing "axé," that is, increasing the power, strength, and vitality to beings so that they can manifest themselves freely and happily, contributing to the construction of a fairer world. In this perspective, Science is reconciled with its concrete practices, which are creative, open-ended, emotional and alive, full of disputes and contractions.

Of course, there is no single way to promote a cosmopolitical pedagogy. One possible path would be to sustain the modernist perspective, cosmovision and language while engaging in dialogue with other collectives. This perspective, however, entails an inherent asymmetry, namely, that Science remains the epistemological yardstick against which all other cultures and natures must be analyzed. Our attempt was to stretch the argument towards a more symmetrical perspective in which scientists would start from a different cosmovision to think about aspects of their own knowledge and practices (de Castro, 2018).

There is a long journey ahead to promote a cosmopolitical (science) education. Firstly, it is important to expand the theorisation about relationships between nature and society so as to include different groups and cosmovisions in this debate. This means expanding conceptual agendas and creating zones of contact between different collectives. In this effort, it is important to stress that all elements brought to a cosmopolitical education must contribute to the construction of a fairer world and to promote life. In this sense, it makes sense to bring different cosmovisions as far as they contribute to preservation of life and to the promotion of justice. A deeper examination of epistemological and metaphysical assumptions involved in different worldviews, informed by several disciplinary and non-disciplinary perspectives, is therefore of utmost importance.

Also it seems relevant to discuss relevant aspects of their educational practices in different collectives, with a particular view to: (i) the way the inseparability between nature and society is effected; (ii) forms of embodied learning that take place and; (iii) ways for enabling the empowerment and emancipation of groups in situations of greater social vulnerability to legitimately act in the world.

Thus, the text is an invitation to think about a research agenda, which is as a radical as necessary, to face up contemporary challenges and dilemmas in Science Education.

Acknowledgments The authors thank the Brazilian Council for Scientific and Technological Development (CNPq) and the Rio de Janeiro's Carlos Chagas Research Foundation (FAPERJ) for financial support.

References

Auler, D. (2018). *Cuidado! Um Cavalo Viciado Tende a Voltar Para o Mesmo Lugar*. Appris.

Auler, D., & Delizoicov, D. (2001). Alfabetização Científico-Tecnológica Para Quê? *ENSAIO – Pesquisa Em Educação Em Ciências, 03*(1), 1–17.

Blaser, M. (2016). Is another Cosmopolitics possible? *Cultural Anthropology, 31*(4), 545–570.

de Castro, E. V. (2018). *Metafísicas Canibais: Elementos Para Uma Antropologia Pós-Estrutural*. Ubu.

de la Bellacasa, M. P. (2011). Matters of Care in Technoscience: Assembling neglected things. *Social Studies of Science, 41*(1), 85–106. https://doi.org/10.1177/0306312710380301

dos Santos, A. B. (2015). *Colonização Quilombos: Modos e Significados*. INCTI/UNB.

Erduran, S. (2022). Trust in Science and Science Education—Part 1. *Science & Education, 31*(5), 1101–1104. https://doi.org/10.1007/s11191-022-00386-4

Freire, P. (2021). *Por Uma Pedagogia Da Pergunta*. Paz e Terra.

Gonzalez, L. (2010). *Primavera Para as Rosas Negras*. Selo Negro Edições.

Haraway, D. J. (1991). *Simians, cyborgs, and women: The reinvention of nature*. Routledge.

Holbraad, M., Pedersen, M. A., & Vi Castro, E. (2014). The politics of ontology: Anthropological positions. *Fieldsights*. 2014. https://culanth.org/fieldsights/the-politics-of-ontology-anthropological-positions

Jasanoff, S., et al. (2021). *Comparative Covid response: Crisis, knowledge, politics*. Harvard Kennedy School. Available at https://compcore.cornell.edu/publications/

Kohn, E. (2015). Anthropology of ontologies. *Annual Review of Anthropology, 44*(1), 311–327. https://doi.org/10.1146/annurev-anthro-102214-014127

Krenak, A. (2019). *Ideias Para Adiar o Fim Do Mundo*. Companhia das Letras.
Latour, B. (1999). *Pandora's hope: Essays on the reality of science studies*. Harvard University Press.
Latour, B. (2004a). *Politics of nature: How to bring science into democracy*. Harvard University Press.
Latour, B. (2004b). Whose cosmos, which cosmopolitics? *Common Knowledge, 10*(3), 450–462. https://muse.jhu.edu/article/171401/summary#info_wrap
Latour, B. (2005). *Reassembling the social: An introduction to actor network theory*. Oxford University Press.
Latour, B. (2013). *Jamais Fomos Modernos*. Editora 34.
Latour, B. (2020). *Onde Aterrar? Como Se Orientar Politicamente No Antropoceno?* Bazar do Tempo.
Lima, N. W. (2023). Cosmopolitical propositions: A historiographic analysis of contemporary anthropological perspectives on sciences. In M. L. Condé & M. Salomon (Eds.), *BT – Handbook for the historiography of science* (pp. 1–21). Springer International Publishing. https://doi.org/10.1007/978-3-030-99498-3_27-1
Lima, N. W., & Guerra, A. (2022). Superando Narciso: Histórias Das Ciências Para Adiar o Fim Do Mundo. *Revista Brasileira de História Da Ciência, 15*(2), 386–399.
Lima, N. W., & Nascimento, M. M. (2021). Aterrando No Sul: Uma Proposta Político-Epistemológica Para a Área de Educação Em Ciências Do Antropoceno. *Ciência & Educação* in press.
Lima, N. W., & Nascimento, M. M. (2022). Not only why but also how to trust science: Reshaping science education based on science studies for a better post-pandemic world. *Science & Education, 31*, 1363–1382. https://doi.org/10.1007/s11191-021-00303-1
Lima, N. W., Vazata, P. A. V., Guerra, A., de Holanda Cavalcanti, C. J., & Ostermann, F. (2019). Science education in post-truth age: Metaphysical reflections from Bruno Latour's science studies. *Revista Brasileira de Pesquisa em Ensino Ciências, 19*, 155–189.
Lima, N. W., Rosa, G. G., Rosa, A. W., & Gois, E. (2023). Em Direção a Uma Pedagogia Cosmopolítica: História e Filosofia No Ensino de Ciências Para Reviver No Antropoceno. In F. Ostermann, I. S. Araujo, & M. M. Nascimento (Eds.), *Cadernos de Pesquisa Do Programa de Pós-Graduação Em Ensino de Física Da UFRGS*. Pimenta Cultural.
Lopes, N., & Simas, L. A. (2021). *Filosofias Africanas: Uma introdução*. Civilização Brasileira.
Mbembe, A. (2018). *Necropolítica*. N-1 Edições.
Mcintyre, L. (2018). *Post-truth*. MIT Press.
Monteiro, B. A. P., Dutra, D., Cassiani, S., Sanchéz, C., & Oliveira, R. (Eds.). (2019). *Decolonialidades na Educação em Ciências*. Livraria da Física.
OXFAM. (2023). *Survival of the richest*. Oxfam International.
Rufino, L. (2019). *Pedagogia Das Encruzilhadas* (Vol. 9, pp. 262–289). Mórula Editorial.
Santos, B. d. S. (2019). *O Fim Do Império Cognitivo – A Afirmação Das Epistemologias Do Sul*. autêntica.
Santos, B. d. S. (2020). *A Cruel Pedagogia Do Vírus*. Almedina.
Santos, B. d. S., & Meneses, M. P. (Eds.). (2014). *Epistemologias Do Sul*. Cortez Editora.
Silva, T. T. (2010). *Documentos de Identidade: Uma Introdução Às Teorias Do Currículo*. Autentica.
Stengers, I. (2007). La Proposition Cosmopolitique. In J. Lolive & O. Soubeyran (Eds.), *L'émergence Des Cosmopolitiques* (1st ed., pp. 45–68). La Découverte.
Stengers, I. (2018). A Proposição Cosmopolítica. *Revista Do Instituto de Estudos Brasileiros, 69*, 442–464.
Tarde, G. (2007). *Monadologia e Sociologia – e Outros Ensaios*. Cosac Naify.
UNO. (2006). *Social justice in an open world: The role of the United Nations*. United Nations.
Whitehead, A. N. (1925). *Science and the modern world*. Pelican Mentor Book.
Whitehead, A. N. (1978). *Process and reality: An essay in cosmology*. The Free Press.

Nathan Willig Lima graduated in Physics with a focus on Medical Physics from Pontifícia Universidade Católica do Rio Grande do Sul (PUCRS) (2014), and has a second degree in Physics from Faculdade Avantis (2017), a master's degree in Materials Engineering and Technology from Pontifícia Universidade Católica do Rio Grande do Sul (PUCRS) (2016), and a PhD in Physics Teaching from Universidade Federal do Rio Grande do Sul (UFRGS) (2018). He is a professor at the Physics Institute of the Federal University of Rio Grande do Sul, a permanent professor in the Graduate Program in Physics Teaching at Universidade Federal do Rio Grande do Sul (UFRGS). He visited the University of Copenhagen, Denmark, participating in a research project on the history of Quantum Theory and implications for teaching. He was the assistant editor of the HPSST Newsletter from 2022 to 2023. His main interests are History, Philosophy, and Physics/Science Teaching.

Isabel Martins is Professor of Science and Health Education at the Federal University of Rio de Janeiro. She taught physics in Rio de Janeiro secondary schools before getting her PhD in Science Education at the University of London. She has also worked at University College London Institute of Education and at the Federal University of Minas Gerais, Brazil. Isabel has led and collaborated in several international funded projects, published in refereed journals, co-edited and co-authored academic books in Science Education. She was President of the Brazilian Science Education Research Association (ABRAPEC) and is editor-in-chief of the electronic teacher education journal Ciência em Tela. Her research interests concern the articulation of discursive, sociopolitical, cultural and epistemological dimensions of science knowledge and its implications for curriculum development, textbook analyses and scientific popularisation.

The manufacturer's authorised representative in the EU is Springer Nature Customer Service Centre GmbH, Europaplatz 3, 69115 Heidelberg, Germany. If you have any concerns regarding our products, please contact ProductSafety@springernature.com

Printed and bound by CPI Group (UK) Ltd, Croydon, CR0 4YY
01/10/2026
02225932-0014